# Waste Management in Climate Change and Sustainability Perspectives

This book addresses the challenges and opportunities associated with the management of lignocellulosic waste in the context of climate change and sustainability. It is divided into three sections exploring the potential of lignocellulosic waste biorefineries, examining waste management strategies in the face of climate change, and providing insightful case studies showcasing innovative waste management systems. It discusses bioprocessing techniques for converting lignocellulosic waste into fuels and chemicals, the application of laccase in lignocellulose degradation, and the utilization of microbes for developing biofertilizers from crop stubble waste. The chapters also examine the influence of climate change and emerging pollutants on agro-residue composting processes and the resulting compost quality. Towards the end, the book presents real-world examples of innovative waste management systems. This book would be useful for a wide range of professionals, researchers, policymakers, educators, and students seeking to deepen their understanding of waste management, climate change mitigation, and sustainable practices in the context of lignocellulosic waste.

**Key Features:**

- Provides comprehensive coverage of lignocellulosic waste management and climate change
- Focuses on climate change resilience and the role of waste management in mitigating environmental impacts
- Presents insights into sustainable practices and circular economy approaches
- Examines in-depth exploration of bioprocessing and bioconversion techniques
- Includes practical strategies for converting waste into valuable resources, including fuels and chemicals

## Sustainable Industrial and Environmental Bioprocesses

This book series aims to provide a comprehensive collection of books focusing on bioprocesses in industrial and environmental biotechnology. The multidisciplinary content encompasses chemical and biochemical engineering, industrial microbiology, and energy biosciences, all with a central theme of sustainable development and circular economy principles. The books align with the Sustainable Development Goals (SDGs) and offer state-of-the-art information and in-depth knowledge on the subject matter. The books in this series also emphasizes the application of emerging tools, such as machine learning and artificial intelligence, for the advancement of bioprocesses. While primarily targeting academicians and researchers, the series is valuable for policy planners and industry professionals, with carefully tailored contents to cater to their specific needs and interests.

*Series Editor: Dr Ashok Pandey*

**Professor Ashok Pandey** is currently Executive Director, Centre for Energy and Environmental Sustainability-India, Lucknow, India and Founder of the Biotech Research Society, India (BRSI) and International Bioprocessing Association (IBA).

*Diatoms and Sustainable Agriculture*
Archana Tiwari and Ashok Pandey

*Biochar and the Circular Economy: Innovations in Production and Applications*
Hrudayanath Thatoi, Shibani Mohapatra, Ashok Pandey, and Lam Su Shiung

*Sustainable Technologies for Value Addition and Management of Plastic Waste*
Abha Kumari, Ashok Pandey, Rohidas Bhoi, and Ahn Tuan Hoang

*Sustainable Biomanufacturing: Innovations for a Circular Bioeconomy and Industry 5.0*
Vinod Kumar Garg, Navish Kataria, and Ashok Pandey

*Waste Management in Climate Change and Sustainability Perspectives: Lignocellulosic Waste*
Sunita Varjani, Carol S.K. Lin, Ashok Pandey, and Xuan-Thanh Bui

For more information about this series, please visit https://www.routledge.com/Sustainable-Industrial-and-Environmental-Bioprocesses/book-series/SIEB

# Waste Management in Climate Change and Sustainability Perspectives

## Lignocellulosic Waste

Edited by
Sunita Varjani, Carol S.K. Lin, Ashok Pandey,
and Xuan-Thanh Bui

CRC Press
Taylor & Francis Group
Boca Raton London New York

CRC Press is an imprint of the
Taylor & Francis Group, an **informa** business

Designed cover image: Shutterstock

First edition published 2027
by CRC Press
2385 NW Executive Center Drive, Suite 320, Boca Raton FL 33431

and by CRC Press
4 Park Square, Milton Park, Abingdon, Oxon, OX14 4RN

*CRC Press is an imprint of Taylor & Francis Group, LLC*



ISBN: 9781032633114 (hbk)
ISBN: 9781032676852 (pbk)
ISBN: 9781032676890 (ebk)

DOI: 10.1201/9781032676890

Typeset in Times
by codeMantra

# Contents

## *SECTION III Case Studies*

# Preface

The world stands at a critical juncture. Our linear "take-make-dispose" model of consumption has led to unprecedented pollution challenges, characterized by the presence of heavy metals, surplus nutrients, and a cocktail of chemicals. Traditional waste management methods have been proven inadequate to address this crisis, failing to eliminate pollutants and often contributing to greenhouse gas (GHG) emissions. This undeniable reality is a driving force for a profound shift within the global scientific and engineering communities. We are moving from a view of waste as a problem of disposal to recognizing it as a valuable resource to be circularly managed—a transition central to the pursuit of genuine sustainable development.

It is in this context that we present the book entitled *Waste Management in Climate Change and Sustainability Perspectives: Lignocellulosic Waste*. This book is born from the conviction that effective waste management is not merely a municipal duty but a powerful lever for climate change mitigation. By adopting integrated strategies—such as the waste to wealth concept and 5R principles (Refuse, Reduce, Reuse, Repurpose, Recycle)—GHG emissions across key sectors (agriculture, energy, manufacturing, etc.) can be reduced.

This book deals with lignocellulosic waste—the most abundant renewable resource having a strong crystalline structure. It comprises cellulose, hemicellulose, and lignin, and this holds immense promise as a feedstock for biofuels and biochemicals. This book delves into the cutting-edge pretreatment technologies—chemical, physical, and biological—required to break down this bio-recalcitrance. We also address a critical subsequent challenge: the generation of inhibitory toxins during pretreatment and the vital detoxification processes needed to enable efficient microbial fermentation and unlock industrial-scale production.

The primary objectives of this book are threefold: to meticulously examine the relationship between waste management and climate change, to evaluate the climate advantages and consequences of various technological solutions, and to provide a comprehensive global analysis of managing lignocellulosic waste. Each chapter offers detailed, updated information intended to serve as a comprehensive guide for students, scientists, and academicians in environmental biotechnology, engineering, and related disciplines.

This publication is the fruit of a collective endeavour. We extend our deepest gratitude to the international team of authors who have generously contributed their expertise, knowledge, and time. Their dedication to addressing these critical issues of sustainability has been instrumental in shaping this book's content. We are also profoundly thankful to the panel of reviewers, whose rigorous evaluation and constructive feedback were indispensable in refining the scientific discussions and enhancing the overall quality of this work.

We sincerely thank Dr Bhavik Sawhney, Senior Publishing Editor, Taylor & Francis, and the CRC Press team for their support in publishing this book. Their expertise, professionalism, and support throughout the publication process have been invaluable in bringing this book to fruition.

We hope that this book will serve as a valuable resource and inspire further innovation in turning the global challenge of waste into a cornerstone for a more sustainable and climate-resilient future.

# About the Editors

**Sunita Varjani** is a professor at UPES in Dehradun, India, and an adjunct professor at Korea University. She also serves as the director of the Institute of Chartered Waste Managers in India. She has worked as a visiting fellow at the City University of Hong Kong and as a scientific officer at the Gujarat Pollution Control Board, Gandhinagar, Gujarat, India. Her research focuses on industrial and environmental (bio)technology, particularly in waste management, wastewater treatment, and bioremediation, with an emphasis on developing circular waste-based biorefineries for sustainable chemical and fuel production. She is an international expert of the European Research Council for Starting, Consolidator, Advanced and Synergy Grant calls.

**Carol S.K. Lin** is a professor at the City University of Hong Kong. She received her PhD in biochemical engineering at the School of Chemical Engineering and Analytical Science at the University of Manchester, England. Her research focuses on technological advancement and development of circular waste-based biorefineries for sustainable production of chemicals, materials, and fuels, which contribute to the reduction of environmental burden of waste disposal and enhancement of resource efficiency. She is an international expert of the Bio4 Project panel on applied biological sciences for the Research Foundation Flanders in Belgium.

**Ashok Pandey** is currently the executive director of the Centre for Energy and Environmental Sustainability-India, Lucknow. He is also the HTBS National Innovation Chair (Biotechnology). His major research and technological development interests include industrial and environmental biotechnology and energy biosciences, focusing on biomass to biofuels and chemicals, waste to wealth and energy, etc.

**Xuan-Thanh Bui** is an associate professor at the Faculty of Environment and Natural Resources, Ho Chi Minh City University of Technology, Vietnam. He obtained a PhD degree in environmental engineering from the Asian Institute of Technology, Thailand. He serves as the chairman of the Department of Water Science & Technology and the head of VNU-HCM Key Laboratory of Advanced Waste Treatment Technology. His research interests are in the area of environmental technology, with a focus on wastewater treatment.

# *Section I*

## *Lignocellulose Biorefineries*

# 1 Waste Biorefineries for a Sustainable and Circular Bioeconomy

*Farah Naaz*[1], *Vivek Suresh Dalvi*[1], *Saurabh Samuchiwal*[1], *Vivek Nair*[2], *and Anushree Malik*[1]

[1]Applied Microbiology Laboratory, Centre for Rural Development and Technology, Indian Institute of Technology Delhi, New Delhi, India

[2]School of Interdisciplinary Research, Indian Institute of Technology Delhi, New Delhi, India

## 1.1 INTRODUCTION

Around 2.01 billion tonnes of solid waste (SW) is being produced every year, and every person on earth does indeed have a stake in the solution to the problem of SW management (SWM). The daily health, productivity, and sanitation of communities are affected by the decisions that people and authorities make for consumption and waste management. The SWM is a costly affair and can be the single largest budget item for many local governments especially in low-income nations, where it accounts for an average of about 20% of total budgets. At least 33% of the 2.01 billion tonnes of SW produced yearly is not managed in an environmentally sound way, and these available data are quite conservative. The amount of SW produced daily across the globe ranges from 0.11 to 4.54 kg/person, with an average of 0.74 kg/person. Industrialization showcases the potential of waste generation, and thereby, high-income developed countries produce 34% of the global waste (683 million tonnes), although they have merely 16% of the global population (Kaza et al., 2018). Although low-income countries collect nearly half of all garbage in urban areas, that number plummets to just 26% in rural areas. Sub-Saharan Africa only manages to collect roughly 44% of the total SW produced, compared to at least 90% in Europe, Central Asia, and North America combined (Vergara and Tchobanoglous, 2012).

The waste stream is not homogeneous across socioeconomic brackets. Dry waste that could be recycled accounts for more than half of all garbage produced in high-income countries. This includes plastic, paper, cardboard, metal, and glass and accounts for only 32% of all SW produced. The percentage of organic waste increases from middle-income to low-income countries, from 53% to 56% for food waste and 56%–58% for agricultural waste. Even if recycling is an option, only 16% of SW in low-income countries is recyclable. There is not much variation in SW streams across regions, besides those that correspond to income. Except European region, major geographical area-occupying regions, such as Central Asia and North America, generate at least half (on average) of their SW as organic.

The widespread belief that technology can solve the issue of growing trash without any human intervention is not always true. When it comes to dealing with SW, technology is rarely the sole deciding factor. Developing nations that want to move away from open dumping as well as other primitive waste management practices have a better chance of success if they choose solutions that are tailored to their specific needs. Currently, landfills are the primary means by which garbage is disposed of around the world. About 37% of trash ends up in a landfill, with only 8% going to sanitary dumps equipped to capture landfill gas. About 33% of organic SW ends up in landfills,

DOI: 10.1201/9781032676890-2

19% is reclaimed through reuse via recycling and composting, and 11% is used for energy generation employing incineration. Only developed countries, i.e., high- and upper-middle-income countries, can afford to have proper waste disposal or treatment systems in place, such as regulated landfills or strictly monitored facilities. Open dumping is commonly used in nations with lower incomes, accounting for 93% ofFFFwaste disposed of, compared to 2% in high-income countries. The landfill percentage is highest (54%) in upper-middle-income countries, which falls to 39% in high-income nations. About 35% of SW is sent to recycling and composting facilities in upper-middle-income countries, and around 22% is being burned in direct air. Till now, incineration is practiced in countries that have high-end technology, high income, and capacity to abate the flue gas produced by adopting incineration. Also, land constraint is an important factor for the same.

As per 2016 data, open dumping of organic SW and disposal in landfills without landfill gas capture systems are the main contributors to an overall 1.6 billion tonnes of carbon dioxide ($CO_2$) equivalent greenhouse gas emissions from SWM and disposal practices. This constitutes about 5% of worldwide emissions. Also, annual emissions are projected to rise to 2.6 billion tonnes of $CO_2$ equivalent by 2050, if no major steps are taken ahead to induce the SWM industry.

Almost 70% of nations have institutions in place that are in charge of developing waste-related policies and supervising regulatory compliance, while most countries delegate SWM operations to local governments. In regard to SWM, over 66% of countries have enacted legislation and rules, while enforcement varies greatly. About 70% of trash services are directly supervised by local public authorities, whereas the central government's engagement in waste service provision is uncommon outside of regulatory control and fiscal transfers. At least fifty percent of all waste management services (including collection, transfer, treatment, and disposal) are provided by government agencies, while 30% are provided through public-private partnerships. Collaboration with the private sector for operations and financing can be successful, but only when the proper reward structures and compliance mechanisms are in place.

This chapter discusses the substitution of petroleum-based resources by biobased resources that would manage the greenhouse gas emissions and carbon release to the atmosphere, contributing to a cleaner and greener environment. Hence, exploitation of biogenic wastes as feedstocks and integration of processes resulting in the generation of high-value products (renewable energy, platform chemicals, biofertilizers, biomaterials, etc.) would serve multiple purposes: bioremediation, resource recovery, and improved investment returns.

## 1.2 TECHNOLOGIES ADOPTED FOR SOLID BIOWASTE MANAGEMENT

Various technologies have been used to convert solid biowaste into new valuable products. The different technologies are broadly classified into three groups based on treatment methods: direct use, biological treatment, and physicochemical treatment, as shown in Table 1.1. The final utilization of generated new compounds is mainly divided into three categories, i.e., agriculture (crop production), food (meat/fish production), and bioenergy (Lohri et al., 2017). In the case of agricultural use, the biowaste-driven products should be rich in carbon and nutrients. With intensive agriculture practices like tillage, sowing, and irrigation on the land, the soil quality degrades with time, leading to a loss in its fertility and soil erosion. Hence, the product formed from biowaste could be used as a soil amendment by providing nutrients and improving soil structure. It could act as a low-cost input for low-middle-income farmers and reduce their need for chemical fertilizer (Smith et al., 2015). Apart from this, the end product could also be used as a food material for animal feed. With the growing population, the demand for meat and milk is increasing and is causing a burden on animal feed. This results in the rise of conventional feed material prices, which further triggers the need to develop low-cost alternative feed sources. Nutrient-rich products like worms, larvae, and organic waste formed from biowaste treatment could be served as alternative feed in animal husbandry (Makkar et al., 2014). The utilization of generated bioenergy

**TABLE 1.1**
**Different Biowaste Treatment Methods Used to Generate Valuable Products**

| Classification | Technology | Products | End Use | Advantages | Disadvantages |
|---|---|---|---|---|---|
| Direct use | Direct land application | | Crop production | Soil fertility, increases crop yield | Trace metal contamination/bioaccumulation; soil pollution |
| | Direct animal feed | | Meat/fish production | Low-cost animal feed | Risk of transfer of pathogens or contaminants |
| | Direct combustion | | Bioenergy | Energy production, complete removal of pathogen | Air pollution, negative environmental impact |
| Biological treatment | Composting and vermicompost | Compost | Crop production | Less energy-consuming, cost-effective, sustainable | Time-consuming process and releases odors |
| | | Worm | Meat/fish production | | |
| | Anaerobic digestion | Digestate | Crop production | Green energy, sustainable, digestate as organic fertilizer | Poor design selection, lack of maintenance, poor quality of feedstock, and lack of demand |
| | | Biogas | Bioenergy | | |
| Physicochemical treatment | Transesterification | Biodiesel | Bioenergy | Mixed with motor vehicle fuels; less harmful | Expensive, difficult to scale up |
| | Densification | Pellets/briquettes | Bioenergy | Alternate to wood, easy to use | Air pollution |
| Thermochemical treatment | Pyrolysis | Char and bio-oil | Bioenergy | Solid fuel, absorbent material, sustainable | Expensive, depending upon the quality of biowaste |
| | Liquefaction | Bio-oil | Bioenergy | Highly valuable bio-oil extraction | Expensive, difficult to scale up |

could be another endpoint use of the solid biowaste treatment. The bioenergy developed from the biowaste can replace the conventional sources for various applications like cooking, transport, and electricity production (Lohri et al., 2016).

### 1.2.1 Direct Use

The direct use of solid biowaste is the simplest and low-cost method for the proper use of discarded biowaste. This "direct use" category includes direct land application, direct animal feed, and direct combustion. However, this method contains certain risks, including contamination, toxic compound formation, and environmental pollution.

#### 1.2.1.1 Direct Land Application

Direct land application refers to the practice of direct spreading of raw organic wastes on the land or agricultural field. In rural areas, this practice is commonly used by people to spread the manure of cattle on their farmland. It improves the nutrient value of soil, making it more fertile for crops. After spreading on the land, the raw organic waste undergoes aerobic biodegradation, leading to the mobilization of the nutrients in the soil. However, the microbial degradation of the organic waste sometimes causes competition for nitrogen with the agricultural crop. The microbes utilize the available nitrogen in the soil for their metabolism process. Smith et al. (2015) reported that the direct use of raw organic waste causes a decrease in nitrogen availability by 66% for the growth of crops. The main benefit of direct land application is the soil fertility amendment. The high organic content in the soil helps in the growth of plants, provides energy to soil microbes, and improves the soil structure and buffering of the soil pH (Hamdi et al., 2019). On the other hand, this method also contains some negative impacts on the environment. The raw biowaste is likely to have a small concentration of trace metals, which would bioaccumulate in the plant or microbes (Olowolafe, 2008; Walsh and McDonnell, 2012). This would result in a harmful impact on human health or can cause pollution in soil or water.

#### 1.2.1.2 Direct Animal Feed

From the beginning of animal domestication, people have been feeding raw organic waste to animals like swine, poultry, and fish (Cheng and Lo, 2016). Organic waste is generally rich in carbohydrates, amino acids, minerals, fibers, and fats. This biowaste is consumed and metabolized by the animals, which helps in their growth. However, the application of biowaste generally depends upon the animal's digestive system. Ruminants have a complex digestive system that can digest cellulosic materials like straw and grass, whereas the digestive system of swine cannot digest them (Lohri et al., 2017). Moreover, some risks may be associated with transferring pathogens in the raw waste to the animals, resulting in animal infections.

#### 1.2.1.3 Direct Combustion

It is commonly known as the open burning of raw biowaste. The main benefits associated with direct combustion are a reduction in the volume of biowaste and the complete removal of pathogens. However, uncontrolled open burning of waste causes the emission of harmful gases into the local environment (Estrellan and Iino, 2010). This method is commonly used by people, but it is not considered as acceptable biowaste management because it poses a considerable threat to the environment by emission of harmful gas and incomplete combustion. In India, stubble burning in the farm field is a major concern because of the increase in the level of air pollution in the surrounding areas (Chawala and Sandhu, 2020).

### 1.2.2 Biological Treatment

Biological treatment refers to the controlled decomposition of biowaste by the living organism. These processes are considered slow compared to thermochemical conversion but are low cost and eco-friendly. There are several types of biological treatment, as discussed below.

#### 1.2.2.1 Composting and Vermicompost

The composting process refers to the aerobic decomposition of biowaste into stable organic compounds called compost or humus (Sayara et al., 2020). This process is driven by a wide range of microorganisms under optimum conditions. Under the ideal optimum composition, the biowaste undergoes three stages of decomposition: mesophilic stage, thermophilic stage, and cooling and maturation stage (Lohri et al., 2017). In the end stage, the inner temperature of compost material is like ambient temperature. The final compost material is stable, with dark brown color, crumbly texture, and earthy smell. It contains various essential plant nutrients such as nitrogen, phosphorus, and potassium (Edgerton, 2009). It is also used to amend the soil structure to improve crop yield or land remediation. However, the quality of compost is highly dependent upon the quality of biowaste. Further, this process is also slow and releases odors. It generally requires 2–3 months for the making of final compost from biowaste. This leads to poor acceptance by the people (Lohri et al., 2017).

Like composting, vermicomposting is also the aerobic decomposition of biowaste using microbes and earthworms under controlled conditions. During the process, the biowaste is partially degraded by the microorganism, and then, earthworm feeds on it and generates casting or vermicompost (Gudeta et al., 2022). Vermicompost contains more nutrients than compost (having 2% more nitrogen than compost). The selection of earthworm species is an important parameter (Adhikary, 2012). It is because of their adaptability for different wastes, feeding rate, digestion, growth, and reproductive rate that would affect the overall vermicomposting process. Widely used earthworm species are *Eisenia fetida*, *Lumbricus rubellus*, *Eudrilus eugeniae*, *Eisenia andrei*, and *Perionyx excavatus* (Lohri et al., 2017). Other parameters affecting the vermicomposting rate are temperature, earthworm density, biowaste composition and feeding rate, incubation time, moisture content, and pH. This process is considered less energy-consuming, cost-effective, sustainable, and economically feasible for biowaste management.

#### 1.2.2.2 Anaerobic Digestion

Anaerobic digestion (AD) is a well-designed biochemical process for the decomposition of biowaste by the range of bacteria under anaerobic or anoxic conditions. Biomethane production is the major output that could be considered green energy (Vögeli et al., 2014). The feedstock for AD includes domestic sewage sludge, animal manure, food industry waste, crop residue, and other organic fractions of urban or rural waste. The anaerobic process must be conducted under moist conditions (more than 60% of water). The efficiency of AD can be controlled by different operational parameters, including pH, moisture content, substrate composition, C/N ratio, feeding rate, temperature, and incubation time (Appels et al., 2011). However, the process efficiency could also be reduced by the acidification of the medium that inhibits the functioning and growth of methanogenic bacteria. The main reason for acidification is the biodegradation of organic matter into small acids or volatile fatty acids (VFAs) that results in the decrease in pH of reactor and inhibits the functioning of methanogenic bacteria. Previous studies reported the composition of bacterial species during the different metabolic stages of AD (Chai et al., 2016; Tyagi et al., 2021). In the initial stages of digestion, i.e., hydrolysis and acidogenesis, bacterial genera of Bacteroides, *Butyrivibrio*, *Eubacterium clostridium*, and *Lactobacillus* were dominant in the microbial community for the transformation of complex organic substances into small fatty acids, sugar, VFAs, alcohols, etc. (Pagliano et al., 2017). It was followed by the acetogenesis stage where bacterial species (such as *Acetogenium* spp. and *Syntrophococcus* spp.) are responsible for the conversion of fatty acids and alcohol into acetate (Wang et al., 2023). The last stage, i.e., methanogenesis, responsible for the transformation of acetate to methane, is led by the methanogenic bacterial community. Bacterial genera such as *Methanobacterium*, *Methanoculleus*, *Methacholine*, *Methanospirillum*, *Methanosphaerula*, *Methanobrevibacter*, and *Methanothermobacter* are examples of methanogenic bacteria (Chai et al., 2016). Moreover, the production of ammonia and sulfide and other heavy metal compounds inhibits the methanogenic bacterial growth (Yenigün and Demirel, 2013; Zhao et al., 2010). The digestate is another byproduct (other than biogas) of the AD process that is rich in nitrogen. It is generally used as an organic fertilizer in the agricultural field to improve crop yield and also help to reduce pressure on chemical fertilizer (Möller and Müller, 2012). However, the utilization of the AD process for biowaste handling is still limited. There are several challenges associated with the efficiency of AD; for instance, poor design selection, lack of maintenance, poor quality of feedstock, and lack of market for biogas are major challenges in the establishment of AD plants.

### 1.2.3 Physicochemical Treatment

Physicochemical treatment refers to the conversion of biowaste into useful compounds using physical, thermal, or chemical processes. It includes densification, transesterification, pyrolysis, liquefaction, and gasification. The densification means the conversion of raw biowaste into pellets and briquettes using physical or mechanical pressure (Karkania et al., 2012). The biowaste used in the densification can be classified into two categories, i.e., crop waste and agro-industrial waste. The crop waste includes the residue of the crop remaining after the harvesting of paddy straw, soya straw, wheat straw, and maize straw, whereas agro-industrial waste includes waste generated during the processing of crops such as rice husk, sawdust, molasses, coffee husk, and other organic waste residues (Felfli et al., 2011). The pellets obtained after the process have a uniform size with high density and low moisture content. It can be used for domestic heating and cooking fuel as an alternative to wood.

Transesterification is a process used for the production of biodiesel using chemical reactions (Knothe, 2005). In this process, liquid biowastes such as animal fats from slaughterhouses, waste vegetable oils, and grease were used as input material. The process starts with the heating of moisture-free input material, followed by adding catalyst and alcohol under continuous mixing in a closed reactor. After a few hours, the mixture is allowed to settle at room temperature. The biodiesel is then collected from the crude mixture. Biodiesel is a yellowish liquid with high energy density (~38–45 MJ/kg). It is generally used either in neat form or mixed with petroleum-based diesel (Guo et al., 2015). However, scaling up this process is still a challenge for the industry, leading to more expenditure than petroleum-based diesel.

Pyrolysis refers to the heating of biowaste in the absence of oxygen, resulting in the production of valuable products like biochar and bio-oil. The biowastes having high carbon, low ash, and moisture content are generally preferred for the pyrolysis process (Zhuang et al., 2019). The conversion process starts with the drying of the biomass (~100°C) to release the moisture of biowaste. It is followed by exothermic dehydration of biowaste (<300°C) with the release of gases like CO, $CO_2$, and water vapor. Afterward, the heating of biowaste at high temperature (300°C–900°C) leads to biochar production. The overall conversion process includes various reactions like dehydration, decarboxylation, depolymerization, and aromatization (Basu, 2018). Literature reported that the efficiency of product yield in the pyrolysis process is influenced by several operating conditions such as temperature, heating time, pressure, chemical composition, and physical properties (size, density, moisture content, ash content, etc.) of the biowaste (Deng et al., 2021; Maroušek et al., 2019). The generated biochar has received attention due to its various applications, such as solid fuel, solid amendment, or absorbent material.

Hydrothermal liquefaction (HTL) is a treatment process that involves the processing of biowaste in hot and high-pressure water environments to break the solid biowaste into liquid components like bio-oil or bio-crude (Elliott et al., 2015). The wet biowaste can directly be used as input material without any energy-consuming drying stage. However, most of the studies reported the demonstration of the HTL process at a bench scale for a short period of time (Lee et al., 2016). Moreover, the economic feasibility of the process is a challenge due to the high cost of the reactor and its operation.

## 1.3 SOLID BIOWASTE-BASED BIOREFINERIES TARGETING CIRCULAR BIOECONOMY

The development of biorefineries that could generate multiple bioproducts from SWs is emerging as an encouraging path to achieve the Sustainable Development Goals (Mohan et al., 2019). The soaring energy demands and the need to conserve biodiversity and environment have attracted the interest of the scientific communities toward exploring SWs as the source of multiple valuable products and thus developing a circular bioeconomy out of waste biorefineries. In the last decade, multiple studies have been reported on the utilization of a variety of biowastes (municipal solid wastes [MSWs], agricultural wastes, agro-industrial wastes) as the feedstock for developing biorefineries, few of which are explained below.

### 1.3.1 Municipal Solid Waste-Based Biorefineries

The MSW broadly comprises biodegradable wastes (food and green wastes) and non-biodegradable wastes (paper, glass, metals, plastics, rubbers, and electronic wastes) that come from households (~70%) and market areas (~30%). The proportions of these wastes vary in developed and developing countries as the developing nations produce more organic waste as compared to the developed ones (Nanda and Berruti, 2021). The countries that are reported to contribute to the maximum

SW generation include the United States, China, and Russia, with the United States producing 265 million tonnes of MSW annually (Dang et al., 2020). The segregation, treatment, and disposal of such vast amounts of wastes are alarming global issues. MSW management has been a major factor affecting a country's economy and its environmental sustainability. The foremost step in waste management is the segregation of waste (based on its post-treatment), which is labor-intensive and labor-expensive even with the use of automated technologies. The most widely used MSW disposal techniques are composting and landfilling, which again have deleterious effects on the environment as on the decomposition of 1 tonne of organic SW results in the release of 50–100 $m^3$ of $CO_2$ and 100–150 $m^3$ of $CH_4$, both of which are the primary global warming gases (Fermoso et al., 2018). Another technique that is used for waste management is incineration or combustion, which, in turn, produces air pollutants such as particulate matter (that causes pulmonary diseases), heavy metals (that cause neurological diseases), toxic chemicals such as dioxins and PFAS (that cause other serious health problems), etc. (Tait et al., 2020). The MSW disposal methods such as landfilling and incineration require large land areas and resources. These uni-product-based systems are economically unfavorable and do not generate return on investment for building and maintaining this facility. Alternative techniques such as waste-to-energy systems and source reduction can reduce the amount of waste sent to landfills, provide renewable energy, and decrease negative environmental impact. In view of this, researchers have investigated the use of MSW as a feedstock for multiple high-value products, which is termed as generation of "wealth out of waste."

Pérez et al. (2020) demonstrated a pilot-scale-integrated biorefinery termed as "URBIOFIN" that involved the valorization of the MSW into biobased products. With the total budget of 15 million euros, 16 European countries worked synergistically on this project that aimed at converting a linear bioeconomy into a circular bioeconomy. MSW was first segregated into organic fraction and non-organic fraction. A part of the organic fraction is hydrolyzed and fermented to produce bioethanol, and the rest is partially converted to VFAs, which are further converted to medium-chain polyhydroxyalkanoates (mc-PHA). Further, the VFA-free digestate is mixed with vinasse (which was generated from the fermentation of organic fraction of the waste) and transformed to produce biogas, to be used as biomethane. So, the three bioproducts of the developed biorefinery are bioethylene, mcl-PHA, and biomethane.

Another interesting work on the development of a biorefinery out of MSW has been reported by Nizami et al. (2017). Every year, millions of worshippers visit the city of Mecca in Saudi Arabia for pilgrimage, and consequently, an enormous amount of MSW (~4,000 tonnes per day) is generated whose management is a serious concern for the concerned authorities. A large fraction of the MSW is food (~50%) and plastic (~17%). The developed biorefinery included AD, pyrolysis, and transesterification that treated around ~80% of the MSW (producing ~2 million barrels of oil and ~11 million MCF of natural gas), and the remaining was recycled. This approach could generate 22 million US dollars from carbon credits with a reduction of ~1.1 million metric $CO_2$ eq. in global warming potential (GWP). Apart from this, if instead of landfills, a biorefinery is developed that produces electricity as one of the products, around 201 million US dollars could be generated.

### 1.3.2 Agricultural Waste-Based Biorefineries

Similar to MSW, the management of waste generated from agricultural practices (indiscriminate dumping and crop residue burning) is also a matter of serious concern as it causes pollution of the environment, which poses a threat to human health. For instance, in the process of stubble burning in Asian countries, where the crop residues are burnt, large amounts of toxic gases (methane, carbon monoxide), carcinogenic polycyclic aromatic hydrocarbons, and particulate matter are emitted into the atmosphere. These gases form toxic white clouds known as smog in winter, cause air pollution, and make it difficult for the people in nearby areas to breathe. The reason for practicing stubble burning is the lack of cost-intensive technologies that can handle the agro-waste in a short span of time (~20–30 days), i.e., between harvesting of one crop and sowing of the next. Such practices of agro-waste management need to be checked. This also calls for developing and implementing cost-effective, eco-friendly, and socially acceptable practices urgently. Agro-waste comprises leaves, roots, bagasse, straw residues, seeds, wood residues, and animal residues, and being rich in lignocellulosic mass, it can be utilized to produce second-generation biofuels (Mafe et al., 2015). Furthermore, agro-waste can also be explored to develop livestock or industrial products (biochemicals, biopesticides, biopromoters), biofertilizers, etc. (Hong et al., 2016; Lehtinen et al., 2014), using physical, thermochemical, and biochemical technologies (Kapoor et al., 2020).

Among the initial studies on waste biorefineries was a study by Nuruddin et al. (2011), who worked on an integrated forest biorefinery concept in which the agricultural wastes (corn stalks/wheat straws/rice straws, etc.) were first fractionated into cellulose, hemicellulose, and lignin using formic acid/performic fatty acid/hydrogen peroxide) giving ~90% of biomass recovery. Further, these fractions were valorized to produce biofuels, biomaterials, chemicals, etc. A very interesting study has been done by Chamarro et al. (2022) where they have reported that the agricultural and industrial wastes generated from the kiwi processing are applicable in pharmaceutics (nutraceuticals/prebiotics), cosmetics (pigments), and food industries (pectins/colorants/coagulants). Another study on waste biorefinery was done by Khounani et al. (2021) that involves the valorization of agro-wastes generated from safflower oil production. The process produced two main products: ethanol and biodiesel. The by-products of the process were biogas, lignin, liquid digestate, seed cake, sodium salts, and glycerol. All these by-products are of high value. The liquid digestate could be a nitrogen fertilizer, sodium salts as agents for manufacturing detergents, and in the kraft process for paper pulping. Furthermore, using thermochemical technique, lignin can be valorized to produce a range of value-added products: ethylbenzene (intermediate for organic synthesis), guaiacol (phenolic resin making), p-hydroxyl acetophenone (cholagogue drug making and perfume synthesis), vanillin (perfume and pharmaceutical intermediate), and carbon materials (supercapacitor and catalyst). The seed cake, on pyrolysis, can be converted to bio-oil/biochar/syngas, and glycerol can directly be utilized in the pharmaceutical, cosmetics, or food industries or used in the manufacturing of commodity chemicals.

Countries like Brazil and India, which are the leaders in producing and exporting agricultural products, have immense stocks of generated wastes that could be reused to develop biorefineries. There are projects with the framework of No Agricultural Waste (NoAW) in the United States; for instance, a project work reported by Gontard et al. (2018) involved the AD to produce biomethane along with the production of polyhydroxy butyrates, polyphenols, and fillers for use in composite materials, out of the agricultural residues (straw/manure/pomace/vine shoots). Such studies on the valorization of agro-wastes prove their potential in developing feasible biorefineries.

### 1.3.3 Food Industries' Waste-Based Biorefineries

Food industries discharge huge amounts of waste/effluents that consist of proteins, lipids, sugars, etc., making them a perfect candidate for being an affordable, renewable, and most easily available source of multiple high-value products like biofuels, specialty chemicals, and biomaterial biofertilizers (Singh et al., 2022). For utilizing the by-products of the process, food industries have become a growing area of interest.

Serrano et al. (2017) conducted a study on the valorization of olive mill solid waste (OMSW). The researchers performed the thermal pre-treatment of the OMSW at 170°C for an hour, followed by the extraction of phenols. This induced organic matter solubilization and increased phenol generation. Furthermore, AD was performed to recover energy in the form of biomethane. For 1 kg of OMSW, around 1,500 mg of hydroxytyrosol was recovered on phenol extraction. This also resulted in improved methane yield (44% increased) as the feedstock was now a de-phenolized liquid. Therefore, such combination of processes could be a sustainable route for managing lignocellulosic residues like OMSW and could help in developing a circular bioeconomy. However, care is needed to deal with the inhibitory components produced during the process that could hamper the viability of the complete system.

Another avenue where waste/residue valorization can be targeted is the dairy industry. For instance, in the cheese industry, millions of tonnes of whey are released in the effluent streams as this fraction is left out after the recovery of coagulated milk, i.e., cheese. While manufacturing 1 kg of cheese, around 10 L of whey is produced, which in small- or medium-scale industries is usually discharged with the effluents (Asunis et al., 2020). With an aim to utilize the proteins/peptides of whey, this whey can be separated and utilized instead of being discharged as dairy waste effluent. Cheese whey being highly loaded with organic content poses severe environmental concerns if released into water bodies without treatment. It can be processed to synthesize curd, cottage, or ricotta cheese along with the remaining secondary cheese whey. An interesting route to valorize the cheese whey is dark fermentation. Here, the high carbohydrate content of the cheese whey plays the main role, which is converted to hydrogen and VFAs (Akhlaghi et al., 2019). This involves the production of glucose and galactose from the hydrolysis of lactose, followed by the conversion of these sugars into lactate. Ultimately, this lactate is converted to hydrogen and VFAs by bacterial fermentation. In this way, dairy waste can be utilized for both biobased product synthesis as well as bioenergy production.

## 1.4 CASE STUDIES ON BIOWASTE-BASED BIOREFINERY MAKING A GLOBAL AND LOCAL IMPACT

The generation of bioenergy and bioproducts through the generated waste from biorefineries and clusters of the biobased industry could play a crucial role in converting a product-based bioeconomy to a circular and biobased one. The utilization of waste organic feedstock for biorefineries could exploit the untapped potential of generated organic waste into value-added products. There are several examples where the biowaste-based biorefinery approach has been applied to convert these biowastes into value-added products. Hence, in the current section, we have discussed a few case studies related to the biobased biorefinery being implemented at ground level.

### 1.4.1 Case Study of "NESTE" Using Food Waste Feedstock

Currently, the aviation industry accounts for 2%–3% of the total $CO_2$ emissions globally. Also, air travel frequency is expected to double in the next 15 years (Sarkar 2012). This would further increase the percentage of global contribution of $CO_2$ through the aviation industry. Hence, in an attempt to decrease the carbon emissions from the aviation industry, a Finnish company named "NESTE" produced sustainable aviation fuel (SAF) and high-quality biodiesel from 100% renewable sources such as waste cooking oil and waste animal and fish fats through their proprietary technology of NEXBTL. In NEXBTL technology, the raw material is first outsourced and pretreated, followed by a catalytic process called hydrodeoxygenation, replacing oxygen with hydrogen atom.

Further, depending on the intended end product, hydrocarbons undergo controlled isomerisation and branching to tailor their physicochemical properties to the specifications required for diesel, jet fuel, gasoline, or other fuels. SAF formed through this process can directly replace the traditional aviation jet fuel and decrease greenhouse gas emissions by up to 80% compared to the conventional jet fuel (NESTE Corporation 2022). Currently, one lakh tonnes of SAF are produced annually, and it is projected to increase to 1.5 million tonnes by the end of 2023 by the company (Jaganathan and Samanta 2019).

### 1.4.2 Case Study of "Novamont" Using Agri-Waste Feedstock

It has been seen that non-edible inescapable agricultural processing food waste also holds significant promise for its valorization into value-added products (Kapoor et al. 2020). "Novamont," a chemical company located in Piemonte, Italy, aims to combine chemistry with agriculture to develop an integrated biorefinery system. They mainly produce bioplastics, biofuels, and eco-friendly cosmetics to supply them in Europe. The product line of "Novamont" offers three products made from agricultural waste, i.e., MATER-BI, MATROL-BI, and CELUS-BI. The MATER-BI product is fully compostable bioplastics developed from cellulose, starch, and vegetable oil combinations. These bioplastics have similar characteristics to those of regular plastic but, at the same time, are biodegradable as per the European and international standards of WIPO. The second product offered by "Novamont" is MATROL-BI, which is classified as the rapidly degrading range of biolubricants and greases designed from the organic waste feedstock. It has fuel properties similar to those of fossils, but due to its rapid biodegradability (unlike fossil-based fuels), it overcomes the hazardous consequences of spilling into the environment. It can be applied to machinery located in ecologically sensitive regions such as coastlines, oceans, agricultural fields, and forests. The MATROL-BI product line consists of four products, i.e., hydraulic fluid, greases, transmission fluid, and dielectric fluids. The third product, CELUS-BI, is a range of eco-friendly and biodegradable cosmetic products such as shampoo, moisturizer, and cream, with high performance and low environmental impact. The CELUS-BI range comprises CELUS-BI esters through which sustainable body care products are developed. Another in this range is CELUS-BI microbeads, which act as exfoliating agents in cosmetics and personal care products by Novamont.

### 1.4.3 Case Study of Anaerobic Digestion-Based Biorefinery in Norway

One of the most widely used strategies for valorizing biowaste and developing a biorefinery is AD (Lytras et al., 2021). On the west coast of Norway at Skogn, Biokraft (an affiliation of a Scandinavian biogas company) installed its biogas plant to treat and valorize the waste generated after the conventional activated sludge treatment of wastewater from the pulp and paper mills to reduce its carbon footprint. Around 20,000 $m^3$ of wastewater used to be generated by the mills daily. The generated wastewater was further treated through the conventional activated sludge system, where excessive aeration was provided to degrade organic matter. Excessive aeration helped increase microbial growth by utilizing nutrients like nitrates and phosphates. These microbes use organic matter and convert it to $CO_2$. Parallelly, excessive microbial biomass growth led to the generation of waste-activated sludge, which again created the problem of sludge disposal (Donnér, 2018). The excess amount of waste-activated sludge is hard to dewater and hence incinerated, which leads to energy loss. So, to minimize the amount of sludge produced, excessive aeration was provided to degrade a large portion of waste-activated sludge along with organic matter. Though this approach solves the problem of excessive sludge generation, on the other hand, it requires higher electricity consumption and also increases the retention period during wastewater treatment. Therefore, a new wastewater treatment system was started by integrating AD with the current treatment to reduce the retention period and energy consumption required for excessive sparging to degrade excessive waste-activated sludge generated after the treatment. For the biogas production during AD, they applied the strategy of co-digestion by mixing the sludge with the fishery waste from Norway's fish aquaculture. The co-digestion approach helped improve the nutrient balance of carbon, nitrogen, and phosphorus that enhanced biogas production by stabilizing the entire digestion process (Kainthola et al. 2019). Also, bioconversion of two different types of wastes into value-added products was achieved through this approach. After digestion, the AD is dewatered, and only a solid fraction is left. Water is again reused in the activated sludge process as a nutrient supplement for microbes, and the solid fraction is used as fertilizers. Overall, through this method, 61,000 tonnes of fish waste, along with the 7.56 million liters of pulp and paper industry wastewater, is handled along with the generation of valuable biogas from them (Feiz et al. 2021). The biogas plant operating to process waste from the mill and organic fish waste can generate 125 GWh of energy per year, equivalent to about 12.5 million liters of diesel (*Pappalardo* 2018). The symbiotic association between a mill and a biogas plant could further lower energy requirements and time for treating the wastewater of mills, thus reducing their carbon footprint and providing the biogas plant with a constant supply of feedstock for biogas production.

### 1.4.4 Case Studies on Food Waste-Based Biorefinery

According to the United Nations Food and Agriculture Organization (FAO) report, one-third of the total food is lost during the agricultural harvesting and supply chain, leading to an estimated loss of 1 trillion USD annually (FAO 2011). Hence, the organic feedstock generated through food waste holds an immense scope for its upscaling and valorizing into value-added products.

Several active biowaste-based biorefinery ventures present globally support sustainability and a circular bioeconomy (Table 1.2). In San Francisco, a company named "ReGrained" used spent grain from the brewery industry to make its product SuperGrain+® *ReGrained.* It is a sustainable way to use the spent grain waste generated in breweries as a protein source. During the brewing process, most of the sugar is processed. However, a significant part of protein and micronutrients still exists in the spent grain, which was utilized to upscale the waste spent grain into protein-rich food products like puffs, nutrition bars, etc. In this regard, another Bengaluru-based company, "Loopworm," adopted a recycling approach to bring food waste into a circular loop. They aim to convert food waste into protein-rich insect feed comprised of black soldier larvae for fish, shrimps, poultry, and pets. Insect-based oil and animal feeds are rich sources of protein and other essential nutrients for the healthy growth of animals *Loopworm.* Another company named "Bio-Beans," located in London, England, industrialized the process of utilizing waste coffee grounds into biofuel and biomass pellets. In 2013, in Cambridgeshire, they built the world's first waste coffee recycling company. The waste coffee ground is collected from nearby coffee factories, offices, and coffee shops. This accumulated coffee waste is further converted into second-generation biofuel and biomass pellets, which are solid dry pellets and are used for burning in local broilers as a renewable heat alternative (*CNBC*, 2015).

## 1.5 SUSTAINABLE WASTE BIOREFINERY AND CIRCULAR BIOECONOMY

As we discussed in the above sections, organic/bio-SWM practices at large scale are expensive and complex, and they must be developed enough to compete for resources with other priority sectors like clean water, education, and health care. Generally, local governments cater the primal responsibility of handling SWM. However, they do not allocate sufficient funds for proper management of MSW and have lesser expertise to plan, manage, and monitor the MSW contracts and operations. Taking account of all these aspects, following sustainable SWM practices is a costly affair especially for low- and middle-income countries parallelly pursuing economic development. Poor waste management has devastating effects that disproportionately affect the poor, who are frequently uninvolved in or have little control over the formal or informal waste disposal processes that take place close to their homes.

As the organic/biowaste constitutes the majority proportion of total waste, the closed-loop biorefinery approach possesses potential for cost recovery and achieves circular bioeconomy. In the coming decades, biorefineries and circular bioeconomy approaches may offer solutions to global problems such as

- Waste separation is possible for rural communities to accomplish sustainable rural development with the support of biowaste since it is a source of energy (methane, cook stove, etc.) or material recovery (fertilizer, animal feed, etc.). Under such acquittance, composting or AD looks highly logical and economically practical when waste is segregated at the point of generation itself and when the other essential requirements are successfully handled, such as financing.
- Utilization is among the most crucial variables that influence activities linked to the creative design, development, and marketing of biomass-based products. The activities involved in the transformation of biomass are coordinated through the supply chain. Through the implementation of plans for SW biorefineries, countries with middle and low incomes will

**TABLE 1.2**
**A Few Companies That Utilize Biowaste to Produce Value-Added Products for Commercial Applications**

| Sr. No. | Company | Waste Types | Biorefinery | Location | Company/Firm |
|---|---|---|---|---|---|
| 1 | PROteINSECT | Food waste | Rearing insects from food waste as an alternative source of protein for animal feed | Scotland | PROteINSECT |
| 2 | CelluComp | Vegetable waste | Utilizing carrots and other root vegetables to develop the product Curran (microfibrillated cellulose) from its carbohydrate | Burntisland, United Kingdom | CelluComp |
| 3 | The Jaw Brewery Limited and Thomas Auld and Sons Ltd. (Auld's Bakery) | Bakery waste | Producing beer from surplus bread | Glasgow, Scotland | Glasgow Live |
| 4 | Bio-Beans | Coffee grounds | In collaboration with Costa Coffee, their 3,000 tonnes of coffee waste is converted into biofuels and biochemical and new solid fuels | London | Bio-Beans |
| 5 | NESTE | Waste cooking oil | Converting waste cooking oil and waste animal and fish fats into biofuels and biodiesel | Finland | NESTE |
| 6 | VEOLIA | Residual sludge from the sewage treatment plant | Producing biogas by the processes of thermal hydrolysis and anaerobic digestion | Paris, France | VEOLIA |
| 7 | Genecis | Food waste | Utilizing bacterial activity to convert food waste to produce bioplastics | Toronto, Canada | Genecis |
| 8 | Loopworm | Food waste | Rearing insects from food waste as an alternative source of protein for animal feed | Bengaluru, India | Loopworm |
| 9 | Barnana | Food waste | Converting unfit, overripe, and bruised bananas, which are left to be rotten at plantation into dehydrated banana snacks | California, United States | Barnana |
| 10 | Arbor Renewable Gas | Forest residue and waste wood | Converting into gasoline, aviation fuel, and hydrogen | Texas, United States | Recycling startups |
| 11 | Beta-renewable | Agricultural waste | Producing second-generation sugars from feedstock that may then be transformed into biofuels or biobased compounds | Italy | Beta-renewable |

continue to be sustained by low collection fees; yet, their SW clearance efficiency is low and that needs to be taken care of.

- Emission of methane to the environment, which is a solid greenhouse gas, is a major concern in the present day, which is a result of uncontrolled/regulated SWM practices. Almost 90% of all solid biowaste emissions, or roughly 18% of global anthropogenic methane emissions, are produced by methane from landfills. This methane, a recoverable form of energy, can be trapped through optimized AD processes in biowaste treatment plants. The optimized anaerobic co-digestion for vitality recovery is important in such cases that improves natural performance while reducing carbon footprint.
- Lack of connection between the production of garbage in cities and the planned consumption of goods from circular economies.

In the near future, two prime workable frameworks for the circular economy system have been proposed earlier for evaluating the impact of application of biowaste biorefineries. The framework includes (1) involving financing niche markets to expand the bioeconomy and (2) lowering transportation costs related to distributing biomass feedstocks. By securing the specialized financing trade and streamlining the transportation, a sustainable yield of products from biowaste is achievable, and certainly in the coming decades, biomass biorefineries may offer a sustainable fixture:

- To convert existing linear economic prototype models to closed-loop circular bioeconomic systems.
- To create varied products that are related to the bioresource/biowaste location in order to facilitate flow and preserve product stream order under the varying effects of bioresources.
- To enhance the biosynthesis of low-quality bioproducts into higher-quality bioproducts, which would reinforce the commercialization function.
- Genetic bioengineering techniques may aid in boosting biomass mixture yields through faster production rates and customized biochemical structures. By transforming it into superior-grade bioproducts and bioenergy, this will considerably increase the economics and viability of biowaste. The regenerative bioindustry creates complex bioproducts like collagen, emulsifiers, surfactants, thickeners, colors, and anticancer medications. These biocosmetics can be produced using microbial materials, biomass, or algal biomass.

## 1.6 CONCLUSIONS AND PERSPECTIVES

Countries worldwide generate significant quantities of biowaste, which can serve as potential feedstock for biorefineries. However, determining if valorizing these wastes is feasible depends on factors like (1) regional availability of waste—the type and quantity of waste vary across regions due to differences in agricultural, industrial, and urban activities; (2) logistics and transportation—the cost of collecting, transporting, and processing waste can make or break the feasibility of a biorefinery (remote or sparsely populated regions may find it economically challenging to transport biowaste to central processing facilities); and (3) waste quality and composition of waste—not all biowaste—are of equal value as different types of waste (agricultural residues, industrial by-products, food waste, etc.) have varying chemical compositions, which influence their suitability for conversion into biobased products. An important role in deciding the development of waste-based biorefineries is played by the national and regional policies on waste management, energy, and environmental protection. Life cycle assessment (LCA) and techno-economic analysis (TEA) of the waste biorefineries are essential to understand their environmental impacts and to evaluate whether the biorefinery is financially viable. While small-scale biorefineries may be easier to manage, they may not achieve the economies of scale needed for economic viability. TEA should consider variables like fluctuating feedstock prices, market demand for bioproducts, and changes in regulatory incentives (e.g., subsidies or carbon credits). In summary, some key queries that need to be addressed while planning to develop a waste biorefinery should be the following: Can the waste valorization technology operate efficiently using the available waste? Does it meet the desired product yield and quality? Is the biorefinery profitable, considering the costs and potential revenues? How will local communities perceive and benefit from the biorefinery? Will it create jobs, and how will it affect public health and safety? What are the risks associated with the proposed technology and how can they be addressed? What are the long-term environmental impacts of the biorefinery? Does it reduce emissions, minimize waste, and conserve resources? How well is the technology aligned with the UN Sustainable Development Goals? To comprehensively answer these questions, a detailed investigation on the technical, environmental, economic, social, and political feasibility of a planned waste biorefinery is essential.

## BIBLIOGRAPHY


Adhikary, S. (2012). Vermicompost, the story of organic gold: A review. *Agricultural Science*. https://doi.org/10.4236/as.2012.37110.

Akhlaghi, M., Boni, M. R., Polettini, A., Pomi, R., Rossi, A., De Gioannis, G., Muntoni, A., & Spiga, D. (2019). Fermentative $H_2$ production from food waste: Parametric analysis of factor effects. *Bioresource Technology*, *276*, 349–360.

Appels, L., Lauwers, J., De Grve, J., Helsen, L., Lievens, B., Willems, K., Van Impe, J., Dewil, R. (2011). Anaerobic digestion in global bio-energy production: Potential and research challenges. *Renewable and Sustainable Energy Reviews*. https://doi.org/10.1016/j.rser.2011.07.121.

Asunis, F., De Gioannis, G., Dessì, P., Isipato, M., Lens, P. N., Muntoni, A., Spiga, D., & Pomi, R. (2020). The dairy biorefinery: Integrating treatment processes for cheese whey valorisation. *Journal of Environmental Management*, *276*, 111240.

Baiano, A. (2014). Recovery of biomolecules from food wastes: A review. *Molecules*, *19*(9), 14821–14842.

Barnana. (2022). Barnana. https://barnana.com/ (accessed on 7 September 2022).

Basu, P. (2018). *Biomass Gasification, Pyrolysis and Torrefaction: Practical Design and Theory*. Elsevier. https://doi.org/10.1016/C2016-0-04056-1.

Beta Renewables. (2022). What is PROESA? https://www.betarenewables.com/it/proesa/che-cos-e-proesa/ (accessed on 7 September 2022).

Bio-Bean Ltd. (2022). Coffee recycling. https://theearthview.in/bio-bean-now-biofuel-from-recycled-coffee-waste/ (accessed on 12 September 2022).

CelluComp Ltd. (2022). Curran. https://www.cellucomp.com/ (accessed on 2 September 2022).

Chai, X., Tonjes, D. J., & Mahajan, D. (2016). Methane emissions as energy reservoir: Context, scope, causes and mitigation strategies. *Progress in Energy and Combustion Science*, *56*, 33–70.

Chamorro, F., Carpena, M., Fraga-Corral, M., Echave, J., Rajoka, M. S. R., Barba, F. J., Chao, H., Xianbo, J., & Simal-Gandara, J. (2022). Valorization of kiwi agricultural waste and industry by-products by recovering bioactive compounds and applications as food additives: A circular economy model. *Food Chemistry*, *370*, 131315.

Chawala, P., & Sandhu, H. A. S. (2020). Stubble burn area estimation and its impact on ambient air quality of Patiala & Ludhiana district, Punjab, India. *Heliyon*, *6*(1), e03095. https://doi.org/10.1016/j.heliyon.2019.e03095.

Cheng, J. Y. K., & Lo, I. M. C. (2016). Investigation of the available technologies and their feasibility for the conversion of food waste into fish feed in Hong Kong. *Environmental Science and Pollution Research*. https://doi.org/10.1007/s11356-015-4668-3.

CNBC. (2015). Grounds for optimism: Turning coffee into fuel. https://www.cnbc.com/2015/01/09/grounds-for-optimism-turning-coffee-into-fuel.html (accessed on 15 September 2022)

Dang, M., Chai, J., Xu, Z., Qin, Y., Cao, J., & Liu, F. (2020). Soil water characteristic curve test and saturated-unsaturated seepage analysis in Jiangcungou municipal solid waste landfill, China. *Engineering Geology*, *264*, 105374.

de Sousa, M. H., da Silva, A. S. F., Correia, R. C., Leite, N. P., Bueno, C. E. G., dos Santos Pinheiro, R. L., de Santana, J. S., da Silva, J. L., Sales, A. T., de Souza, C. C., Aquino, K. A. S., de Souza, R. B., Pinheiro, I. O., Henríquez, J. R., Schuler, A. R. P., Sampaio, E. V. S. B., Dutra, E. D., & Menezes, R. S. C. (2022). Valorizing municipal organic waste to produce biodiesel, biogas, organic fertilizer, and value-added chemicals: An integrated biorefinery approach. *Biomass Conversion and Biorefinery*, *12*(3), 827–841.

Deng, B., Yuan, X., Siemann, E., Wang, S., Fang, H., Wang, B., Gao, Y., Shad, N., Liu, X., Zhang, W., Guo, X., & Zhang, L. (2021). Feedstock particle size and pyrolysis temperature regulate effects of biochar on soil nitrous oxide and carbon dioxide emissions. *Waste Management*. https://doi.org/10.1016/j.wasman.2020.11.015.

Donnér, D. (2018). Impact on carbon emissions applying the sustainable EffiSludge wastewater treatment concept to the Nordic pulp- and paper industry (Dissertation). https://urn.kb.se/resolve?urn=urn:nbn:se:liu:diva-148610.

Edgerton, M. D. (2009). Increasing crop productivity to meet global needs for feed, food, and fuel. *Plant Physiology*. https://doi.org/10.1104/pp.108.130195.

Elliott, D. C., Biller, P., Ross, A. B., Schmidt, A. J., & Jones, S. B. (2015). Hydrothermal liquefaction of biomass: Developments from batch to continuous process. *Bioresource Technology*. https://doi.org/10.1016/j.biortech.2014.09.132.

Estrellan, C. R., & Iino, F. (2010). Toxic emissions from open burning. *Chemosphere*. https://doi.org/10.1016/j.chemosphere.2010.03.057.

Feiz, R., Larsson, M., Ekstrand, E., Hagman, L., Ometto, F., & Tonderski, K. (2021). The role of biogas solutions for enhanced nutrient recovery in biobased industries—Three case studies from different industrial sectors. *Resources, Conservation and Recycling*, *175*, 105897. https://doi.org/10.1016/j.resconrec.2021.105897.

Felfli, F. F., Mesa P. J. M., Rocha, J. D., Filippetto, D., Luengo, C. A., & Pippo, W. A. (2011). Biomass briquetting and its perspectives in Brazil. *Biomass and Bioenergy*. https://doi.org/10.1016/j.biombioe.2010.08.011.

Fermoso, F. G., Serrano, A., Alonso-Farinas, B., Fernandez-Bolanos, J., Borja, R., & Rodriguez-Gutierrez, G. (2018). Valuable compound extraction, anaerobic digestion, and composting: A leading biorefinery approach for agricultural wastes. *Journal of Agricultural and Food Chemistry*, *66*(32), 8451–8468.

Genecis Bioindustries Inc. (2022). Genecis. https://genecis.co/ (accessed on 18 September 2022).

Glasgow Live. (2017). Award win for Glasgow's pioneering project. https://www.glasgowlive.co.uk/news/glasgow-news/award-win-glasgows-pioneeringproject-12461326/ (accessed on 12 September 2022).

Gontard, N., Sonesson, U., Birkved, M., Majone, M., Bolzonella, D., Celli, A., Hénaut, A., & Sebök, A. (2018). A research challenge vision regarding management of agricultural waste in a circular bio-based economy. *Critical Reviews in Environmental Science and Technology*, *48*(6), 614–654.

Gudeta, K., Bhagat, A., Julka, J. M., Sinha, R., Verma, R., Kumar, A., Kumari, S., Ameen, F., Bhat, S. A., Amarowicz, R., & Sharma, M. (2022). Vermicompost and its derivatives against phytopathogenic fungi in the soil: A review. *Horticulturae*, *8*. https://doi.org/10.3390/horticulturae8040311.

Guo, M., Song, W., & Buhain, J. (2015). Bioenergy and biofuels: History, status, and perspective. *Renewable and Sustainable Energy Reviews*. https://doi.org/10.1016/j.rser.2014.10.013.

Gustavsson, J., Cederberg, C., & Sonesson, U. (2011). *Global Food Losses and Food Waste: Extent, Causes and Prevention*. Food and Agriculture Organization of the United Nations.

Hamdi, H., Hechmi, S., Khelil, M. N., Zoghlami, I. R., Benzarti, S., Mokni-Tlili, S., Hassen, A., & Jedidi, N. (2019). Repetitive land application of urban sewage sludge: Effect of amendment rates and soil texture on fertility and degradation parameters. *Catena*. https://doi.org/10.1016/j.catena.2018.08.015.

Hong, J., Ren, L., Hong, J., & Xu, C. (2016). Environmental impact assessment of corn straw utilization in China. *Journal of Cleaner Production*, *112*, 1700–1708.

Jaganathan, J., & Samanta, K. (2019, July 31). Finland's Neste expands Singapore refinery as it taps renewable growth. *Reuters*. https://www.reuters.com/article/us-singapore-neste-interviewidUSKCN1UQ0OW.

Kainthola, J., Kalamdhad, A. S., & Goud, V. V. (2019). Enhanced methane production from anaerobic codigestion of rice straw and hydrilla verticillata and its kinetic analysis. *Biomass and Bioenergy*, *125*, 8–16.

Kapoor, R., Ghosh, P., Kumar, M., Sengupta, S., Gupta, A., Kumar, S. S., Vijay, V. K., Kumar, V., & Pant, D. (2020). Valorization of agricultural waste for biogas based circular bioeconomy in India: A research outlook. *Bioresource Technology*, *304*, 123036.

Karkania, V., Fanara, E., & Zabaniotou, A. (2012). Review of sustainable biomass pellets production: A study for agricultural residues pellets' market in Greece. *Renewable and Sustainable Energy Reviews*. https://doi.org/10.1016/j.rser.2011.11.028.

Kaza, S., Yao, L., Bhada-Tata, P., & Van Woerden, F. (2018). *What a Waste 2.0: A Global Snapshot of Solid Waste Management to 2050*. World Bank Publications.

Khounani, Z., Hosseinzadeh-Bandbafha, H., Nazemi, F., Shaeifi, M., Karimi, K., Tabatabaei, Mourtaza, A., & Lam, S. S. (2021). Exergy analysis of a whole-crop safflower biorefinery: A step towards reducing agricultural wastes in a sustainable manner. *Journal of Environmental Management*, *279*, 111822.

Knothe, G. (2005). The history of vegetable oil-based diesel fuels. In G. Knothe, J. Van Gerpen, & J. Krahl (Eds.), *The Biodiesel Handbook* (pp. 5–15). AOCS Press. https://doi.org/10.1201/9781003040262

Lee, A., Lewis, D., Kalaitzidis, T., & Ashman, P. (2016). Technical issues in the large-scale hydrothermal liquefaction of microalgal biomass to biocrude. *Current Opinion in Biotechnology*. https://doi.org/10.1016/j.copbio.2016.01.004

Lehtinen, T., Schlatter, N., Baumgarten, A., Bechini, L., Krüger, J., Grignani, C., ... & Spiegel, H. (2014). Effect of crop residue incorporation on soil organic carbon and greenhouse gas emissions in European agricultural soils. *Soil Use and Management, 30*(4), 524–538.

Lohri, C. R., Diener, S., Zabaleta, I., Mertenat, A., & Zurbrügg, C. (2017). Treatment technologies for urban solid biowaste to create value products: A review with focus on low- and middle-income settings. *Reviews in Environmental Science and Biotechnology, 16*, 81–130. https://doi.org/10.1007/s11157-017-9422-5.

Lohri, C. R., Rajabu, H. M., Sweeney, D. J., & Zurbrügg, C. (2016). Char fuel production in developing countries - A review of urban biowaste carbonization. *Renewable and Sustainable Energy Reviews*. https://doi.org/10.1016/j.rser.2016.01.088.

Loopworm Pvt. Ltd. (2022). About Loopworm. https://loopworm.in/ (accessed on 21 September 2022).

Lytras, G., Lytras, C., Mathioudakis, D., Papadopoulou, K., & Lyberatos, G. (2021). Food waste valorization based on anaerobic digestion. *Waste and Biomass Valorization, 12*(4), 1677–1697.

Mafe, O. A., Davies, S. M., Hancock, J., & Du, C. (2015). Development of an estimation model for the evaluation of the energy requirement of dilute acid pretreatments of biomass. *Biomass and Bioenergy, 72*, 28–38.

Makkar, H. P. S., Tran, G., Heuzé, V., & Ankers, P. (2014). State-of-the-art on use of insects as animal feed. *Animal Feed Science and Technology*. https://doi.org/10.1016/j.anifeedsci.2014.07.008.

Maroušek, J., Strunecký, O., & Stehel, V. (2019). Biochar farming: Defining economically perspective applications. *Clean Technologies and Environmental Policy*. https://doi.org/10.1007/s10098-019-01728-7.

Mohan, S. V., Dahiya, S., Amulya, K., Katakojwala, R., & Vanitha, T. K. (2019). Can circular bioeconomy be fueled by waste biorefineries: A closer look. *Bioresource Technology Reports, 7*, 100277.

Möller, K., & Müller, T. (2012). Effects of anaerobic digestion on digestate nutrient availability and crop growth: A review. *Engineering in Life Sciences*. https://doi.org/10.1002/elsc.201100085.

Nanda, S., & Berruti, F. (2021). A technical review of bioenergy and resource recovery from municipal solid waste. *Journal of Hazardous Materials, 403*, 123970.

Neste Corporation (2022a). First flight in history with 100% sustainable aviation fuel on a regional commercial aircraft. https://www.neste.com/releases-and-news/renewable-solutions/first-flight-history-100-sustainable-aviation-fuel-regional-commercial-aircraft/ (accessed on 21 September 2022).

Neste Corporation. (2022b). Sustainable aviation fuel (SAF). https://www.neste.com/products/all-products/saf#590cbc04/ (accessed on 21 September 2022).

Nizami, A. S., Shahzad, K., Rehan, M., Ouda, O. K. M., Khan, M. Z., Ismail, I. M. I., Almeelbi, T., & Demirbas, A. (2017). Developing waste biorefinery in Makkah: A way forward to convert urban waste into renewable energy. *Applied Energy, 186*, 189–196.

Novamont S.p.A. (2022). Novamont. https://www.novamont.com/ (accessed on 2 September 2022).

Nuruddin, M., Chowdhury, A., Haque, S. A., Rahman, M., Farhad, S. F., Jahan, M. S., & Quaiyyum, A. (2011). Extraction and characterization of cellulose microfibrils from agricultural wastes in an integrated biorefinery initiative. *Biomaterials, 3*, 5–6.

Olowolafe, E. A. (2008). Effects of using municipal waste as fertilizer on soil properties in Jos area, Nigeria. *Resources, Conservation & Recycling*. https://doi.org/10.1016/j.resconrec.2008.07.005

Pagliano, G., Ventorino, V., Panico, A., & Pepe, O. (2017). Integrated systems for biopolymers and bioenergy production from organic waste and by-products: A review of microbial processes. *Biotechnology for Biofuels, 10*(1), 113.

Pappalardo (2018, 18 July). Effisludge: Biogas from the pulp and paper industry. *BESustainable*. https://www.besustainablemagazine.com/cms2/effisludge-biogas-from-paper-and-pulp-industry/.

Pérez, V., Pascual, A., Rodrigo, A., Torreiro, M. G., Latorre-Sánchez, M., Lozano, C. C., Fernández, C., & Muñoz, R. (2020). Integrated innovative biorefinery for the transformation of municipal solid waste into biobased products. In R. Chandra (Ed.), *Waste Biorefinery* (pp. 41–80). Elsevier.

PROteINSECT. (2022). *PROteINSECT*. https://www.proteinsect.eu/ (accessed on 2 September 2022).

Recycling Startup. (2022). Top bio-waste startups. https://www.recyclingstartups.org/top/bio-waste/ (accessed on 2 September 2022).

Regrained. (2022). SuperGrain+ page. https://www.regrained.com/pages/supergrain-page/ (accessed on 21 September 2022).

Sarkar, A. N. (2012). Evolving green aviation transport system: A holistic approach to sustainable green market development. *Journal of Transport Literature, 6*(3), 143–167.

Sayara, T., Basheer-Salimia, R., Hawamde, F., & Sánchez, A. (2020). Recycling of organic wastes through composting: Process performance and compost application in agriculture. *Agronomy*. https://doi.org/10.3390/agronomy10111838.

Serrano, A., Fermoso, F. G., Alonso-Fariñas, B., Rodríguez-Gutierrez, G., Fernandez-Bolaños, J., & Borja, R. (2017). Olive mill solid waste biorefinery: High-temperature thermal pre-treatment for phenol recovery and biomethanization. *Journal of Cleaner Production*, *148*, 314–323.

Singh, T. A., Sharma, M., Sharma, M., Sharma, G. D., Passari, A. K., & Bhasin, S. (2022). Valorization of agro-industrial residues for production of commercial biorefinery products. *Fuel*, *322*, 124284.

Smith, J. U., Fischer, A., Hallett, P. D., Homans, H. Y., Smith, P., Abdul-Salam, Y., Emmerling, H. H., & Phimister, E. (2015). Sustainable use of organic resources for bioenergy, food and water provision in rural Sub-Saharan Africa. *Renewable and Sustainable Energy Reviews*. https://doi.org/10.1016/j.rser.2015.04.071.

Tait, P. W., Brew, J., Che, A., Costanzo, A., Danyluk, A., Davis, M., Khalaf, A., McMahon, K., Watson, A., & Bowles, D. (2020). The health impacts of waste incineration: A systematic review. *Australian and New Zealand Journal of Public Health*, *44*(1), 40–48.

Tengerdy, R. P., & Szakacs, G. (2003). Bioconversion of lignocellulose in solid substrate fermentation. *Biochemical Engineering Journal*, *13*(2–3), 169–179.

Tyagi, L., Dhodapkar, R., & Patil, M. (2025). Exploring the potential of fruit peels and sheep feces-derived garbage enzymes for bio-hydrogen production from sewage sludge. *Biomass Conversion and Biorefinery*, *15*(21), 28239–28252.

Veolia. (2022). Veolia group profile. https://www.veolia.com/en (accessed on 15 September 2022).

Vergara, S. E., & Tchobanoglous, G. (2012). Municipal solid waste and the environment: a global perspective. *Annual Review of Environment and Resources*, *37*, 277–309.

Vögeli, Y., Riu, C., Gallardo, A., Diener, S., & Zurbrügg, C. (2014). *Anaerobic Digestion of Biowaste in Developing Countries: Practical Information and Case Studies*. Department of Water and Sanitation in Developing Countries.

Walsh, E., & McDonnell, K. P. (2012). The influence of added organic matter on soil physical, chemical, and biological properties: A small-scale and short-time experiment using straw. *Archives of Agronomy and Soil Science*. https://doi.org/10.1080/03650340.2012.697999.

Wang, K., Wu, Z., Du, J., Liu, Y., Zhu, Z., Feng, P., ... & Tan, T. (2023). Metabolic engineering of saccharomyces cerevisiae for conversion of formate and acetate into free fatty acids. *Fermentation*, *9*(11), 984.

Xiong, X., Iris, K. M., Tsang, D. C., Bolan, N. S., Ok, Y. S., Igalavithana, A. D., Park, J. H., Kim, K.-H., & Vikrant, K. (2019). Value-added chemicals from food supply chain wastes: State-of-the-art review and future prospects. *Chemical Engineering Journal*, *375*, 121983.

Yenigün, O., & Demirel, B. (2013). Ammonia inhibition in anaerobic digestion: A review. *Process Biochemistry*. https://doi.org/10.1016/j.procbio.2013.04.012.

Zhao, T., Zhang, L., & Zhao, Y. (2010). Study on the inhibition of methane production from anaerobic digestion of biodegradable solid waste. *Waste Management & Research*. https://doi.org/10.1177/0734242X09351180.

Zhuang, X., Zhan, H., Song, Y., Yin, X., & Wu, C. (2019). Structure-reactivity relationships of biowaste-derived hydrochar on subsequent pyrolysis and gasification performance. *Energy Conversion and Management*. https://doi.org/10.1016/j.enconman.2019.112014

# 2 Bioprocessing of Lignocellulosic Waste for Production of Fuels and Chemicals

*Runze Pan, Lu Mou, Min Qiu, Hao Gao, Yujia Jiang, and Fengxue Xin*
State Key Laboratory of Materials-Oriented Chemical Engineering, College of Biotechnology and Pharmaceutical Engineering, Nanjing Tech University, Nanjing, P.R. China

## 2.1 INTRODUCTION

Due to the rapid economic development, global energy consumption (including renewable and non-renewable energy) has increased by about 200% from 1990 to 2016. To date, oil remains the most dominant fuel source, accounting for 32.9% of global energy consumption. Coal supplies about 40% of the world's electricity. The increase in energy demand caused by urbanization, the development of electric drive technology, and transportation is taking a serious toll, and non-renewable energy reserves such as oil and coal are rapidly being depleted. However, the widespread use of these non-renewable energy sources has led to a growing threat to the environment. The development of the global economy in different sectors, such as energy, food, and agriculture, is under unprecedented pressure to address global climate issues and reduce dependence on fossil fuels. The rising cost of fuel and the overuse of fossil fuels have triggered a global challenge to find alternatives to fossil resources to produce renewable biofuels (Islam et al., 2020).

As the most abundant renewable biomass resource on Earth, lignocellulose biorefining methods have recently emerged and are currently used in a wide range of industrial processes (Abraham et al., 2020; Figure 2.1). Importantly, lignocellulose is a potential substrate for biotransformation due to its rich carbohydrate content. Advanced biofuels and useful chemicals derived from lignocellulosic biomass are expected to significantly reduce greenhouse gas emissions and increase domestic energy production in countries with sufficient biomass resources or suitable for growing energy crops (Baral et al., 2019). Studies have shown that the world produces about 181.5 billion tons of lignocellulosic biomass each year. About 8.2 billion tons of biomass is currently used: 7 billion tons from forests, agriculture, and pasture, and the remaining 1.2 billion tons from agricultural waste (Dahmen et al., 2019). In the past, this biomass was mainly used for cooking, heating, building materials, and the paper industry. However, the focus of recent research has shifted to the production of renewable biofuels and other value-added chemical production, such as organic acids, biochemicals, and bioplastics (Usmani et al., 2021). The main sources of lignocellulosic biomass are forest residues, agricultural waste, dedicated energy crops, organic municipal solid waste, and industrial waste (Roy et al., 2021). Lignocellulose generally consists of cellulose (35%–37%), hemicellulose (23%–25%), and lignin (18%–22%) (Stefanidis et al., 2014). For cellulose and hemicellulose, since most microorganisms that can synthesize chemicals cannot directly utilize these polysaccharides, it is necessary to hydrolyze lignocellulose into fermentable sugars by cellulase and hemicellulase.

 DOI: 10.1201/9781032676890-3

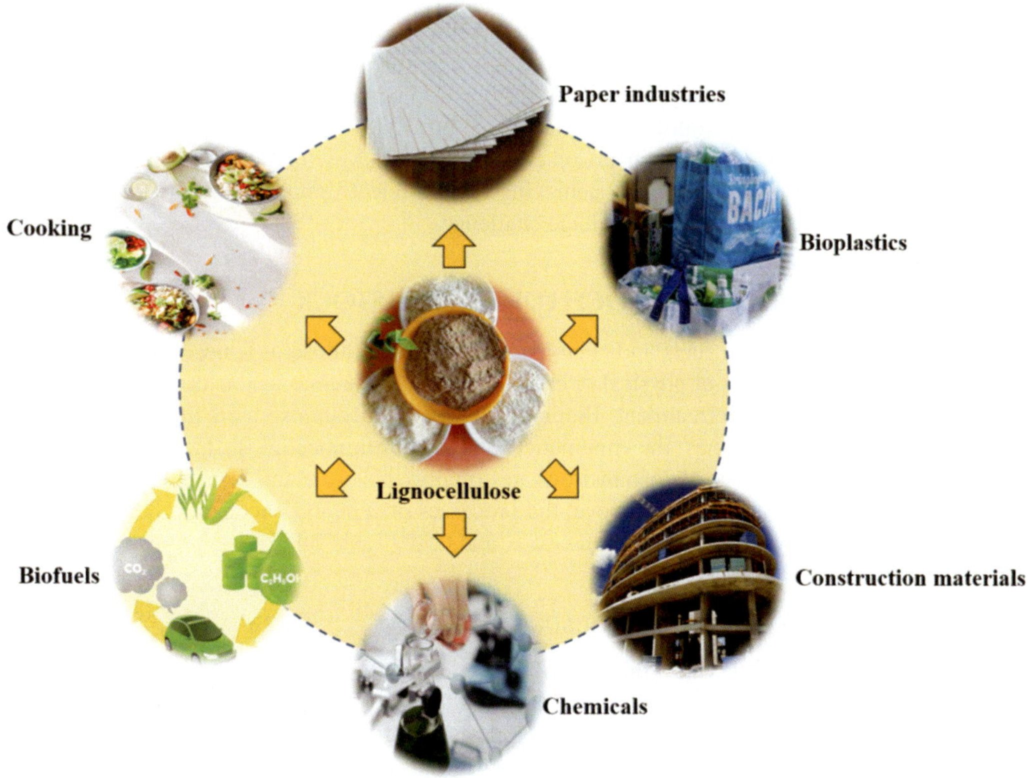


**FIGURE 2.1** Application of lignocellulosic waste.

However, lignin's complex structure and high mechanical strength hinder its biotransformation. At present, lignin is mainly used as an additive in epoxy resins, rubber, and thermoplastics, and some researchers have also begun to synthesize lignin biopolymers into many products, such as lignin micro-nanocapsules, biosensors, platform chemicals, and so on (Garlapati et al., 2020). Depending on their type, these polymers are organized into complex, non-uniform three-dimensional spatial structures that vary in relative degree and composition. The crystallinity of cellulose, the hydrophobicity of lignin, and the coating of cellulose by a lignin–hemicellulosic matrix tightly bound by covalent and hydrogen bonds help lignocellulose maintain its stable structure (Woiciechowski et al., 2020). In addition, due to the complex structure of the lignin polymer, it is difficult for microbial degradation and enzymatic hydrolysis of cellulose into sugar, resulting in low yield (Sun et al., 2016; Sheng et al., 2021a). As a result, researchers have developed a variety of pretreatment techniques to break down the complex structure of lignocellulose for further value-added chemical production. In the pretreatment technology, the removal of lignin helps to change the structure of the biomass, making the hydrolysis process easier to obtain a high yield. Various pretreatment techniques, including physical, chemical, biological, and physicochemical methods, have been used to degrade lignocellulose into fermentable sugars (Mankar et al., 2021).

Several biological processes for the production of fuels and chemicals from lignocellulosic waste have been developed, mainly including four fermentation processes, namely separate hydrolysis and fermentation (SHF), simultaneous saccharification and fermentation (SSF), simultaneous saccharification and co-fermentation (SSCF), and consolidated bioprocessing (CBP) (Lu et al., 2021). Over the past decade, SHF, SSF, SSCF, and CBP have been used to produce biofuels and biochemicals that can serve as sustainable alternatives to petrochemicals. However, SHF, SSF, and SSCF have a common disadvantage, that is, a large amount of hydrolase needs to be added during the fermentation process, which greatly increases the production cost and limits large-scale industrial production (Akhtar & Idris, 2017). Therefore, integrated bioprocessing is a promising strategy, in which hydrolase production, lignocellulosic hydrolysis, and microbial fermentation are carried out simultaneously in a single reactor (Choi et al., 2019). Compared with bioprocesses that require a separate step to produce hydrolase, CBP can reduce process-related equipment investment and operating costs, while reducing the conversion cost of lignocellulosic organisms by approximately 78% (Lynd et al., 2005). Simultaneously achieving both efficient lignocellulosic degradation and chemical synthesis in one system is also a huge challenge.

## 2.2 BIOPROCESSING OF LIGNOCELLULOSE FEEDSTOCK

In order to efficiently produce fuels and chemicals from lignocellulose, it is necessary to pretreat lignocellulose through a series of physical or chemical methods, such as acid hydrolysis, ultrasonic treatment, and ionic liquid pretreatment. Bioprocessing of lignocellulose biorefinery involves the production of hydrolytic enzymes, the enzymatic hydrolysis of lignocellulosic biomass, as well as the microbial fermentation of cellulose hydrolysates (mainly hexoses) and hemicellulose hydrolysates (mainly pentoses). As described above, the processes of fuel and chemical production from lignocellulose can be divided into SHF, SSF, SSCF, and CBP (Table 2.1).

### 2.2.1 Separate Hydrolysis and Fermentation

SHF includes two separated steps, in which lignocellulose is firstly hydrolyzed by hydrolytic enzymes, and then the hydrolyzed sugars are fermented into different target products by various functional microbes (Tavva et al., 2016). The advantage of SHF is that hydrolysis and fermentation can be carried out under their most suitable conditions, respectively (Nguyen et al., 2018).

**TABLE 2.1**
**Summary of Fuel and Chemical Production Process from Lignocellulosic Biomass**

| Substrates | Methods | Strains | Products | Titer (g/L) or Yield (g/g) | References |
|---|---|---|---|---|---|
| Waste paper and newspaper | SHF | *S. cerevisiae* | Ethanol | 11.15 and 6.65 g/L | Annamalai et al. (2020) |
| Sugarcane bagasse | SHF | *S. cerevisiae* | Ethanol | 33 g/L | Méndez et al. (2021) |
| Corn stover hydrolysate | SHF | *A. succinogenes* | Succinic acid | 56.4 g/L | Li et al. (2011) |
| Arundo donax hydrolysate | SHF | *B. succiniciproducens* | Succinic acid | 7.4 g/L | Ventorino et al. (2017) |
| Oil palm empty fruit bunches | SSF | *A. succinogenes* | Succinic acid | 42.9 g/L | Akhtar et al. (2020) |
| Sugar beet pulp | SSF | LAB | Lactic acid | 30 g/L | Berlowska et al. (2018) |
| Sweet sorghum bagasse | SSF | *S. cerevisiae* | Ethanol | 53.1 g/L | Darkwah et al. (2016) |
| Sawdust | SSCF | *Z. mobilis* | Ethanol | 0.069 g/g | Chen et al. (2017) |
| Glucose/xylose | SSCF | *S. cerevisiae* | Ethanol | 30.1 g/L | Tran et al. (2020) |
| Cellobiose | CBP | *E. coli* | Succinic acid | 26.5 g/L | Dong et al. (2017) |
| Cellobiose | CBP | *M. thermophila* | Malic acid | 79.7 g/L | Li et al. (2020) |
| Cellulose and xylose | CBP | *T. reesei*, LAB, and *C. butyricum* | Butyric acid | 0.35 g/g | Shahab et al. (2020) |

Obtaining chemicals and fuels from different lignocellulose through SHF has been widely studied (Slathia et al., 2020). A recent study related to the ethanol production from wastepaper and newspaper is a successful example by using SHF (Annamalai et al., 2020). Specifically, the wastepaper was pretreated by $H_2O_2$ (0.5% v/v) at 121°C for 30 minutes to effectively remove lignin. After that, 24.5 g/L of sugar was achieved through the enzymatic pretreatment, and the hydrolysis efficiency is 91.8%. Further, the ethanol was produced by using *Saccharomyces cerevisiae* DSM 70449 from the hydrolysate, and the ethanol production and productivity reached 11.15 g/L and 0.32 g ethanol/(L·h), respectively. Similarly, pretreated sugarcane bagasse was degraded by the enzyme mixture produced by *Trichoderma harzianum*, *Penicillium funiculosum*, and *Aspergillus niger*, and the obtained fermentable hydrolysate produced 33 g/L of ethanol by *S. cerevisiae* JP1 with 11 g/(L·h) of productivity (Méndez et al., 2021). The production of organic acids from lignocellulose by SHF has also been widely studied. For example, an industrial fermentation system using corn stover hydrolysate, economical succinic acid (SA) production by *Actinobacillus succinogenes* NJ113 was developed (Li et al., 2011). The final titer of SA was 56.4 g/L with a yield of 0.73 g/g. These results indicate that SHF is a promising strategy to achieve high productivities for fuel and chemical production.

However, due to the increase of fermentable sugar concentration in the hydrolysis stage, the feedback inhibition of the product may decrease the enzymatic reaction and result in the uncompleted hydrolyzed (Akhtar & Idris, 2017; Sharma et al., 2023). Therefore, improving the tolerance of hydrolase to high concentrations of products is the key to increasing the synthesis efficiency of products through SHF.

### 2.2.2 Simultaneous Saccharification and Fermentation

Both SHF and SSF processes need exogenous hydrolase, and SHF needs to complete saccharification and production procedures in different reactors. Compared with SHF, SSF integrates two steps in the same bioreactor, including saccharification of lignocellulose and biotransformation of hydrolyzed sugar into fuels and chemicals. It means that the feedback inhibition can be eliminated, and a higher titer and yield of the product can be obtained. At the same time, the SSF processes are carried out in the same reactor, which can reduce the equipment investment cost (Valles et al., 2020).

In several studies, it has been shown that lignocellulose can be efficiently converted into target products by SSF (Maslova et al., 2019). For example, through the optimization of SSF process parameters, the temperature, pH value, and enzyme dosage were 36°C, pH 5, and 39.7 FPU/g, respectively. The final production and yield of SA were 42.9 g/L and 0.61 g/g, respectively, by using oil palm empty as the substrate (Akhtar et al., 2020). Importantly, the SSF process commonly shows lower processing cost and higher productivity than SHF. A study showed that only a portion of soluble sugars released by the enzymes from the sugar beet pulp was assimilated by the lactic acid (LA) bacteria during SHF. By adding the substrate either continuously or at scheduled intervals during fed-batch SSF, the viscosity of the fermentation medium can be maintained at a low value as cellulosic biomass is continuously degraded into glucose and fermented into the target product (Rudolf et al., 2005). The sweet sorghum bagasse pretreated with 5% (w/w) acetic acid at an accumulated solid concentration of 20% (w/v) during the 96-h fed-batch SSF achieved a maximum ethanol concentration of 53.1 g/L with 88.7% yield, compared with 25.7 g/L for the 96-h batch SSF at a solid concentration of 10% (w/v), respectively (Darkwah et al., 2016).

However, it should be noticed that due to the different temperature requirements between the optimal temperature of enzymatic hydrolysis and fermentation, enzymatic hydrolysis usually cannot be carried out under its optimal hydrolysis temperature, which greatly limits its further application. Therefore, the key to further optimizing SSF is to screen and edit genes to construct strains with higher temperature tolerance (Lu et al., 2021).

### 2.2.3 Simultaneous Saccharification and Co-Fermentation

SSCF is developed based on SSF, enabling the saccharification of the lignocellulosic waste to react simultaneously with the co-fermentation of hexose and pentose by microorganisms, effectively increasing the final product titer and productivity by reducing end-product inhibition and affording higher solid loading (Qin et al., 2017). Gene editing technology is usually applied to construct microbial strains with the capability of simultaneous pentose and hexose utilization. Taking ethanol production as an example, wood dust has been effectively used to obtain reducing sugars (glucose, xylose, and other by-products) through the use of either a supercritical fluid extraction (SFE) or steam explosion (SE) pretreatment step. At the same time, *Zymomonas mobilis* can utilize a variety of fermentable sugars, which effectively improves the titer of the product. Using 5% (w/v) pretreated sawdust as the substrate, *Z. mobilis* was immobilized onto alginate beads, and a new SSCF bioreactor was used to produce ethanol. The final yields of ethanol were 0.069 g/g for SFE pretreatment and 0.049 g/g for SE pretreatment (Chen et al., 2017). In addition, by developing a powerful mixed sugar co-fermentation *S. cerevisiae* strain XUSEA, the transformation ability of xylose was improved in the process of simultaneous glucose/xylose co-fermentation. By further increasing the fermentation temperature, the xylose catabolism ability of the engineering strain was effectively improved, and the co-fermentation time was significantly shortened by half. Eventually, *S. cerevisiae* XUSEA achieved simultaneous co-fermentation of lignocellulosic hydrolysates composed of 39.6 g/L glucose and 23.1 g/L xylose, and 30.1 g/L ethanol, with a yield of 0.48 g/g, was obtained within 24 hours (Tran et al., 2020). Generally, owing to the efficient co-fermentation of both pentose and hexose, SSCF shows a higher yield of chemical and fuel production than the other two strategies (Sharma et al., 2021).

### 2.2.4 Consolidated Bioprocessing

CBP can complete the multi-step processes in one reactor at the same time, including hydrolase production, substrate hydrolysis, and microbial fermentation, which is the ultimate goal of biorefining of inferior biomass such as lignocellulose (Zheng et al., 2012). Chemical and fuel production directly from lignocellulose is expected to be much less expensive than a process where hydrolytic enzyme production, cellulose saccharification, and microbial fermentation are accomplished separately (Xin et al., 2019). Current research aims to endow cellulolytic microorganisms with the ability to produce target products or to endow a product producer with the ability to degrade lignocellulose. For example, a β-glucosidase encoding gene *bglA* was first cloned from *Paenibacillus* species M1, and enzymatic characterization was identified. Furthermore, the metabolically engineered *E. coli* suc260(pTbglA) was constructed by overexpression of *bglA*, which could directly utilize cellobiose as the substrate for SA production (Dong et al., 2017). Finally, 26.50 g/L SA was produced by strain suc260(pTbglA) from cellobiose with a corresponding yield of 88.30%. An efficient CBP microbial strain should not only hydrolyze lignocellulosic biomass, but also produce the target products with high titer, yield, and productivity (Olson et al., 2012). However, prokaryotes such as bacteria cannot be used to synthesize chemicals or fuels directly due to their lack of organelles and enzyme production capacity, as well as poor tolerance to the adverse environment.

Various fungi can effectively secrete cellulases; for example, *T. reesei* and *Penicillium* have been shown to produce abundant cellulases (Peterson & Nevalainen, 2012). Therefore, cellulose-degrading fungi can be used as an important platform to build a CBP platform for chemical and fuel production. For example, *Myceliophthora thermophila* is an efficient cellulose- and hemicellulose-degrading strain. Its genome encodes a large number of lignocellulose-decomposing enzymes, and it can efficiently utilize various monosaccharides produced by hydrolysis. However, the malic acid titer of the wild strain is very low. By overexpressing the malic acid export gene (*Aomae*) and the malic acid transporter encoding gene (*Aopyc*) in *M. thermophila*, the engineered strain JG207 obtained a stronger ability to synthesize malic acid than the wild-type strain (Li et al., 2020). In shaken-flask cultures, strain JG207 obtained 63.8 g/L malic acid from 75 g/L glucose, corresponding to an overall yield of 0.85 g/g carbon source. When 75 g/L cellobiose was provided as the substrate, the total titer and yield of malic acid reached 79.7 g/L and 1.06 g/g, respectively.

However, the most successful cases of using a single strain to produce chemicals and fuels from lignocellulose is only using oligo-polysaccharide compounds as the substrate. The combination of excellent lignocellulose degradation and target products biosynthesis capability into one single microbe is a challenging issue. Compared with single-strain culture, microbial communities can complete the biotransformation of complex compounds more effectively because of its inherent advantages, such as reducing metabolic burden by division of labor (Lu & Qu, 2019). In a recent study, a microbial community was established to synthesize chemicals directly from lignocellulose by co-culturing a cellulose-degrading strain and a product-synthesizing strain (Shahab et al., 2020). The microbial community consists of the aerobic fungus *T. reesei*, which secretes cellulolytic enzymes and thus provides soluble sugars, and facultative anaerobic LA bacteria, which possess excellent capability for the LA production. Furthermore, a stable three-strain system was established by adding *Clostridium butyricum* into the microbial community, in which *C. butyricum* could co-consume LA and acetic acid in a molar ratio of 3: 1 to produce butyric acid. With a mixture of cellulose and xylose as the model substrate, 0.35 g of butyric acid per gram of carbohydrate was obtained.

In nature, few natural strains possess the abilities of both lignocellulose hydrolysis and product generation. Hence, metabolic engineering and setting up a microbial co-cultivation system have become two main strategies for direct fuel and chemical generation from lignocellulose through CBP. Metabolic engineering and microbial-consortia-based CBP are promising trends in biomass biorefinery, but still face challenges in terms of complicated microbial structure, low conversion efficiency, and operational instability (Du et al., 2022).

## 2.3 FUEL AND CHEMICAL PRODUCTION FROM LIGNOCELLULOSIC WASTE

### 2.3.1 Fuels

With rapid economic growth and an increasing population in modern society, the global energy demand takes on a drastically rising trend. As reported, global energy consumption has reached 418 EJ in 2019 and is forecasted to increase further by 23% to reach 516 EJ by 2040. However, more than 80% of fuel and chemical output still relies on fossil energy, which has caused a series of serious environmental issues. Moreover, based on the current reserve-to-production ratio, depletion of natural gas and oil is expected to occur in about 50 years. Thus, with the advantages of being non-edible, abundant, low-cost, and environmentally friendly, lignocellulose is considered a potential alternative to petroleum energy and has been developed for the production of several fuels and chemicals through bioconversion.

#### 2.3.1.1 Ethanol

Ethanol has been widely used as a clean fuel and solvent and was developed in a very early stage since the 1970s due to oil crises and adverse ecological effects of fossil fuels (Ko et al., 2016). Nowadays, bioethanol is usually derived from lignocellulose instead of cereal feedstocks, avoiding competition with people for food. Among all bioethanol producers, yeast is the most popular platform for conventional large-scale ethanol production. As reported, the annual production of alcohol worldwide is over 100 billion liters, with yeast being the predominantly used industrial microorganism for ethanol production (da Silva Fernandes et al., 2022).

However, using lignocellulose as a fermentation substrate brings an issue in that many yeasts cannot naturally utilize xylose, which is a large part of biomass. And efficient co-fermentation of glucose and xylose is vital to realizing the bioconversion of lignocellulosic biomass to ethanol in a more economical way. However, even in several strains with intrinsic xylose assimilation ability, their utilization efficiency was quite low. As reported, three xylose-utilizing yeasts were newly isolated from oil palm waste, and their ethanol production only ranged from 0.85 to 1.34 g/L (Kusumawati et al., 2023). Thus, to overcome this issue of converting lignocellulosic feedstocks into bioethanol more economically, the cellular system of yeasts was modified, such as by introducing a xylose-utilizing pathway. For example, xylose isomerase-expressing *S. cerevisiae* SXA-R2P-E produced 50 g/L of ethanol with a yield of 0.43 g/g at 72 hours by using 70 g/L of glucose and 40 g/L of xylose as substrates. Moreover, the engineered strain obtained 18–21 g/L ethanol by fermenting hydrolysates of rice straw and hardwood (Ko et al., 2016). Similarly, a xylose assimilation pathway consisting of xylose reductase (XR), xylitol dehydrogenase (XDH), and xylulokinase (XK) was introduced to *S. cerevisiae*. And 0.26 g/g of ethanol was achieved by the genetically modified strain *S. cerevisiae* NBRC1440/B-EC3-X ΔPHO13 from Napier grass, which was based on 76% of carbohydrate content. However, the introduction of xylose utilization pathway brought out cofactor imbalance, resulting in by-product generation. To enhance substrate utilization and target production at the same time, NADH oxidase was introduced to xylose-assimilating engineered *S. cerevisiae* JX123, and the resulting strain (JX123_noxE) successfully produced 55.5 g/L of ethanol along with reduced xylitol production from the silver grass hydrolysates (Lee et al., 2017).

In industrial bioethanol production, the presence of inhibitors in biomass hydrolysates is another hurdle that hampers the bioconversion efficiency among microbial biorefinery platforms, such as acetic acid, phenols, and furfural. And several strategies have been used to overcome it. For example, through adaptive laboratory evolution, *S. cerevisiae* XUSAE57 efficiently converted xylose to ethanol with high yields of 0.43–0.50 g ethanol/g xylose under 2–5 g/L of acetic stress, and the ethanol yield was twice that of the initial one (Ko et al., 2020). Based on the role of nitroreductase in coping with oxidative stress response, its overexpression can effectively enhance the tolerance of *Kluyveromyces marxianus* to lignocellulose-derived inhibitors (phenols, furfural, and acetic acid), and its ethanolic fermentation productivity was improved by 18.75% (Zhang et al., 2023).

For the intricate structure of lignocellulose, appropriate pretreatments also play an important role. For example, a laccase-mediator pretreatment system improved glucose yield and ethanol production to 157 and 72.4 g/L, respectively, realizing 74% glucan conversion of high-solids-loading sugarcane bagasse (Raj & Krishnan, 2019). Moreover, polyethylene glycol (PEG6000) addition plays a significant role in wheat straw enzymatic hydrolysis, and the maximum ethanol concentration achieved was 97.1 g/L at 96 hours with 2.0% PEG addition, achieving a 17.1% increase than the control group (Kadhum et al., 2019).

#### 2.3.1.2 Acetoin

Acetoin, known as 3-hydroxy-2-butanone or acetyl methyl carbinol, is a four-carbon hydroxyl-keto compound that naturally exists in fruits, vegetables, flours, or fermented foods such as vinegar, butter, and wine. For its diverse applications in the food, cosmetic, and pharmaceutical sectors, acetoin has been included in the 30 most important sugar-derived platform chemicals by the US Department of Energy (DOE) (Cui et al., 2018). *Bacillus* strains are considered the main chassis for acetoin production for its food safety. However, the generation of by-products lowers substrate utilization and targets formation efficiency.

Hence, several genetic modifications were conducted. For example, a series of genes coding for by-products were knocked out to reduce carbon source waste, including *spoA*, *sigE*, and *sigF*, and the recombinant strain BSMAY-4-$P_{srfA}$ could produce 64.3 g/L of acetoin with the productivity of 0.765 g/(L·h) in a 5-L bioreactor from sugarcane bagasse hydrolysate (Wang et al., 2022). Similarly, Yan et al. deleted *bdhA* and *acoA* to avoid target shunts and incorporated gene xylose isomerase (XI) and XK for utilizing pentose sugar. Finally, the engineered *B. subtilis* strain BSK814 produced up to 23.3 g/L acetoin, which was 39% higher than that of the original one from 50 g/L xylose (Yan et al., 2018).

#### 2.3.1.3 2,3-Butanediol

2,3-Butanediol (2,3-BD) is a promising bio-based chemical having various industrial applications. In particular, 2,3-BD was considered as an ideal fuel additive for its advantage of having a higher heating value (27.2 kJ/g) than other liquid fuels such as methanol (22.1 kJ/g). What is more, except used as fuel additives, 2,3-BD also has great potential in the manufacturing of printing inks, perfumes, synthetic rubber, antifreeze agents, foods, and pharmaceuticals (Ge et al., 2016). The global market of 2,3-BD is expected to reach a value of $220 million by 2027, growing at a CAGR of 3% from 2019 to 2027.

However, low 2,3-BD production efficiency is a main obstacle hindering its industry development. Hence, several explorations have been conducted to improve the productivity of bio-2,3-BD. For example, the *pflB* and *ptsG* deletions in *Enterobacter aerogenes* had an effect on 2,3-BD production, improving the production by 40% and 70%, respectively (Um et al., 2017). The two genes were involved in phosphotransferase system (PTS), which has an effect on reducing carbon catabolite repression (CCR) and making the industrial strains to utilize glucose and xylose more economically. In addition, obtaining high-yield strains via mutagenesis technology is also an effective strategy. For example, a mutant *Enterobacter cloacae* M22 generated from atmospheric and room temperature plasma (ARTP) produced 24.32 g/L 2,3-BD, with an improvement of 114% than that of the wild type (Wu et al., 2019). Similarly, a newly isolated thermophilic *B. licheniformis* strain X10 produced a high titer of 74.0 g/L 2,3-BD from corn stover hydrolysate with a productivity of 2.1 g/L/h and a yield of 94.6% (Li et al., 2014).

### 2.3.2 Organic Acids

Besides biofuels, organic acids are another group of bulk chemicals derived from biomass that have wide applications in the food, pharmaceutical, cosmetic, detergent, polymer, and textile industries (Lu et al., 2020). According to the organic acid market analysis report, the global organic acid market is expected to reach $12.54 billion by 2026, and these biochemicals provide significant relief for environment concern and energy depletion (Kim et al., 2018). Here, several typical organic acid production from lignocellulose sources were listed, and challenges on manufacture as well as solutions were also discussed and summarized.

#### 2.3.2.1 Lactic Acid

LA has attracted a great deal of attention due to its wide applications in the food, cosmetic, and pharmaceutical industries. What is more, the escalating call for environmental protection drives the market expansion for polylactic acid (PLA) for its biodegradable and biocompatible characteristics, directly increasing the demand for LA (Murariu & Dubois, 2016). Although the technology for bio-LA production is quite mature, decreasing competition with food supplies and cutting down cost due to its production are still what the society exactly searches for (Shan et al., 2023). As a cheap and abundant resource, forestry and agricultural residues are a promising alternative to LA production. Yet, inhibitors generated during the pretreatment process of lignocellulose doses have a toxic effect on LA producers, resulting in low conversion efficiency and LA productivity (Shan et al., 2023). Therefore, several approaches have been taken to improve the tolerance robustness of LA strains to overcome these inhibitors' effect.

One of these efforts is to screen new inhibitor-tolerance strains. For instance, a newly isolated *Bacillus* sp. P38 with high 2-furfural tolerance produced 180 g/L LA with a productivity of 2.4 g/L/h from corn stover hydrolysate in a fed-batch fermentation (Peng et al., 2013). In another study, the isolated *B. coagulans* strain IPE22 also showed good tolerance to some inhibitors (such as furans, acetate, and sulfuric acid) from wheat straw hydrolysate treated by dilute sulfuric acid, resulting in 46.12 g LA production from 100 g dry wheat straw via SSF (Zhang et al., 2014). Zhao et al. also demonstrated a high tolerance of temperature and lignocellulose-derived toxics LA producer

*Pediococcus acidilactici* DQ2, whose LA titer, yield, and productivity reached 101.9 g/L, 77.2%, and 1.06 g/L/h, respectively (Zhao et al., 2013).

Metabolic engineering on strains to accelerate toxic inhibitors conversion into less toxic intermediates is another common strategy for enhancing LA production. For example, a short-chain dehydrogenase CGS9114_RS09725 from *Corynebacterium glutamicum* was expressed in *P. acidilactici* to enhance vanillin degradation rate, and it finally produced 115 g/L D-LA from biodetoxified corn stover with the productivity of 1.6 g/L/h and overall yield of 61.1% (Qiu et al., 2020a). Similarly, overexpression of an oxidoreductase in *P. acidilactici* achieved a rapid accumulation of 123.8 g/L D-LA from corn stover as a feedstock by converting the p-benzoquinone into less toxic hydroquinone (Qiu et al., 2020b). Besides, mutagenesis, adaptive laboratory evolution, and genome shuffling have also worked (Ling et al., 2014; Wang et al., 2019; Yan et al., 2018).

#### 2.3.2.2 Succinic Acid

SA, which is a kind of dicarboxylic acid with a C4 structure, has been identified as one of the 12 value-added bio-based platform chemicals by the US DOE. Due to its basic chemical structure, SA has extensive applications in food, medicine, surfactants, biodegradable plastics, and other industries (Lu et al., 2021). And the market potential of SA and its direct derivatives is estimated to be as high as 245,000 tons every year, while the market size of SA polymers is estimated to be 25 million tons every year (Bozell & Petersen, 2010). With the depletion of fossil feedstocks and the demand for a sustainable global economy, a shift to SA biosynthesis was encouraged and the current yield of bio-SA even exceeds 130 g/L (Wang et al., 2016). However, using glucose as a substrate was not economic and practical at the industrial scale. Hence, making full use of cheaper raw lignocellulosic materials is a feasible way (Zhou et al., 2023).

Several pretreatments have been developed to render the biomass more accessible for the uptake of microbes since the key point affecting the utility of biomass in fermentation is the recalcitrant polysaccharides hindering sugar release (Zhou et al., 2023). For example, ionic liquid (IL) 1-allyl-3-methylimidazolium chloride (AmimCl) pretreatment improved the enzymatic hydrolysis rate of pinewood and corn stover. Scanning electron microscopy indicated that AmimCl was effective in extracting cellulose from pinewood and in degrading pinewood into a uniform pulp, and finally, *A. succinogenes* 130Z produced 20.7 g/L from AmimCl-pretreated pinewood with an average yield of 0.37 g/g (Wang et al., 2014). Similarly, corn stover was pretreated by concentrated-alkali under steam-assistant (CASA) conditions, and CASA treatment resulted in a higher sugar yield as well as a lower black liquid, which is a more environmentally acceptable strategy. And finally, 64.16 g/L of SA was obtained from 150 g/L CASA-pretreated corn stover by *C. glutamicum* (Li et al., 2023).

The yield degradation cost can also be reduced by optimization of the fermentation process. For example, a two-stage pH regulation strategy successfully improved SA production to 54.4 g/L with 0.44 g/g yield from 42.2 g/L with 0.38 g/g yield. Meanwhile, the gradual adjustment of pH from 6 to 5.5 resulted in 40% reduction of NaOH consumption, greatly lowering the cost (Stylianou et al., 2021). Similarly, a cell-recycled continuous fermentation (CRCF) mode was carried out in order to enhance SA productivity, and the productivity was strengthened to 3.86 g/L/h, which was an approximately 5.1-fold increase compared with the parallel one. What is more, a 24% increment in SA production was also achieved by this cost-effectiveness process. Besides, CBP is another kind of practical process optimization to reduce the cost. For example, the microbial alliance consisting of a hemicellulase-producing bacterium *Thermoanaerobacterium thermosaccharolyticum* M5 and a SA production specialist *A. succinogenes* 130Z can directly produce 32.50 g/L of SA with a yield of 0.39 g/g from xylan, and 12.51 g/L of SA was directly produced from 80 g/L of untreated corn cob.

#### 2.3.2.3 Fumaric Acid

Fumaric acid (FA), another essential intermediate in the tricarboxylic acid cycle (TCA cycle), is a common metabolite that can be synthesized in a wide range of microbial species (Lu et al., 2020).

Meanwhile, FA is the simplest unsaturated carboxylic acid, containing an unsaturated double bond and two carboxylic groups. Hence, as an important kind of raw material for fine chemicals, FA was considered one of 12 platform compounds, having applications in food, paint, resin, plasticizers, pharmaceuticals, and other fields (Guo et al., 2020). As reported, the current market size stands at USD 660.9 million per annum and is expected to grow at an annual rate of 5.5% for the period of 2021–2026 (Swart et al., 2022). Traditionally, FA was a by-product produced from petrochemicals, namely butane and benzene; thus, the drive toward renewable feedstocks for the chemical production has grown. Among a variety of natural and genetically modified FA producer, *Rhizopus* spp., has been found to be the most effective workhorse. However, one main challenge facing the lignocellulosic hydrolysate utilization was the much lower consumption efficiency than that of glucose despite *Rhizopus* spp. having the ability to assimilate xylose (Sebastian et al., 2019). Hence, efforts have been made to improve the production efficiency.

Strain modification was one of the practical methods. For instance, combining furfural domestication and UV mutagenesis, Wu et al. obtained an improved *R. oryzae* WHT5 strain, whose FA titer reached 49.05 g/L from corncob hydrolytes in a 7-L bioreactor (Wu et al., 2018). Similarly, a 2-deoxyglucose-resistant mutant *R. oryzae* DG-3, which was bred through nitrogen ion ($N^+$) treatment, produced about 39.80 g/L FA from cornstarch, which is 1.28-fold as much as that of the parent. Further, a higher FA production (44.10 g/L) was achieved with the initial total sugar concentration of cornstarch improved to 100 g/L (Deng et al., 2012).

Besides strain improvement, conducting an optimization on the fermentation process also has an effect on FA production. Through a two-stage utilization strategy, *R. oryzae* produced 27.79 g/L of FA from corn straw hydrolysates with a productivity of 0.33 g/L/h (Xu et al., 2010). Besides, a further FA yield could be improved to 0.735 g/g, which was an increase of 67.4% than pre-optimization by maintaining a continuous low feed rate when using a synthetic lignocellulosic hydrolysate as substrate (Swart et al., 2022).

#### 2.3.2.4 Malic Acid

Malic acid, a dicarboxylic acid which is currently chemically manufactured using fossil resources, has a wide variety of applications in industries such as food and beverages, pharmaceuticals, oil, wastewater treatment, metal cleaning, and textile finishing (Kövilein et al., 2020). As one of the most important platform chemicals considered by the US DOE, the global malic acid market size in 2022 reached US$ 211.7 million and is expected to reach US$ 283.7 million by 2028, exhibiting a compound annual growth rate of 5.1% during 2023–2028 according to the report of IMARC Group. However, the low productivity of bio-malic acid is still the main challenge, and several efforts have been made to solve this matter.

One is mainly focused on strain modifications. For example, a modified strain *T. fusca* muC was obtained through expressing an exogenous pyruvate carboxylase derived from *C. glutamicum* ATCC 13032, and the engineered strain achieved 62.76 g/L malic acid from 100 g/L cellulose. What is more, it even converted milled corn stover to malic acid with the highest titer of 21.47 g/L (Deng et al., 2016). Similarly, by introducing a malate transporter gene SpMAE1 to an acid-tolerant yeast *Pichia kudriavzevii*, the malic acid titer was improved to 43.8 g/L from below 3 g/L (Xi et al., 2023). Mutagenesis is another common method to obtain a strain with desirable characteristics. For instance, a high malic acid-producing mutant *A. oryzae* FMME218–37 was bred through combined mutagenesis methods, including atmospheric and room temperature plasma (ARTP), 60Co-γ, and diethyl sulfate (DES) iterative irradiation, with a morphological characterization screening system. And the mutant boosted L-malic acid production from 30.73 to 78.12 g/L (Ding et al., 2018).

## 2.4 CONCLUSIONS AND PERSPECTIVES

Lignocellulose biomass has great potential for valuable biochemical and biofuel production, and choosing the appropriate host is the key to achieving high-yield production. *E. coli* is a good host for the biosynthesis of various organic acids because of its complete gene editing tools and clear metabolic pathways. By endowing it with the ability to hydrolyze lignocellulose, the direct biotransformation from lignocellulose to chemicals can be realized. However, because it can't tolerate low pH, it is necessary to add extra alkali liquor to stabilize the pH during fermentation. In comparison, yeast has strong tolerance and good stress resistance to low pH and high product concentrations and is widely used in the production of chemicals and fuels such as ethanol. In addition, most molds have the potential to produce chemicals because of their natural ability to degrade lignocellulose. However, due to the lack of complete gene editing tools, its research is limited. In the future, the development of efficient and well-tolerated chassis cells is an effective means to biotransform lignocellulosic wastes into fuels and chemicals. Lignocellulose is the most abundant renewable biomass resource on Earth. However, most of the biomass resources are not well utilized. Hemicellulose is combined with cellulose and lignin as a molecular binder, and lignin has a three-dimensional network aromatic ring structure, which surrounds and strengthens cellulose and hemicellulose as a supporting skeleton. It is precisely because of the dense three-dimensional reticular aromatic ring structure of lignin and the high crystallinity of cellulose that lignocellulose is difficult to degrade by enzymes or microorganisms. Therefore, the removal of lignin is the key to realizing the efficient utilization of lignocellulose.

In recent years, obtaining chemicals and fuels from lignocellulose by biotransformation has been widely studied. SHF divides the conversion process of lignocellulose into two stages, so it can provide the best conditions for hydrolysis and microbial fermentation. However, a high concentration of hydrolyzed sugar will inhibit the enzymatic reaction, resulting in a decrease in the reaction rate. In contrast, the main advantage of SSF is the use of a single reaction vessel and reduced the fermentation period. However, due to the difference in optimum temperature and pH between hydrolysis and fermentation, SSF may lead to lower efficiency and lower production yield. Further, SSCF is based on SSF; the productivity is effectively improved by endowing microorganisms with the ability to co-ferment hexose and pentose. Moreover, CBP is a promising development trend in the field of lignocellulose biorefinery, characterized by hydrolytic enzyme production, enzymatic hydrolysis, and microbial fermentation in a single step, which largely decreases the cost and complexity of the process. Constructing a production strain capable of producing cellulase or introducing a cellulase-producing gene into the production strain, and realizing the direct transformation of a single strain from lignocellulose to products. However, complex and lengthy metabolic pathways often lead to low production efficiency. Therefore, microbial community has been widely studied to replace a single strain for biotransformation. Microbial community is a promising substitute for complex biotransformation because of its inherent advantages, such as the distribution of metabolic burden by division of labor, and it can transform complex substrates more effectively. However, due to the incompatibility and growth competition between strains, the microbial community is fragile and unstable. Therefore, regulating the proportion of strains and building strains with stronger stress resistance through genetic engineering to establish a more stable microbial community for production is the direction of future research.

## REFERENCES

Abraham, A., Mathew, A.K., Park, H., Choi, O., Sindhu, R., Parameswaran, B., Pandey, A., Park, J.H., Sang, B.I. (2020). Pretreatment strategies for enhanced biogas production from lignocellulosic biomass. *Bioresource Technology*, **301**, 122725.

Akhtar, J., Hassan, N., Idris, A., Ngadiman, N.H.A. (2020). Optimization of simultaneous saccharification and fermentation process conditions for the production of succinic acid from oil palm empty fruit bunches. *Journal of Wood Chemistry and Technology*, **40**(2), 136–145.

Akhtar, J., Idris, A. (2017). Oil palm empty fruit bunches a promising substrate for succinic acid production via simultaneous saccharification and fermentation. *Renewable Energy*, **114**, 917–923.

Annamalai, N., Al Battashi, H., Anu, S.N., Al Azkawi, A., Al Bahry, S., Sivakumar, N. (2020). Enhanced bio-ethanol production from waste paper through separate hydrolysis and fermentation. *Waste and Biomass Valorization*, **11**(1), 121–131.

Baral, N.R., Sundstrom, E.R., Das, L., Gladden, J., Eudes, A., Mortimer, J.C., Singer, S.W., Mukhopadhyay, A., Scown, C.D. (2019). Approaches for more efficient biological conversion of lignocellulosic feedstocks to biofuels and bioproducts. *ACS Sustainable Chemistry & Engineering*, **7**(10), 9062–9079.

Berlowska, J., Cieciura-Wloch, W., Kalinowska, H., Kregiel, D., Borowski, S., Pawlikowska, E., Binczarski, M., Witonska, I. (2018). Enzymatic conversion of sugar beet pulp: a comparison of simultaneous saccharification and fermentation and separate hydrolysis and fermentation for lactic acid production. *Food Technology and Biotechnology*, **56**(2), 188–196.

Bozell, J.J., Petersen, G.R. (2010). Technology development for the production of biobased products from biorefinery carbohydrates-the US Department of Energy's "Top 10" revisited. *Green Chemistry*, **12**(4), 539–554.

Chen, W.C., Lin, Y.C., Ciou, Y.L., Chu, I.M., Tsai, S.L., Lan, J.C.W., Chang, Y.K., Wei, Y.H. (2017). Producing bioethanol from pretreated-wood dust by simultaneous saccharification and co-fermentation process. *Journal of the Taiwan Institute of Chemical Engineers*, **79**, 43–48.

Choi, K.R., Jang, W.D., Yang, D., Cho, J.S., Park, D., Lee, S.Y. (2019). Systems metabolic engineering strategies: integrating systems and synthetic biology with metabolic engineering. *Trends in Biotechnology*, **37**(8), 817–837.

Cui, X.K., Zhao, X.B., Liu, D.H. (2018). A novel route for the flexible preparation of hydrocarbon jet fuels from biomass-based platform chemicals: a case of using furfural and 2,3-butanediol as feedstocks. *Green Chemistry*, **20**(9), 2018–2026.

Da Silva Fernandes, F., de Souza É.S., Carneiro, L.M., Alves Silva, J.P., de Souza, J.V.B., da Silva Batista, J. (2022). Current ethanol production requirements for the yeast *Saccharomyces cerevisiae*. *International Journal of Microbiology*, **2022**, 7878830.

Dahmen, N., Lewandowski, I., Zibek, S., Weidtmann, A. (2019). Integrated lignocellulosic value chains in a growing bioeconomy: status quo and perspectives. *Global Change Biology Bioenergy*, **11**(1), 107–117.

Darkwah, K., Wang, L., Shahbazi, A. (2016). Simultaneous saccharification and fermentation of sweet sorghum after acid pretreatment. *Energy Sources Part A-Recovery Utilization and Environmental Effects*, **38**(10), 1485–1492.

Deng, W.P., Feng, Y.C., Fu, J., Guo, H.W., Guo, Y., Han, B.X., Jiang, Z.C., Kong, L.Z., Li, C.Z., Liu, H.C., Nguyen, P.T.T., Ren, P.N., Wang, F., Wang, S., Wang, Y.Q., Wang, Y., Wong, S.S., Yan, K., Yan, N., Yang, X.F., Zhang, Y.B., Zhang, Z.R., Zeng, X.H., Zhou, H. (2023). Catalytic conversion of lignocellulosic biomass into chemicals and fuels. *Green Energy & Environment*, **8**(1), 10–114.

Deng, Y., Mao, Y., Zhang, X.J. (2016). Metabolic engineering of a laboratory-evolved *Thermobifida fusca* muC strain for malic acid production on cellulose and minimal treated lignocellulosic biomass. *Biotechnology Progress*, **32**(1), 14–20.

Deng, Y.F., Li, S., Xu, Q., Gao, M., Huang, H. (2012). Production of fumaric acid by simultaneous saccharification and fermentation of starchy materials with 2-deoxyglucose-resistant mutant strains of Rhizopus oryzae. *Bioresource Technology*, **107**, 363–367.

Ding, Q., Luo, Q.L., Zhou, J., Chen, X.L., Liu, L.M. (2018). Enhancing l-malate production of FMME218–37 by improving inorganic nitrogen utilization. *Applied Microbiology and Biotechnology*, **102**(20), 8739–8751.

Dong, W.L., Xue, M.L., Zhang, Y., Xin, F.X., Wei, C., Zhang, W.M., Wu, H., Ma, J.F., Jiang, M. (2017). Characterization of a β-glucosidase from *Paenibacillus* species and its application for succinic acid production from sugarcane bagasse hydrolysate. *Bioresource Technology*, **241**, 309–316.

Du, R., Li, C., Lin, W.C., Lin, C.S.K., Yan, J.B. (2022). Domesticating a bacterial consortium for efficient lignocellulosic biomass conversion. *Renewable Energy*, **189**, 359–368.

Garlapati, V.K., Chandel, A.K., Kumar, S.P.J., Sharma, S., Sevda, S., Ingle, A.P., Pant, D. (2020). Circular economy aspects of lignin: towards a lignocellulose biorefinery. *Renewable & Sustainable Energy Reviews*, **130**, 109944.

Ge, Y.S., Li, K., Li, L.X., Gao, C., Zhang, L.J., Ma, C.Q., Xu, P. (2016). Contracted but effective: production of enantiopure 2,3-butanediol by thermophilic and GRAS. *Green Chemistry*, **18**(17), 4693–4703.

Guo, F., Wu, M., Dai, Z.X., Zhang, S.J., Zhang, W.M., Dong, W.L., Zhou, J., Jiang, M., Xin, F.X. (2020). Current advances on biological production of fumaric acid. *Biochemical Engineering Journal*, **153**, 107398.

Islam, M.K., Wang, H.M., Rehman, S., Dong, C.Y., Hsu, H.Y., Lin, C.S.K., Leu, S.Y. (2020). Sustainability metrics of pretreatment processes in a waste derived lignocellulosic biomass biorefinery. *Bioresource Technology*, **298**, 122492.

Kadhum, H.J., Mahapatra, D.M., Murthy, G.S. (2019). A comparative account of glucose yields and bioethanol production from separate and simultaneous saccharification and fermentation processes at high solids loading with variable PEG concentration. *Bioresource Technology*, **283**, 67–75.

Kim, H.M., Park, J.H., Choi, I.S., Wi, S.G., Ha, S., Chun, H.H., Hwang, I.M., Chang, J.Y., Choi, H.J., Kim, J.C., Park, H.W. (2018). Effective approach to organic acid production from agricultural kimchi cabbage waste and its potential application. *Plos One*, **13**(11), e0207322.

Ko, J.K., Enkh-Amgalan, T., Gong, G., Um, Y., Lee, S.M. (2020). Improved bioconversion of lignocellulosic biomass by *Saccharomyces cerevisiae* engineered for tolerance to acetic acid. *Global Change Biology Bioenergy*, **12**(1), 90–100.

Ko, J.K., Um, Y., Woo, H.M., Kim, K.H., Lee, S.M. (2016). Ethanol production from lignocellulosic hydrolysates using engineered *Saccharomyces cerevisiae* harboring xylose isomerase-based pathway. *Bioresource Technology*, **209**, 290–296.

Kövilein, A., Kubisch, C., Cai, L., Ochsenreither, K. (2020). Malic acid production from renewables: a review. *Journal of Chemical Technology & Biotechnology*, **95**(3), 513–526.

Kusumawati, N., Sumarlan, S.H., Zubaidah, E., Wardani, A.K. (2023). Isolation of xylose-utilizing yeasts from oil palm waste for xylitol and ethanol production. *Bioresources and Bioprocessing*, **10**(1), 71.

Lee, Y.G., Jin, Y.S., Cha, Y.L., Seo, J.H. (2017). Bioethanol production from cellulosic hydrolysates by engineered industrial. *Bioresource Technology*, **228**, 355–361.

Li, J., Zheng, X.Y., Fang, X.J., Liu, S.W., Chen, K.Q., Jiang, M., Wei, P., Ouyang, P.K. (2011). A complete industrial system for economical succinic acid production by *Actinobacillus succinogenes*. *Bioresource Technology*, **102**(10), 6147–6152.

Li, J.G., Lin, L.C., Sun, T., Xu, J., Ji, J.X., Liu, Q., Tian, C.G. (2020). Direct production of commodity chemicals from lignocellulose using *Myceliophthora thermophila*. *Metabolic Engineering*, **61**, 416–426.

Li, K., Li, C., Zhao, X.Q., Liu, C.G., Bai, F.W. (2023). Engineering for efficient production of succinic acid from corn stover pretreated by concentrated-alkali under steam-assistant conditions. *Bioresource Technology*, **378**, 128991.

Li, L.X., Li, K., Wang, K., Chen, C., Gao, C., Ma, C.Q., Xu, P. (2014a). Efficient production of 2,3-butanediol from corn stover hydrolysate by using a thermophilic *Bacillus licheniformis* strain. *Bioresource Technology*, **170**, 256–261.

Li, X., Liu, Y., Yang, Y., Zhang, H., Wang, H., Wu, Y., Zhang, M., Sun, T., Cheng, J., Wu, X., Pan, L., Jiang, S., Wu, H. (2014b). High levels of malic acid production by the bioconversion of corn straw hydrolyte using an isolated *Rhizopus delemar* strain. *Biotechnology and Bioprocess Engineering*, **19**(3), 478–492.

Ling, H., Teo, W.S., Chen, B.B., Leong, S.S.J., Chang, M.W. (2014). Microbial tolerance engineering toward biochemical production: from lignocellulose to products. *Current Opinion in Biotechnology*, **29**, 99–106.

Liu, G., Qu, Y. (2019). Engineering of filamentous fungi for efficient conversion of lignocellulose: tools, recent advances and prospects. *Biotechnology Advances*, **37**(4), 519–529.

Liu, R.M., Liang, L.Y., Chen, K.Q., Ma, J.F., Jiang, M., Wei, P., Ouyang, P.K. (2012). Fermentation of xylose to succinate by enhancement of ATP supply in metabolically engineered *Escherichia coli*. *Applied Microbiology and Biotechnology*, **94**(4), 959–968.

Lu, J.S., Li, J.W., Gao, H., Zhou, D.W., Xu, H.X., Cong, Y.X., Zhang, W.M., Xin, F.X., Jiang, M. (2021). Recent progress on bio-succinic acid production from lignocellulosic biomass. *World Journal of Microbiology & Biotechnology*, **37**(1), 169.

Lu, J.S., Lv, Y., Jiang, Y.J., Wu, M., Xu, B., Zhang, W.M., Zhou, J., Dong, W.L., Xin, F.X., Jiang, M. (2020a). Consolidated bioprocessing of hemicellulose-enriched lignocellulose to succinic acid through a microbial cocultivation system. *ACS Sustainable Chemistry & Engineering*, **8**(24), 9035–9045.

Lu, J.S., Lv, Y., Qian, X.J., Jiang, Y.J., Wu, M., Zhang, W.M., Zhou, J., Dong, W.L., Xin, F.X., Jiang, M. (2020b). Current advances in organic acid production from organic wastes by using microbial co-cultivation systems. *Biofuels, Bioproducts and Biorefining*, **14**(2), 481–492.

Lynd, L.R., van Zyl, W.H., McBride, J.E., Laser, M. (2005). Consolidated bioprocessing of cellulosic biomass: an update. *Current Opinion in Biotechnology*, **16**(5), 577–583.

Mankar, A.R., Pandey, A., Modak, A., Pant, K.K. (2021). Pretreatment of lignocellulosic biomass: a review on recent advances. *Bioresource Technology*, **334**, 125235.

Maslova, O., Stepanov, N., Senko, O., Efremenko, E. (2019). Production of various organic acids from different renewable sources by immobilized cells in the regimes of separate hydrolysis and fermentation (SHF) and simultaneous saccharification and fermentation (SFF). *Bioresource Technology*, **272**, 1–9.

Méndez, J., Passos, D.D., Wischral, D., Modesto, L.F., Pereira, N. (2021). Second-generation ethanol production by separate hydrolysis and fermentation from sugarcane bagasse with cellulose hydrolysis using a customized enzyme cocktail. *Biofuels-UK*, **12**(10), 1225–1231.

Murariu, M., Dubois, P. (2016). PLA composites: from production to properties. *Advanced Drug Delivery Reviews*, **107**, 17–46.

Nguyen, T.H., Ra, C.H., Sunwoo, I.Y., Sukwong, P., Jeong, G.T., Kim, S.K. (2018). Bioethanol production from soybean residue via separate hydrolysis and fermentation. *Applied Biochemistry and Biotechnology*, **184**(2), 513–523.

Olson, D.G., McBride, J.E., Shaw, A.J., Lynd, L.R. (2012). Recent progress in consolidated bioprocessing. *Current Opinion in Biotechnology*, **23**(3), 396–405.

Peng, L.L., Wang, L.M., Che, C.C., Yang, G., Yu, B., Ma, Y.H. (2013). *Bacillus* sp. strain P38: an efficient producer of L-lactate from cellulosic hydrolysate, with high tolerance for 2-furfural. *Bioresource Technology*, **149**, 169–176.

Peterson, R., Nevalainen, H. (2012). *Trichoderma reesei* RUT-C30-thirty years of strain improvement. *Microbiology-SGM*, **158**, 58–68.

Qin, L., Li, X., Liu, L., Zhu, J.Q., Guan, Q.M., Zhang, M.T., Li, W.C., Li, B.Z., Yuan, Y.J. (2017). Dual effect of soluble materials in pretreated lignocellulose on simultaneous saccharification and co-fermentation process for the bioethanol production. *Bioresource Technology*, **224**, 342–348.

Qiu, Z.Y., Fang, C., Gao, Q.Q., Bao, J. (2020a). A short-chain dehydrogenase plays a key role in cellulosic D-lactic acid fermentability. *Bioresource Technology*, **297**, 122439.

Qiu, Z.Y., Fang, C., He, N.L., Bao, J. (2020b). An oxidoreductase gene ZMO1116 enhances the p-benzoquinone biodegradation and chiral lactic acid fermentability of *Pediococcus acidilactici*. *Journal of Biotechnology*, **323**, 231–237.

Raj, K., Krishnan, C. (2019). Improved high solid loading enzymatic hydrolysis of low-temperature aqueous ammonia soaked sugarcane bagasse using laccase-mediator system and high concentration ethanol production. *Industrial Crops and Products*, **131**, 32–40.

Roy, S., Dikshit, P.K., Sherpa, K.C., Singh, A., Jacob, S., Rajak, R.C. (2021). Recent nanobiotechnological advancements in lignocellulosic biomass valorization: a review. *Journal of Environmental Management*, **297**, 113281.

Rudolf, A., Alkasrawi, M., Zacchi, G., Lidén, G. (2005). A comparison between batch and fed-batch simultaneous saccharification and fermentation of steam pretreated spruce. *Enzyme and Microbial Technology*, **37**(2), 195–204.

Sebastian, J., Hegde, K., Kumar, P., Rouissi, T., Brar, S.K. (2019). Bioproduction of fumaric acid: an insight into microbial strain improvement strategies. *Critical Reviews in Biotechnology*, **39**(6), 817–834.

Shahab, R.L., Brethauer, S., Davey, M.P., Smith, A.G., Vignolini, S., Luterbacher, J.S., Studer, M.H. (2020). A heterogeneous microbial consortium producing short-chain fatty acids from lignocellulose. *Science*, **369**(6507), 1073–1078.

Shan, W.W., Yan, Y.L., Li, Y.D., Hu, W., Chen, J.H. (2023). Microbial tolerance engineering for boosting lactic acid production from lignocellulose. *Biotechnology for Biofuels and Bioproducts*, **16**(1), 90.

Sharma, A., Dogra, S., Thakur, B., Yadav, J., Soni, R., Soni, S.K. (2023). Separate hydrolysis and fermentation of kitchen waste residues using multi-enzyme preparation from *Aspergillus niger* P-19 for the production of biofertilizer formulations. *Sustainability*, **15**(12), 9584.

Sharma, S., Nair, A., Sarma, S.J. (2021). Biorefinery concept of simultaneous saccharification and co-fermentation: challenges and improvements. *Chemical Engineering and Processing: Process Intensification*, **169**, 108607.

Sheng, Y.Q., Tan, X., Gu, Y.J., Zhou, X., Tu, M.B., Xu, Y. (2021). Effect of ascorbic acid assisted dilute acid pretreatment on lignin removal and enzyme digestibility of agricultural residues. *Renewable Energy*, **163**, 732–739.

Slathia, P.S., Raina, N., Kiran, A., Kour, R., Bhagat, D., Sharma, P. (2020). Dilute acid pretreatment of pine needles of *Pinus roxburghii* by response surface methodology for bioethanol production by separate hydrolysis and fermentation. *Biomass Conversion and Biorefinery*, **10**(1), 95–106.

Stefanidis, S.D., Kalogiannis, K.G., Iliopoulou, E.F., Michailof, C.M., Pilavachi, P.A., Lappas, A.A. (2014). A study of lignocellulosic biomass pyrolysis via the pyrolysis of cellulose, hemicellulose and lignin. *Journal of Analytical and Applied Pyrolysis*, **105**, 143–150.

Stylianou, E., Pateraki, C., Ladakis, D., Damala, C., Vlysidis, A., Latorre-Sánchez, M., Coll, C., Lin, C.S.K., Koutinas, A. (2021). Bioprocess development using organic biowaste and sustainability assessment of succinic acid production with engineered *Yarrowia lipolytica* strain. *Biochemical Engineering Journal*, **174**, 108099.

Sun, S.N., Sun, S.L., Cao, X.F., Sun, R.C. (2016). The role of pretreatment in improving the enzymatic hydrolysis of lignocellulosic materials. *Bioresource Technology*, **199**, 49–58.

Sun, X., Wu, H., Zhao, G., Li, Z., Wu, X., Liu, H., Zheng, Z. (2018). Morphological regulation of *Aspergillus niger* to improve citric acid production by *chsC* gene silencing. *Bioprocess and Biosystems Engineering*, **41**(7), 1029–1038.

Swart, R.M., Brink, H., Nicol, W. (2022). *Rhizopus oryzae* for fumaric acid production: optimising the use of a synthetic lignocellulosic hydrolysate. *Fermentation-Basel*, **8**(6), 278.

Tavva, S., Deshpande, A., Durbha, S.R., Palakollu, V.A.R., Goparaju, A.U., Yechuri, V.R., Bandaru, V.R., Muktinutalapati, V.S.R. (2016). Bioethanol production through separate hydrolysis and fermentation of *Parthenium hysterophorus* biomass. *Renewable Energy*, **86**, 1317–1323.

Tran, P.H.N., Ko, J.K., Gong, G.T., Um, Y., Lee, S.M. (2020). Improved simultaneous co-fermentation of glucose and xylose by *Saccharomyces cerevisiae* for efficient lignocellulosic biorefinery. *Biotechnology for Biofuels and Bioproducts*, **13**(1), 183.

Um, J., Kim, D.G., Jung, M.Y., Saratale, G.D., Oh, M.K. (2017). Metabolic engineering of *Enterobacter aerogenes* for 2,3-butanediol production from sugarcane bagasse hydrolysate. *Bioresource Technology*, **245**, 1567–1574.

Usmani, Z., Sharma, M., Awasthi, A.K., Lukk, T., Tuohy, M.G., Gong, L., Nguyen-Tri, P., Goddard, A.D., Bill, R.M., Nayak, S.C., Kumar, V. (2021). Lignocellulosic biorefineries: the current state of challenges and strategies for efficient commercialization. *Renewable & Sustainable Energy Reviews*, **148**, 111316.

Valles, A., Alvarez-Hornos, F.J., Martínez-Soria, V., Marzal, P., Gabaldón, C. (2020). Comparison of simultaneous saccharification and fermentation and separate hydrolysis and fermentation processes for butanol production from rice straw. *Fuel*, **282**, 118747.

Ventorino, V., Robertiello, A., Cimini, D., Argenzio, O., Schiraldi, C., Montella, S., Faraco, V., Ambrosanio, A., Viscardi, S., Pepe, O. (2017). Bio-based succinate production from Arundo donax hydrolysate with the new natural succinic acid-producing strain *Basfia succiniciproducens* BPP7. *Bioenergy Research*, **10**(2), 488–498.

Wang, C.X., Yan, D.J., Li, Q., Sun, W., Xing, J.M. (2014). Ionic liquid pretreatment to increase succinic acid production from lignocellulosic biomass. *Bioresource Technology*, **172**, 283–289.

Wang, J., Yang, L., Wang, D., Dong, L., Chen, R. (2016). Enhanced succinic acid productivity by expression of mgtCB gene in *Escherichia coli* mutant. *Journal of Industrial Microbiology & Biotechnology*, **43**(4), 505–516.

Wang, Q., Zhang, X., Ren, K.X., Han, R.M., Lu, R.Q., Bao, T., Pan, X.W., Yang, T.W., Xu, M.J., Rao, Z.M. (2022). Acetoin production from lignocellulosic biomass hydrolysates with a modular metabolic engineering system. *Biotechnology for Biofuels and Bioproducts*, **15**(1), 58.

Wang, S.Z., Sun, X.X., Yuan, Q.P. (2018). Strategies for enhancing microbial tolerance to inhibitors for biofuel production: a review. *Bioresource Technology*, **258**, 302–309.

Wang, W.T., Wu, B., Qin, H., Liu, P.T., Qin, Y., Duan, G.W., Hu, G.Q., He, M.X. (2019). Genome shuffling enhances stress tolerance of *Zymomonas mobilis* to two inhibitors. *Biotechnology for Biofuels and Bioproducts*, **12**(1), 139.

Woiciechowski, A.L., Neto, C.J.D., Vandenberghe, L.P.D., Neto, D.P.D., Sydney, A.C.N., Letti, L.A., Karp, S.G., Torres, L.A.Z., Soccol, C.R. (2020). Lignocellulosic biomass: acid and alkaline pretreatments and their effects on biomass recalcitrance: conventional processing and recent advances. *Bioresource Technology*, **304**, 122982.

Wu, J., Zhou, Y.J., Zhang, W., Cheng, K.K., Liu, H.J., Zhang, J.A. (2019). Screening of a highly inhibitor-tolerant bacterial strain for 2,3-BDO and organic acid production from non-detoxified corncob acid hydrolysate. *AMB Express*, 9(1), 153.

Wu, X., Liu, Q., Deng, Y., Chen, X., Zheng, Z., Jiang, S., Li, X. (2018). Production of fumaric acid by bioconversion of corncob hydrolytes using an improved *Rhizopus oryzae* strain. *Applied Biochemistry and Biotechnology*, **184**(2), 553–569.

Xi, Y.Y., Xu, H.T., Zhan, T., Qin, Y., Fan, F.Y., Zhang, X.L. (2023). Metabolic engineering of the acid-tolerant yeast *Pichia kudriavzevii* for efficient L-malic acid production at low pH. *Metabolic Engineering*, **75**, 170–180.

Xiao, Z., Lu, J.R. (2014). Strategies for enhancing fermentative production of acetoin: a review. *Biotechnology Advances*, **32**(2), 492–503.

Xin, F.X., Dong, W.L., Zhang, W.M., Ma, J.F., Jiang, M. (2019). Biobutanol production from crystalline cellulose through consolidated bioprocessing. *Trends in Biotechnology*, **37**(2), 167–180.

Xu, Q., Li, S., Fu, Y., Tai, C., Huang, H. (2010). Two-stage utilization of corn straw by *Rhizopus oryzae* for fumaric acid production. *Bioresource Technology*, **101**(15), 6262–6264.

Yan, P.P., Wu, Y.Q., Yang, L., Wang, Z.W., Chen, T. (2018). Engineering genome-reduced *Bacillus subtilis* for acetoin production from xylose. *Biotechnology Letters*, **40**(2), 393–398.

Zhang, N.N., Wang, F., Nwamba, M.C., Wang, D.M., Hong, J. (2023). Enhancing tolerance of *Kluyveromyces marxianus* to lignocellulose-derived inhibitors and its ethanol production from corn cob via overexpression of a nitroreductase gene. *Industrial Crops and Products*, **203**, 117136.

Zhang, Y.M., Chen, X.R., Luo, J.Q., Qi, B.K., Wan, Y.H. (2014). An efficient process for lactic acid production from wheat straw by a newly isolated strain *Bacillus coagulans* IPE22. *Bioresource Technology*, **158**, 396–399.

Zhao, C., Chen, S.L., Fang, H. (2018). Consolidated bioprocessing of lignocellulosic biomass to itaconic acid by metabolically engineering. *Applied Microbiology and Biotechnology*, **102**(22), 9577–9584.

Zhao, K., Qiao, Q.G., Chu, D.Q., Gu, H.Q., Dao, T.H., Zhang, J., Bao, J. (2013). Simultaneous saccharification and high titer lactic acid fermentation of corn stover using a newly isolated lactic acid bacterium *Pediococcus acidilactici* DQ2. *Bioresource Technology*, **135**, 481–489.

Zheng, P., Fang, L., Xu, Y., Dong, J.J., Ni, Y., Sun, Z.H. (2010). Succinic acid production from corn stover by simultaneous saccharification and fermentation using *Actinobacillus succinogenes. Bioresource Technology*, **101**(20), 7889–7894.

Zheng, Z.B., Chen, T., Zhao, M.N., Wang, Z.W., Zhao, X.M. (2012). Engineering *Escherichia coli* for succinate production from hemicellulose via consolidated bioprocessing. *Microbial Cell Factories*, **11**, 49.

Zhou, P.-P., Meng, J., Bao, J. (2017). Fermentative production of high titer citric acid from corn stover feedstock after dry dilute acid pretreatment and biodetoxification. *Bioresource Technology*, **224**, 563–572.

Zhou, S.Z., Zhang, M.M., Zhu, L.Y., Zhao, X.L., Chen, J.Y., Chen, W., Chang, C. (2023). Hydrolysis of lignocellulose to succinic acid: a review of treatment methods and succinic acid applications. *Biotechnology for Biofuels and Bioproducts*, **16**(1), 94.

# 3 Application of Laccase in Lignocellulose Degradation

*Kavya Dashora, Gyan Datta Tripathi,*
*Zoya Javed, and Gattupalli Meghana*
Centre for Rural Development and Technology,
Indian Institute of Technology Delhi, New Delhi, India

## 3.1 INTRODUCTION

Laccases are a type of multinuclear copper-containing oxidase that is capable of interacting with a large number of aromatic/non-aromatic compounds and are regarded as an excellent candidate for use in a variety of bio-technological and environmental fields due to their extensive substrate specificity. In the field of biotechnology, they are known as green catalysts or green tools. Since the enzyme laccase plays a very crucial role in environmental management, identifying the enzyme source, structure, mechanism of reaction, application, etc., for future prospects will pave the way for sustainable environment and industrial management. This chapter gives a brief account of the enzymes used in the degradation of biomass with special reference to laccases. Thus, this chapter will aid in the significance of laccases as well as areas that do not receive enough attention and need it. Although in the past, a lot of research work was performed on laccases, still new discoveries and enhancing diverse industrial applications are continuing. Fungal laccases are intriguing enzymes that remain only partially explored and are not yet fully understood. Apart from industrial applications, laccases and certain fungal enzymes from white-rot basidiomycetes are considered to be instrumental in degrading persistent organic pollutants. Attempts have been made at both research and industrial levels to develop processes reusing immobilized laccases in special reactors for the degradation of water pollutants.

Millions of tonnes of agro-industrial wastes are produced every year by agro-industries and are generally discarded in the open environment, resulting in an excessive build-up of organic material (Naraian & Gautam, 2018). Fungal enzymes are also employed for the bioremediation of such chemically complex and heavily polluted soils with large indigenous microbial communities. Lignocellulosic biomass is predominantly constituted of cellulose, carbohydrate polymers, hemicellulose, and lignin along with non-carbohydrate phenolic polymers (Abdel-Hamid et al., 2013). Among these, the degradation of lignin is very challenging due to its recalcitrant nature wherein cellulose fibrils are very closely integrated that prevents the transformation of structural polysaccharides into fermentable sugars, which is a major challenge in addressing lignocellulosic waste. Therefore, the need for pretreatment is an essential step for lignocellulosic degradation. Physical and chemical pretreatments have been shown in recent studies to rapidly break down lignin into lignocellulosic biomass (Hideno, 2017). These processes, however, have considerable limitations. Biological pretreatment is thought to be a safe and superior technique for the delignification of lignocellulosic biomass, offering a lower energy cost alternative to chemical pretreatment for reducing recalcitrance to cellulolytic enzymes (Bule et al., 2016).

In the biological pretreatment and/or degradation of lignocellulosic biomass, the microorganisms and enzymes are the main actors. However, the fungi particularly from the white-rot fungus class are the most efficient and are known to have additional indirect effects on rhizosphere and edaphic ecology and species composition. The fungi *Basidiomycetes* enzymatically attack the complex-structured lignocellulose polymer. Based on the ability to degrade lignin with or without cellulose

 DOI: 10.1201/9781032676890-4

and hemicellulose, wood decay has traditionally been divided into white rot and brown rot mainly exerted by basidiomycetes and soft rot mainly performed by Ascomycetes. The brown-rot fungi attack lignin and chemically modify it into a brown oxidized form, which facilitates the enzymatic oxidative depolymerization of cellulose. It was earlier indicated that white-rot fungi release highly specialized class II peroxidase enzymes having high oxidation potential contrary to the recent studies involving the decay along with the genomes of some selected basidiomycetes establishing that white-rot fungi perform independently of class II peroxidase enzymes.

Cellulases, hemicellulases, and lignin-modifying enzymes (LMEs) are the terms for enzymes that depolymerize cellulose, hemicellulose, and lignin, respectively. They are mostly extracellular in location and have large molecular sizes. These enzymes likely diverged in different species to deal with the numerous harmful lignin metabolites and related compounds in the production of humus. In addition to depolymerizing lignin molecules, they also play a significant role in primary and secondary metabolism. These microorganisms have a depolymerizing enzyme system comprising laccase (*p*-diphenol:dioxygen oxidoreductase), manganese peroxidase (MnP), and lignin peroxidase (LiP) (Plácido & Capareda, 2015). The discovery of several white-rot basidiomycetes that can break down lignin or lignin model compounds in the absence of LiP and/or MnP has sparked interest in the role of laccases in lignin degradation (Heap et al., 2014). Laccase-producing microbial strains can break down lignin effectively as they produce a potent extracellular lignin-degrading enzyme system (Wang et al., 2019).

Laccase (EC 1.10.3.2) is an extracellular copper-containing enzyme composed of tetrameric, monomeric, and dimeric glycoproteins. It is found mostly in microorganisms such as bacteria, actinomycetes, and fungi (Janusz et al., 2017). Laccases have a wide range of functions such as lignin production and degradation, pigment generation in fungal spores, plant pathology, and other processes (Christopher et al., 2014). Hence, in this chapter, the significance of laccase in lignocellulose degradation, challenges for mass production, exploitation of laccases as biocatalysts, and its bioprocess technology for industrial applications will be emphasized.

## 3.2 LACCASES ALONG THE TEMPORAL SCALE OF EVOLUTION

Laccase was discovered by Yoshida in 1883 in the latex of Chinese or Japanese lacquer trees (*Rhus* sp.) and has been one of the first enzymes to be identified. Studies carried out during the past 140 years have shown the presence of this copper-containing protein with unusual catalytic properties in all life forms (Janusz et al., 2020). Initially, peroxidase was thought to be the only plant enzyme involved in the oxidative polymerization of lignin precursors. Later, laccase was also found to possess delignification properties (Bagewadi et al., 2017). For example, laccase III attacks polymeric lignin, partially breaking the framework structure, and introduces more hydrophilic groups, leading to the formation of water-soluble material (Christopher et al., 2014). Since laccases have a very high catalytic function, they form an essential industrial enzyme for delignification. It is particularly used in environmental and electrobiochemistry applications, biological fuel cells, and biosensors, which open up new avenues for applications ranging from industrial to health care (Fernández-Fernández et al., 2013) in the degradation of xenobiotic compounds, bioremediation, textile industries, biosensors, biomedical industries, etc. (Gonçalves et al., 2015). To maximize the potential use of laccase immobilized systems, the search for low-cost assistance and the recovery of activity during the immobilization process should be addressed. It exists in a different molecular arrangement like monomeric, heteromeric, heterodimeric, and multimeric. As they are found in a wide range of microbes and higher plants, their molecular weight also differs from 50 to 130 kDa varying in different organisms (Jaiswal et al., 2015).

Recent studies have facilitated a better understanding of strains and their potential, as well as the success of the production on a small scale. Laccases have a very interesting primary structure consisting of a Greek key β-barrel topology that consists of three tiers or domains where nearly

500 amino acid residues are arranged, respectively. There are 150 initial amino acids in the first domain; in the second domain, around 300 amino acids are arranged, and subsequently, in the third domain, about 300–500 amino acids are arranged. The disulphide bonds present between domains I and II and domains I and III provide a stable structure to laccase (Agrawal et al., 2018). The study by Hakulinen et al. (2002) found an exception to this rule in *Melanocarpus albomyces*, which has three disulphide bridges—one within domain I, one between domains I and III, and one between domains II and III.

## 3.3 MICROBIAL SOURCES OF LACCASES

Most commonly, higher plants and fungi produce laccase; however, some bacteria have also recently been identified with some percentage of this enzyme as shown in Figure 3.1. The first Ascomycetes laccase with peroxidase activity was identified as *Monocillium indicum*. While *Pycnoporus sanguineus* produces laccase as a phenol oxidase, *Pycnoporus cinnabarinus* produces laccase as a ligninolytic enzyme (Shekher et al., 2011).

Fungi produce laccases more frequently than higher plants. The significant laccase-producing fungi are shown in Table 3.1. Laccases have also been found abundantly in many *Trichoderma* species, including *T. atroviride*, *T. harzianum*, and *T. longibrachiatum* (Hölker et al., 2002; Velázquez-Cedeño et al., 2004).

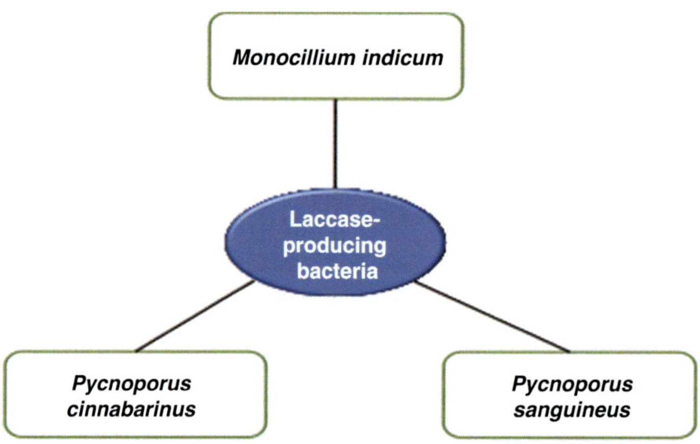


**FIGURE 3.1** Significant laccase-producing bacteria.

**TABLE 3.1**
**Significant Laccase-Producing Fungi**

| Laccase-Producing Fungal Group | Name of the Fungi | References |
|---|---|---|
| Basidiomycetes | *Phanerochaete chrysosporium*<br>*Theiophora terrestris*<br>*Lenzites betulina* | Kiiskinen et al. (2004), Viswanath et al. (2008) |
| White-rot fungi | *Phlebia radiata*<br>*Pleurotus ostreatus*<br>*Trametes versicolor* | Baldrian (2006), Viswanath et al. (2014) |

## 3.4 MOLECULAR IDENTIFICATION OF NEW LACCASE SOURCES

Many laccase genes produce protein isoforms that range in size from 520 to 550 amino acids and have an N-terminal secretion signal peptide of 20 amino acids. In the laccase active site, one cysteine and ten histidine residues comprise four conserved areas that act as a cage-like habitat for chelating ligands to bind to copper ions (Ece et al., 2017). According to Hoegger et al. (2004), numerous laccase isozymes that are produced by the majority of known filamentous fungi are encoded by four, five, or even eight genes. Multicopper oxidases were divided into laccases based on a phylogenetic investigation, with basidiomycete fungi housing the biggest category (Lettera et al., 2010). It is unclear if these enzymes developed in accordance with the lifestyle of the relevant fungus species or some other unidentified reason because the phylogeny of laccase gene does not closely track the phylogeny of species (Hoegger et al., 2006). The higher similarity rate between the sequences of the amino acids in the corresponding proteins and the identical structure of intron and exon was observed in the genes such as pox1, pox4, and poxc that suggest a close relationship between these genes (Pezzella et al., 2009). The placement of introns along the gene appears to be a crucial factor in the conservation of the bulk of the sequences (introns) in the laccase found in Ascomycetes and Basidiomycetes. Amino acid sequences are important factors that are used to classify laccase enzymes (Janusz et al., 2013).

TATA sequences required for the initiation of transcription and at least one CAAT motif key to overall transcription control can be found in every laccase promoter region that has been identified. Different fungal laccase genes have different TATA and CAAT element placements. These two motifs might be connected to the conserved areas of the genes. As the expression of laccase is frequently triggered by a variety of physiological factors, its promoter region may include components involved in the oxidative stress response, temperature shock response, the xenobiotic response mechanism, the fungal metal response mechanism, and the XRE (xenobiotic responsive element) and ARE (antioxidant response element) (Okamoto et al., 2003). T/GCG/AT/CGC/G and NGAAN sequences repeated in either orientation for HSE in eukaryotes were found to be less consistent when known XRE sequences were aligned. The consensus TGCRCNC region is adhered to by all laccase metal-responsive element (MRE) sequences. The quantity (up to 21) and selective presence of HSE in various laccase genes may determine an organism's evolutionary success. Multiple copies of the cis-acting MREs, which are necessary for metal-inducible transcription, have been found in the *Saccharomyces cerevisiae* metallothionein promoter. One MRE in the promoter sequence essentially serves as a marker for how well the organism can respond to heavy metals. Heavy metal exposure boosted laccase synthesis in *Trametes pubescens*, which may indicate the physiological importance of MRE in the lap2 promoter (Galhaup et al., 2002; Levasseur et al., 2010).

## 3.5 DEGRADATION OF LIGNIN AND OTHER MACROMOLECULES

The process of production of industrial laccases is a fairly cost-intensive process. The option of exploring natural and biological sources for the degradation of phenolic compounds is being researched and reviewed very aggressively globally. Additionally, laccase-mediated multi-enzyme co-immobilization and laccase mediator systems utilized in the degradation of phenolic compounds are not thoroughly studied. It is crucial to examine the shortcomings of immobilization strategies for laccase and find cost-effective immobilization techniques in order to improve and optimize laccase stability and activity in order to ensure the effect of degradation of phenolic compounds is applicable to various ecosystems (Chen et al., 2022).

According to Zofair et al. (2022), laccase is engaged in numerous interesting environmental applications like the biodegradation of hazardous chemicals, the bio-bleaching of paper pulp, and the decolorization of wastewater. Laccase is known to perform various industrially important functions, yet the key challenge remains its scaled production. The free laccase production involves many capital assets, and still, the enzymatic instability remains as a major challenge with the laccase enzyme (Ispas et al., 2009). However, the problem is somewhat addressed by the immobilization of the laccase enzyme in the large-scale production.

## 3.6 MECHANISMS OF DEGRADATION OF LIGNIN BY LACCASE

Laccases are produced by different microbial sources and are able to detoxify a range of organic pollutants and degrade various complex polymers like lignin (Kumar & Sonkar, 2013). The catalytic site of laccases contains four copper atoms, is present on the surface, and forms active site environments which directly interact with either lignin or a mediator. In the case of fungal laccases, the active site is termed as T1 (first Cu ion), which directly interacts with the substrate followed by the T2 site (second Cu ion) as well as the third and fourth (T3 site) and forms a clustered structure. In the catalytic process, these four copper ions play a sequential role in the redox mechanisms. The T1 site is the site where the substrate binds and the electrons are transferred to the next copper-clustered sites for further processing. Most importantly, pH plays a significant role in the entire process, and it has been observed that a higher pH may initially enhance the substrate's oxidation to the T1 site but further might reduce the electron transfer to the T2/T3 site due to the presence of hydroxyl anions in its active sites (Curran et al., 2021). The $O_2$ molecule connects with the T2/T3 cluster and prevents the oxidizing molecules to enter to the process. In short, the lignin degradation process happens because of the cleavage of the α- and β-carbon bond because of an oxidation proceeded by phenoxy free radicals. The production of the free radicals (oxygen centred) is further converted into quinone by the second enzyme catalytic reaction, and a further different reaction process takes place. Further, the nonphenolic laccase activity may be achieved by the addition of the mediators. Mediators are the smaller (low molecular weight) organic molecules which can be oxidized by the laccases (Shekher et al., 2011).

## 3.7 POTENTIAL CHALLENGES IN INDUSTRIAL BIOPROCESSING OF LACCASES

The lignocellulose-degrading enzymes are effective inside their native environment or cellular environment of their host, but it is very difficult to replicate their activities in a different environment. Due to the diversity in the producers and their growth patterns, it becomes very difficult to optimize fermentation processes such as effective growth medium, high recovery through downstream process, etc., through bioprocess technology. Not only the downstream processes but also the upstream processes like availability of suitable strain or suitable expression system are yet to be intensely explored by researchers.

Genetic engineering has provided some hope in the development of heterologous expression system for the production of laccases. There are several model organisms like *E. coli*, *Pichia pastoris*, and *Saccharomyces cerevisiae*, which have been exploited by researchers for heterologous production. Several fungi like *Aspergillus niger*, *Aspergillus oryzae*, and *Trichoderma atroviride* have also been considered for genetic improvement because of their capability to produce higher production of laccase (Curran et al., 2021; Piscitelli et al., 2011). In the production process, an effective fermentation medium for the growth and production of the enzymes is also important, but due to the diversity of microbial species and the unavailability of suitable model organisms, it becomes very difficult to optimize the medium. There are several other factors like nitrogen and carbon sources and conditions like pH, temperature, reactor running conditions, and type, which are also important for an effective bioprocess operation of the laccases.

The downstream processing is an important part of the product recovery, and it adds to the cost of the products and makes them reusable, and several methods have been reported for the immobilization and stability of the laccases for suitable industrial production as discussed below.

## 3.8 METHODS FOR LACCASE IMMOBILIZATION

The stability of the free enzyme is a big constraint in using it in large-scale application and can be potentially overcome by immobilization of laccase. Entrapping the enzyme in beads is a further method of immobilizing enzymes. Since laccase is a copper-dependent enzyme, immobilizing it in copper alginate beads may allow for greater activity retention than alternative techniques. The most common methods for immobilizing laccase are either physical or chemical methods. Teerapatsakul et al. (2008) reported that using $CuSO_4$ as the cross-linking solution increased the immobilization yield and enzyme activity compared to $CaCl_2$. Some of the common methods for immobilization of laccase for the degradation of phenolic compounds are (1) ionic absorption, (2) physical adsorption, (3) entrapment, (4) self-immobilization, and (5) covalent binding.

### 3.8.1 Physical Adsorption

Physical adsorption is also known as the method of reversible immobilization and is used to rapidly bind or adsorb an enzyme to a carrier substance (Atiroğlu et al., 2021). In this process, the adhering of enzyme to carrier molecule is done through weak forces such as hydrogen bonds, hydrophobic interactions, and non-specific van der Waals forces (Mehdi et al., 2018). Though this process offers a cost advantage, it is not fully exploited as laccase has a weak adsorption affinity to the carrier and it gets displaced very easily (Yang et al., 2021). Physical adsorption is frequently employed to immobilize laccase. Effective adsorption inside the pores only takes place when the second broad dimension of the adsorbate molecule is at least 1.2 times or 1.7 times larger than the average pore size of the adsorbent. The secondary conformation of the adsorbed laccase experiences a sizable conformational shift, and its laccase activity is increased.

### 3.8.2 Ionic Adsorption

#### 3.8.2.1 Covalent Bonding

Covalent bonds connect an enzyme to amino, imidazole, thiol, hydroxyl, and guanidyl groups present on carriers. The availability of the functional groups on the tyrosine phenol ring, the carboxyl group present at the C-terminus of enzymes, and the indole ring of tryptophan are all implicated in the covalent bonding with the support (Brady & Jordaan, 2009). Numerous supports and carriers, including cellulose, agarose, collagen, silica, glass, metallic nanoparticles, e.g. silver and gold, and glutaraldehyde and other cross-linkers, are most frequently utilized. Despite the benefits of covalent linkage, such as irreversible binding and great stability, the conformational shift of the enzyme during binding places a severe restriction on covalent linkage (Fernandez-Fernandez et al., 2013).

#### 3.8.2.2 Entrapment

Due to its simplicity of use, lack of structural modifications to the enzyme, and ability to reduce enzyme leakage into the solution, the entrapment method of immobilization is most frequently used in industries. Polymers like gelatine, alginate, carrageenan, hydrogel structures, microcapsules of alginate-chitosan, and metallic-linked beads of enzymes are the most frequently used support materials for entrapment (Lu et al., 2007).

## 3.9 BIOENGINEERING OF ENZYMATIC HYDROLYSIS OF CELLULOSE AND LIGNIN

The most crucial procedure for the effectiveness of bioengineering was identified to be biological pretreatment in conjunction with moderate chemical and/or physical pretreatments. Bioconversion of biomass into bioproducts such as biogas, biocomposites, and bioethanol has experienced some success Ghaffar et al. (2015). However, there are a lot more limitations associated with it.

Compared to LiP and MnP, laccase is the most promising ligninolytic enzyme for industrial delignification applications. It is also easier to control and more widely available. Most agricultural wastes are abundant in soluble carbohydrates and include laccase synthesis inducers, resulting in significant laccase production. The highest laccase activity was seen when agro-wastes, particularly those high in cellulose, were used as a support substrate (Bharathiraja et al., 2017).

In order to provide hydrolytic enzymes access to the holocellulose fraction to form streams of cellulose-derived glucose and a combination of hemicellulose-derived pentoses and hexoses, a pretreatment step is frequently necessary using different physical, chemical, and biological approaches (Champreda et al., 2019).

Pretreatments with laccases and laccase mediators using common redox mediators like ABTS cause fibre changes that may be seen using spectroscopic techniques (Barsberg, 2002). An initial crust of lignin was visible in the SEM images of the pretreated fibres, but it was completely removed by the enzyme pretreatment. It also indicates that laccase pretreatment loosens the surface crust because the lignin on the fibre's surface is sensitive to laccase reactivity (Hüttermann et al., 2001).

Direct enzymatic saccharification of cellulose into glucose is difficult due to the intricate connectivity of cellulose, hemicellulose, and lignin. It is not necessary to multiply or decrease the amount of the lignin fraction; rather, it should be designed so that it can be deconstructed chemically or enzymatically under milder conditions than are currently required. For instance, by including a limited number of linkages and precursor units. As a result, lignin bioengineering may attempt to increase biomass saccharification yield while also facilitating enhanced lignin valorization through additional catalytic treatment (Rinaldi et al., 2016). The removal of cell wall components results in the formation of surface apertures or interior slits, gaps, or spaces, which increase the direct physical contact between the enzymes and the substrate and raise cellulase reactivity (Brijwani & Vadlani, 2011).

The catalytic reactions by laccase are of two types: direct substrate oxidation and indirect substrate oxidation as shown in Figure 3.2. Direct oxidation refers to the oxidation of a substrate to a comparable radical as a result of direct contact with a copper cluster. Certain features of mediators must be demonstrated in order for a reaction to occur without a barrier, as the T1 copper ion has a low redox potential, and direct oxidation is not possible in several processes. The indirect substrate oxidation process is a two-step process in which the first enzyme catalyses the mediator and subsequently the mediator oxidizes the substrate (Janusz et al., 2017).

## 3.10 CONCLUSIONS AND PERSPECTIVES

With the growing demand for laccases in various fields, it is required to find newer industrial production and purification strategies. Researchers are continuously working on finding novel production strategies, but there are several challenges like medium, sources, and recovery that are still a challenge. Fungal sources seem to be more helpful in degrading lignocellulosic biomass, but it varies from case to case. Among them, white-rot fungal organisms are traditionally used for the production of laccases due to their potential to degrade lignocellulose. Additionally, fermentation strategies, for example submerged fermentation, have been reported in multiple studies, but due to high operational and downstream processing costs, it is more challenging to adopt by industries. It is observed that the pretreatment and depolymerization are beneficial in the degradation of lignocellulose. So, in the future, there is a vast scope of technological advancement, which will benefit the improvement in the degradation activity of the laccases. Interestingly, it has been reported that some species of mushrooms (e.g. *Agaricus*) utilize various lignocellulosic biomasses for their growth, which can be explored as a source of laccases because of the cost effectiveness, minimal pre-processing of substrates, and higher fold of enzyme production. Overall, applying a meaningful scientific approach in the selection of strain, utilization of substrates with minimal processing, and effective fermentation strategies with simpler and cheaper downstream processing is key to producing the laccases and ultimately playing a role in lignocellulose degradation.

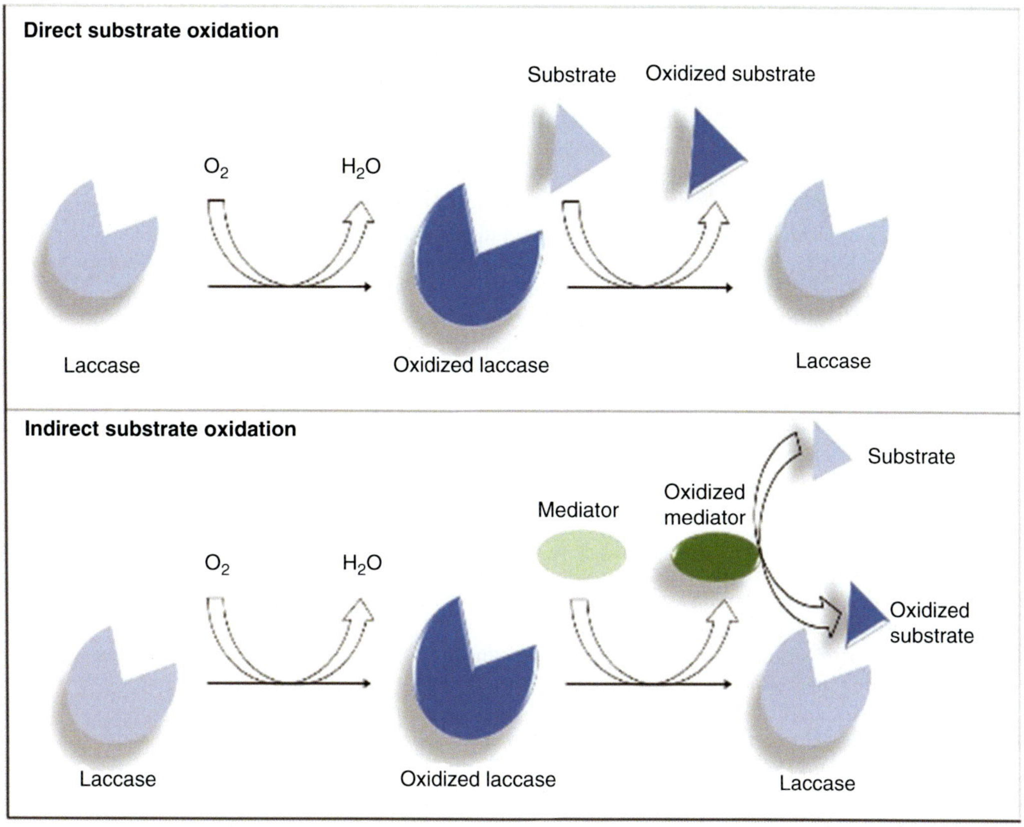


**FIGURE 3.2** Representing direct and indirect substrate oxidation.

## REFERENCES

Abdel-Hamid, A. M., Solbiati, J. O., & Cann, I. K. O. (2013). Insights into lignin degradation and its potential industrial applications. In S. Sariaslani & G. M. Gadd (Eds.), *Advances in Applied Microbiology* (Vol. 82, pp. 1–28). Academic Press. https://doi.org/10.1016/B978-0-12-407679-2.00001-6.

Agrawal, K., Chaturvedi, V., & Verma, P. (2018). Fungal laccase discovered but yet undiscovered. *Bioresources and Bioprocessing*, *5*(1), 1–12.

Atiroğlu, V., Atiroğlu, A., & Özacar, M. (2021). Immobilization of α-amylase enzyme on a protein@ metal–organic framework nanocomposite: a new strategy to develop the reusability and stability of the enzyme. *Food Chemistry*, *349*, 129127.

Bagewadi, Z. K., Mulla, S. I., & Ninnekar, H. Z. (2017). Optimization of laccase production and its application in delignification of biomass. *International Journal of Recycling of Organic Waste in Agriculture*, *6*(4), 351–365. https://doi.org/10.1007/s40093-017-0184-4.

Baldrian, P. (2006). Fungal laccases – occurrence and properties. *FEMS Microbiology Reviews*, *30*(2), 215–242. https://doi.org/10.1111/j.1574-4976.2005.00010.x.

Barsberg, S. (2002). Modification phenomena of solid-state lignin caused by electron-abstracting oxidative systems. *Archives of Biochemistry and Biophysics*, *404*(1), 62–70.

Bharathiraja, S., Suriya, J., Krishnan, M., Manivasagan, P., & Kim, S.-K. (2017). Production of enzymes from agricultural wastes and their potential industrial applications. In S.-K. Kim & F. Toldrá (Eds.), *Marine Enzymes Biotechnology: Production and Industrial Applications, Part III: Application of Marine Enzymes* (Vol. 80, pp. 125–148). Academic Press. https://doi.org/10.1016/bs.afnr.2016.11.003.

Brady, D., & Jordaan, J. (2009). Advances in enzyme immobilisation. *Biotechnology Letters*, *31*(11), 1639–1650.

Brijwani, K., & Vadlani, P. V. (2011). Cellulolytic enzymes production via solid-state fermentation: effect of pretreatment methods on physicochemical characteristics of substrate. *Enzyme Research*, 2011, 860234.

Bule, M. V., Chaudhary, I., Gao, A. H., & Chen, S. (2016). Effects of extracellular proteome on wheat straw pretreatment during solid-state fermentation of Phlebia radiata ATCC 64658. *International Biodeterioration & Biodegradation*, *109*, 36–44. https://doi.org/10.1016/j.ibiod.2015.12.002.

Champreda, V., Mhuantong, W., Lekakarn, H., Bunterngsook, B., Kanokratana, P., Zhao, X.-Q., Zhang, F., Inoue, H., Fujii, T., & Eurwilaichitr, L. (2019). Designing cellulolytic enzyme systems for biorefinery: from nature to application. *Journal of Bioscience and Bioengineering*, *128*(6), 637–654. https://doi.org/10.1016/j.jbiosc.2019.05.007.

Chen, Z., Oh, W.-D., & Yap, P.-S. (2022). Recent advances in the utilization of immobilized laccase for the degradation of phenolic compounds in aqueous solutions: a review. *Chemosphere*, *296*, 135824.

Christopher, L. P., Yao, B., & Ji, Y. (2014). Lignin biodegradation with laccase-mediator systems. *Frontiers in Energy Research*, *2*. https://doi.org/10.3389/fenrg.2014.00012.

Curran, L. M. C. L. K., Sale, K. L., & Simmons, B. A. (2021). Review of advances in the development of laccases for the valorization of lignin to enable the production of lignocellulosic biofuels and bioproducts. *Biotechnology Advances*, *51*, 107809.

Ece, S., Lambertz, C., Fischer, R., & Commandeur, U. (2017). Heterologous expression of a Streptomyces cyaneus laccase for biomass modification applications. *AMB Express*, *7*(1), 86. https://doi.org/10.1186/s13568-017-0387-0.

Fernandez-Fernandez, M., Sanromán, M. Á., & Moldes, D. (2013). Recent developments and applications of immobilized laccase. *Biotechnology Advances*, *31*(8), 1808–1825.

Galhaup, C., Goller, S., Peterbauer, C. K., Strauss, J., & Haltrich, D. (2002). Characterization of the major laccase isoenzyme from Trametes pubescens and regulation of its synthesis by metal ions The GenBank accession numbers for the lap2 and lap1a genes reported in this paper are AF414807 and AF414808, respectively. *Microbiology*, *148*(7), 2159–2169.

Ghaffar, S. H., Fan, M., & McVicar, B. (2015). Bioengineering for utilisation and bioconversion of straw biomass into bio-products. *Industrial Crops and Products*, *77*, 262–274. https://doi.org/10.1016/j.indcrop.2015.08.060.

Gonçalves, I., Silva, C., & Cavaco-Paulo, A. (2015). Ultrasound enhanced laccase applications. *Green Chemistry*, *17*(3), 1362–1374.

Hakulinen, N., Kiiskinen, L.-L., Kruus, K., Saloheimo, M., Paananen, A., Koivula, A., & Rouvinen, J. (2002). Crystal structure of a laccase from Melanocarpus albomyces with an intact trinuclear copper site. *Nature Structural Biology*, *9*(8), 601–605.

Heap, L., Green, A., Brown, D., van Dongen, B., & Turner, N. (2014). Role of laccase as an enzymatic pretreatment method to improve lignocellulosic saccharification. *Catalysis Science & Technology*, *4*(8), 2251–2259. https://doi.org/10.1039/C4CY00046C.

Hideno, A. (2017). Short-time alkaline peroxide pretreatment for rapid pulping and efficient enzymatic hydrolysis of rice straw. *Bioresource Technology*, *230*, 140–142.

Hoegger, P. J., Kilaru, S., James, T. Y., Thacker, J. R., & Kües, U. (2006). Phylogenetic comparison and classification of laccase and related multicopper oxidase protein sequences. *The FEBS Journal*, *273*(10), 2308–2326.

Hoegger, P. J., Navarro-Gonzalez, M., Kilaru, S., Hoffmann, M., Westbrook, E. D., & Kües, U. (2004). The laccase gene family in Coprinopsis cinerea (Coprinus cinereus). *Current Genetics*, *45*(1), 9–18.

Hölker, U., Dohse, J., & Höfer, M. (2002). Extracellular laccases in ascomycetes Trichoderma atroviride and Trichoderma harzianum. *Folia Microbiologica*, *47*(4), 423–427.

Hüttermann, A., Mai, C., & Kharazipour, A. (2001). Modification of lignin for the production of new compounded materials. *Applied Microbiology and Biotechnology*, *55*(4), 387–394.

Ispas, C., Sokolov, I., & Andreescu, S. (2009). Enzyme-functionalized mesoporous silica for bioanalytical applications. *Analytical and Bioanalytical Chemistry*, *393*(2), 543–554.

Jaiswal, N., Pandey, V. P., & Dwivedi, U. N. (2015). Purification of a thermostable alkaline laccase from papaya (Carica papaya) using affinity chromatography. *International Journal of Biological Macromolecules*, *72*, 326–332.

Janusz, G., Kucharzyk, K. H., Pawlik, A., Staszczak, M., & Paszczynski, A. J. (2013). Fungal laccase, manganese peroxidase and lignin peroxidase: gene expression and regulation. *Enzyme and Microbial Technology*, *52*(1), 1–12.

Janusz, G., Pawlik, A., Sulej, J., Świderska-Burek, U., Jarosz-Wilkołazka, A., & Paszczyński, A. (2017). Lignin degradation: microorganisms, enzymes involved, genomes analysis and evolution. *FEMS Microbiology Reviews*, *41*(6), 941–962. https://doi.org/10.1093/femsre/fux049.

Janusz, G., Pawlik, A., Świderska-Burek, U., Polak, J., Sulej, J., Jarosz-Wilkołazka, A., & Paszczyński, A. (2020). Laccase properties, physiological functions, and evolution. *International Journal of Molecular Sciences*, *21*(3). https://doi.org/10.3390/ijms21030966.

Kiiskinen, L., Rättö, M., & Kruus, K. (2004). Screening for novel laccase-producing microbes. *Journal of Applied Microbiology*, *97*(3), 640–646.

Kumar, V., & Sonkar, P. (2013). Laccases: sources and their environmental application. *International Journal of Bioassays*, *2*(6), 909–911.

Lettera, V., Piscitelli, A., Leo, G., Birolo, L., Pezzella, C., & Sannia, G. (2010). Identification of a new member of Pleurotus ostreatus laccase family from mature fruiting body. *Fungal Biology*, *114*(9), 724–730.

Levasseur, A., Saloheimo, M., Navarro, D., Andberg, M., Pontarotti, P., Kruus, K., & Record, E. (2010). Exploring laccase-like multicopper oxidase genes from the ascomycete Trichoderma reesei: a functional, phylogenetic and evolutionary study. *BMC Biochemistry*, *11*(1), 1–10.

Lu, L., Zhao, M., Zhang, B.-B., Yu, S.-Y., Bian, X.-J., Wang, W., & Wang, Y. (2007). Purification and characterization of laccase from Pycnoporus sanguineus and decolorization of an anthraquinone dye by the enzyme. *Applied Microbiology and Biotechnology*, *74*(6), 1232–1239.

Mehdi, W. A., Mehde, A. A., Özacar, M., & Özacar, Z. (2018). Characterization and immobilization of protease and lipase on chitin-starch material as a novel matrix. *International Journal of Biological Macromolecules*, *117*, 947–958.

Naraian, R., & Gautam, R. L. (2018). Penicillium enzymes for the saccharification of lignocellulosic feedstocks. In V. K. Gupta & S. Rodriguez-Couto (Eds.), *New and Future Developments in Microbial Biotechnology and Bioengineering* (pp. 121–136). Elsevier. https://doi.org/10.1016/B978-0-444-63501-3.00006-5.

Okamoto, K., Ito, Y., Shigematsu, I., Yanagi, S. O., & Yanase, H. (2003). Cloning and characterization of a laccase gene from the white-rot basidiomycete Pleurotus ostreatus. *Mycoscience*, *44*(1), 11–17.

Pezzella, C., Giardina, P., Piscitelli, A., Sannia, G., & Faraco, V. (2009). The Pleurotus ostreatus laccase multi-gene family: isolation and heterologous expression of new family members. *Current Genetics*, *55*(1), 45–57.

Piscitelli, A., Del Vecchio, C., Faraco, V., Giardina, P., MacEllaro, G., Miele, A., Pezzella, C., & Sannia, G. (2011). Fungal laccases: versatile tools for lignocellulose transformation. *Comptes Rendus Biologies*, *334*(11), 789–794. https://doi.org/10.1016/j.crvi.2011.06.007.

Plácido, J., & Capareda, S. (2015). Ligninolytic enzymes: a biotechnological alternative for bioethanol production. *Bioresources and Bioprocessing*, *2*(1), 23.

Rinaldi, R., Jastrzebski, R., Clough, M. T., Ralph, J., Kennema, M., Bruijnincx, P. C. A., & Weckhuysen, B. M. (2016). Paving the way for lignin valorisation: recent advances in bioengineering, biorefining and catalysis. *Angewandte Chemie International Edition*, *55*(29), 8164–8215.

Shekher, R., Sehgal, S., Kamthania, M., & Kumar, A. (2011). Laccase: microbial sources, production, purification, and potential biotechnological applications. *Enzyme Research*, *2011*, 217861.

Teerapatsakul, C., Bucke, C., Parra, R., Keshavarz, T., & Chitradon, L. (2008). Dye decolorisation by laccase entrapped in copper alginate. *World Journal of Microbiology and Biotechnology*, *24*(8), 1367–1374.

Velázquez-Cedeño, M. A., Farnet, A. M., Ferre, E., & Savoie, J. M. (2004). Variations of lignocellulosic activities in dual cultures of Pleurotus ostreatus and Trichoderma longibrachiatum on unsterilized wheat straw. *Mycologia*, *96*(4), 712–719.

Viswanath, B., Chandra, M. S., Pallavi, H., & Reddy, B. R. (2008). Screening and assessment of laccase producing fungi isolated from different environmental samples. *African Journal of Biotechnology*, *7*(8), 1129–1133.

Viswanath, B., Rajesh, B., Janardhan, A., Kumar, A. P., & Narasimha, G. (2014). Fungal laccases and their applications in bioremediation. *Enzyme Research*, *2014*, 163242. https://doi.org/10.1155/2014/163242.

Wang, F., Xu, L., Zhao, L., Ding, Z., Ma, H., & Terry, N. (2019). Fungal laccase production from lignocellulosic agricultural wastes by solid-state fermentation: a review. *Microorganisms*, *7*(12), 665. https://doi.org/10.3390/microorganisms7120665.

Yang, B., Tang, K., Wei, S., Zhai, X., & Nie, N. (2021). Preparation of functionalized mesoporous silica as a novel carrier and immobilization of laccase. *Applied Biochemistry and Biotechnology*, *193*(8), 2547–2566.

Zofair, S. F. F., Ahmad, S., Hashmi, M. A., Khan, S. H., Khan, M. A., & Younus, H. (2022). Catalytic roles, immobilization and management of recalcitrant environmental pollutants by laccases: significance in sustainable green chemistry. *Journal of Environmental Management*, *309*, 114676.

# 4 Application of Waste-Derived Biochar for Resource Recovery to Promote Circular Economy and Climate Change Mitigation

*Pranjal Tripathi*[1], *Sunita Varjani*[2,3,4], *Tasmia Falaque*[2], *and Ram Sharan Singh*[1]

[1]Department of Chemical Engineering and Technology, Indian Institute of Technology, Varanasi, Uttar Pradesh, India

[2]School of Engineering, University of Petroleum and Energy Studies, Dehradun, Uttarakhand, India

[3]School of Health Sciences and Technology, University of Petroleum and Energy Studies, Dehradun, Uttarakhand, India

[4]KU-KIST, Korea University, Seoul, Republic of Korea

## 4.1 INTRODUCTION

The world is dealing with various water pollution challenges, characterized by the widespread presence of several pollutants, heavy metals, surplus nutrients, and unexpected chemicals (Samuel Olugbenga et al., 2024). Traditional water treatment methods consistently fail to adequately eliminate these pollutants. The need for sustainable development, especially the circular use of resources, along with the inefficiency of current wastewater treatment methods, has prompted a shift in the scientific community's approach to wastewater solutions (Varjani et al., 2023). In recent years, there has been a shift from the removal of pollutants in wastewater treatment plants (WWTPs) to the recovery of nutrients in water resource recovery facilities (WRRFs). Nowadays, we consider nutrient contamination as a valuable resource to promote circular economy (CE) (Wu & Vaneeckhaute, 2022). The CE is characterized by minimizing the use of non-renewable resources and curtailing the production of waste (Samoraj et al., 2022).

The global demand for fertilizers, including nitrogen (N), phosphorus (P), and potassium (K), is predicted to be 195 million tons for the 2023–2024 cycle. The CE approach in the fertilizer industry is characterized by the extraction of biogenic components such as N, P, and K from the nutrient-rich wastewater. Municipal and domestic wastewater contains high levels of phosphorus (P) and nitrogen, which are discharged into aquatic water bodies. Phosphorus (P) is handled cautiously in WWTPs as it causes eutrophication in aquatic environments (Buss et al., 2020; Nardis et al., 2021). Phosphorus, a non-renewable mineral derived from phosphate rock mining, is widely used as an agricultural fertilizer. Several authors predict depletion in the short to medium term (50–100 years) (Shakoor et al., 2021; Sun et al., 2018). While nitrogen (N) can be traditionally treated from municipal wastewater by converting $NH_4^+$, $NO_2^-$, and $NO_3^-$ to $N_2$ via the nitrification–denitrification process in activated sludge systems, there are no such technologies available for P recovery. Consequently, there is an immediate necessity to create a sustainable method for recovering P from nutrient-rich wastewater.

Various conventional techniques have been utilized by researchers for the recovery of resources from wastewater, such as adsorption, chemical precipitation, biological technologies, electrochemical

 DOI: 10.1201/9781032676890-5

technologies, and microbial electrochemical technologies (Deng et al., 2023; Liu et al., 2021; Puyol et al., 2017). Among all of these, adsorption is the most desired and feasible strategy for resource recovery because of its various advantages, such as ease of operation, minimal waste creation, and cost-effectiveness (Saheed et al., 2022). Biochar is produced by pyrolyzing organic materials like biomass and waste from forests and agriculture under oxygen-limiting circumstances, resulting in carbon-rich charcoal-like compounds (Agarwal et al., 2022). It is a cost-effective and sustainable multifunctional adsorbent with unique qualities such as high specific surface area, pH buffering capability, and ion exchange ability (Huang et al., 2020). Considering its unique characteristics, biochar has been researched for a variety of uses, including carbon sequestration, waste management, adsorption, soil amendment, and energy storage (Allohverdi et al., 2021; Luo et al., 2023a; Patel et al., 2020; Rawat et al., 2023).

In recent studies, several attempts have been made to enhance the P and N adsorption by biochar via surface modification with metal ($Mg^{+2}$, $Fe^{+3}$, and $Al^{+3}$) impregnation (Luo et al., 2023b; Shin et al., 2020). Pristine biochar has a limited capacity to bind anionic nutrients ($NO_3^-$, $PO_4^{-3}$) due to its negative surface charge and insufficient adsorption sites. Surface modification with metal enhances adsorption and can be a viable alternative to pristine biochar for soil amendment (Marcińczyk et al., 2022). This nutrient-rich biochar could be further utilized as a slow-release fertilizer (SRF) to promote CE. Application of SRF can improve nutrient synergy and reduce stress and toxicity associated with conventional fertilizers (Marcińczyk & Oleszczuk, 2022a).

This chapter aims to critically examine the use of waste-derived biochar for nutrient recovery to promote CE. Several existing data were collected on the application of biochar for the resource recovery from wastewater and its application to use as SRFs. In addition to that, this chapter also explores the potential of engineered biochar for enhanced resource recovery. This chapter also discusses the possible co-applications of biochar-based fertilizers with synthetic fertilizers.

## 4.2 BIOCHAR

Biochar is a carbon-rich and highly porous substance formed through the thermal decomposition of biomass under low oxygen conditions (Wang et al., 2024). The potential biomass sources are agricultural residue, sewage waste, food waste, forest residue, animal waste, etc. Its physicochemical property and efficiency mainly depend on biomass type, pyrolysis environment (temperature, time, and carrier gas), and synthesis conditions. Pyrolysis can be categorized as slow, rapid, or flash according to heating rates and temperature specifications. Slow pyrolysis results in biochar synthesis, while rapid and flash pyrolysis mostly produces bio-oil (Wang & Yao, 2023). Biochar is a low-cost, environmentally sustainable substance, which offers greater potential for wastewater treatment due to its mesoporous structure. This structure offers high surface area, and the presence of a variety of functional groups on its surface makes it a suitable adsorbent for wastewater treatment (Xiao et al., 2020; Li et al., 2022). Due to their extremely porous structure and relatively high surface area, biochar can enhance soil structure and improve their water retention capacity (Nguyen, 2021).

In the past few years, biochar has garnered considerable interest in political and academic circles for its potential as a carbon storage mechanism and a soil amendment, providing environmental advantages for land use (Yao et al., 2013; Vo et al., 2022). Due to their unique structure and adsorption capacity, biochar can be functionalized to further improve its nutrient removal characteristics. Functionalized biochar is pristine biochar that has been treated with metals, nanomaterials, microbes, or hydroxides. This functionalization alters its surface properties, increases the surface area, and reduces the electrostatic repulsion between the biochar and negative ions (Zhang et al. 2023). Recent research efforts have focused on improving the adsorption ability of biochar for nutrient removal through surface modifications and impregnation with various metals such as iron (Fe), magnesium (Mg), calcium (Ca), and aluminium (Al) (Shakoor et al., 2021).

## 4.3 NUTRIENT SOURCES

It is well established that municipal and domestic wastewater contains high concentrations of nutrients, particularly P and N, which are eventually discharged into aquatic ecosystems. In aquatic ecosystems, N and P exist in the form of ammonium ($NH_4^+$) and phosphate ($PO_4^{-3}$), which causes eutrophication. As a result, nutrient recovery from wastewater has gained significant interest. The macronutrients (N, P, and K) and micronutrients present in this wastewater could be utilized as fertilizers for soil amendment. The worldwide data for wastewater generation is estimated to be 380 billion m$^3$/year, of which 48% is discharged into the aquatic environment without proper treatment (Jones et al., 2021). The recovery of nutrients from this untreated wastewater could fulfill 10%–15% of global P demand (Jing et al., 2019).

Municipal wastewater, especially black water, is a great source of N and P. Human urine can potentially be utilized as a nutrient source. Food wastewater generated during food processing because of dewatering or anaerobic digestion is also abundant in nutrients. Massive volumes of N- and P-rich products are processed and consumed for industrial, agricultural, and domestic use; excess amounts could be released into aquatic environments. However, it is important to recognize that only nutrients obtained from reliable sources, such as municipal wastewater and fermentation residues, can serve as viable alternatives to conventional sources (Maroušek & Gavurová, 2022). In the CE paradigm, it is required to develop an appropriate technique for recovering nutrients from wastewater and reusing them as fertilizers.

## 4.4 APPLICATION OF BIOCHAR FOR RESOURCE RECOVERY: CIRCULAR ECONOMY PARADIGM

### 4.4.1 Biochar Production

Thermochemical methods, including pyrolysis, gasification, hydrothermal carbonization, and torrefaction, have been employed for biochar production (Figure 4.1). Torrefaction, or moderate heating, is a thermal process that transpires between 200°C and 300°C, usually at ambient conditions and in a restricted or absent oxygen medium (Thengane et al., 2020). Torrefied biomass exhibits improved hydrophobic properties, increased grindability, higher energy density, and resistance to biodegradation, making it well suited for a range of applications. Torrefaction is considered more economically beneficial than pyrolysis and gasification for producing biochar as bio-coal, as it yields a higher amount of solid carbon compared with other methods. The disadvantage of the torrefaction process for biochar formation is its resultant inadequate pore size and reduced surface area.

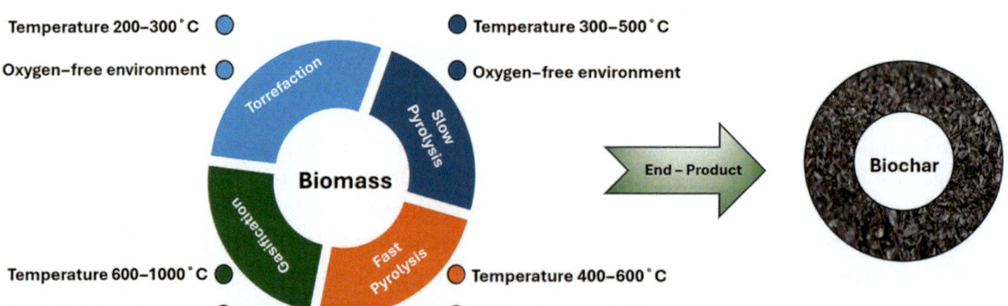


**FIGURE 4.1** Thermochemical conversion technologies for biochar synthesis from biomass.

Pyrolysis appears to be a better alternative to torrefaction when biochar is employed as a bio-adsorbent. The biochar produced by pyrolysis is generally characterized by high surface area, larger pore size, and strong zeta potential (Sharma et al., 2023). Pyrolysis may be classified as slow and rapid pyrolysis according to heating rate, pressure, and residence time. Slow pyrolysis is a decomposition process that takes place at low temperatures (ranging from 350°C to 500°C), allowing biomass sufficient time to generate vapor and optimize secondary cracking processes. Therefore, the biomass that contains cellulose, hemicellulose, and lignin generates a greater quantity of biochar than fast pyrolysis or gasification (Kumari et al., 2023). Fast pyrolysis is defined by a fast heating rate (20°C–200°C/s), wherein biomass is converted into bio-oil instead of biochar. The process shortens the contact time and promotes thermal cracking by heating the biomass above its thermal stability threshold.

Gasification is the decomposition of carbonaceous substances into gaseous products, including syngas, which comprises CO, $CO_2$, $H_2$, $CH_4$, and other hydrocarbons, within a reactor that utilizes oxygen and steam at a higher temperature (Kumari et al., 2023). Post-experiment, a solid residue including a blend of carbonaceous material, tar, ash, and a minor quantity of oil was observed. The biochemical composition of biomass significantly affects its quality and quantity. Gasification offers the benefit of high energy recovery efficiency; however, its drawbacks include complex technology and significant investment and operational costs.

### 4.4.2 Nutrient Adsorption by Pristine Biochar

Adsorption methods are regarded as effective for nutrient recovery from wastewater. This method has several benefits, such as being easy to implement, using inexpensive sorbents, and being economically viable (Marcińczyk et al., 2022). The pristine biochar produced from direct thermochemical processing of biomass feedstocks typically lacks nutrients essential for fertilization. Water contains two principal forms of nitrogen (N): ammonia nitrogen ($NH_4^+$ N) and nitrate nitrogen ($NO_3^-$ N); recovering N from wastewater can simultaneously address water pollution and resource scarcity. Activated carbon and other carbonaceous substances have demonstrated their efficacy in N recovery; thus, biochar is a viable choice for this purpose (Yang et al., 2020).

Numerous adsorption studies have been conducted to assess the removal effectiveness of ammonium nitrogen ($NH_4^+$ N) using biochar derived from various feedstocks and production temperatures. Gao et al. (2015) have evaluated the ability of biochar for ammonium ion adsorption. The authors have prepared three types of biochar from peanut shells (PS), corncobs (CC), and cotton stalks (CS) at various temperatures ranging from 300 to 600°C. Among them, the CS biochar which pyrolyzed at 300°C has shown maximum adsorption of $NH_4^+$ N ($Q_{max}$ = 17.6 mg/g).

In addition to $NH_4^+$ N, nitrate nitrogen ($NO_3^-$ N) also exists in the aqueous solutions. The ions' adsorption onto the biochar surface may be hindered by the electrostatic repulsion between the biochar's negatively charged surface and $NO_3^-$ N. Compared with $NH_4^+$ N, the biochar produced at higher temperatures exhibits relatively higher adsorption of $NO_3^-$ N (Marcińczyk et al., 2022). Biochar synthesized at higher temperatures generally possesses a high specific surface area, resulting in a substantial number of active sites on its surface. Alsewaileh et al. (2019) have reported that biochar produced at 700°C resulted in higher $NO_3^-$ N adsorption compared with the biochar produced at 300° C. As the pyrolysis temperature rises, the quantity of oxygen-containing functional groups diminishes, resulting in a loss in electrostatic repulsion between biochar and the $NO_3^-$ N ion.

Phosphate ions, a crucial macronutrient for plant development, exist in aqueous solution in many forms, such as $H_2PO_4^-$ and $PO_4^{-3}$, relying on ambient pH. The $PO_4^{-3}$ adsorption by unmodified biochar is often impeded due to the repulsion of anions and the negatively charged surface of the biochar. P adsorption on biochar is determined by the amount and availability of cations in the inorganic fraction. Several authors have suggested that the key parameter for $PO_4^{-3}$ adsorption depends on the Mg and Ca contents of the biochar (Marcińczyk et al., 2022). Jung et al. (2015) reported that biochar with a low Mg/P and Ca/P ratio resulted in poor adsorption of $PO_4^{-3}$. The adsorption of $PO_4^{-3}$ P was shown to depend on the pyrolysis temperature used in the production of biochar. Increased pyrolysis temperature correlates with elevated $PO_4^{-3}$ P adsorption values.

### 4.4.3 Nutrient Adsorption by Surface-Modified Biochar

Pristine biochar has several disadvantages, including reduced nutrient adsorption and a highly negatively charged surface, which have led to enhancements through surface modification. Modifications to biochar enhance its cation exchange capacity (CEC) and provide an additional functional group, significantly influencing the processes governing nutrient adsorption. The main mechanisms involved in nutrient adsorptions on biochar surface are electrostatic attraction, ion exchange, surface precipitation, etc. as illustrated in Figure 4.2. Various researchers have modified the biochar surface through physical, chemical, and biological modifications and recovered nutrients as presented in Table 4.1.

Surface-modified biochar may effectively eliminate $NH_4^+$ N from wastewater; yet, post-adsorption, it can serve as a fertilizer, providing a source of $NH_4^+$ N for soil. Amendments to the biochar surface enhance its CEC and include extra functional groups, significantly influencing the mechanisms governing $NH_4^+$ N adsorption (Marcińczyk et al., 2022). The incorporation of minerals

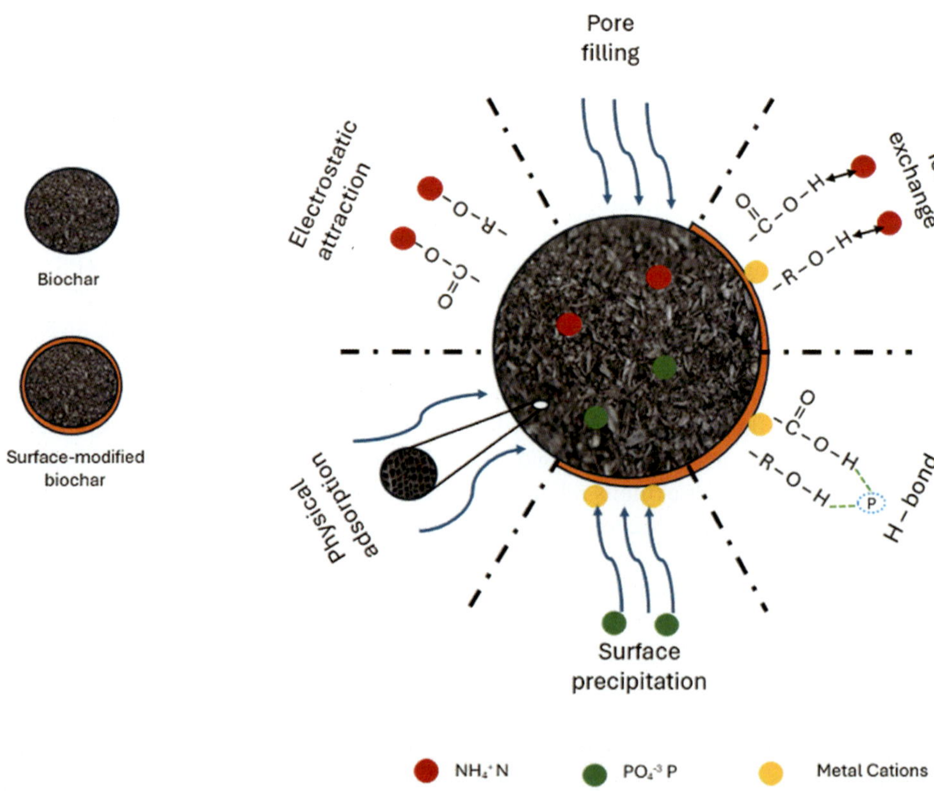


**FIGURE 4.2** Mechanism of nutrient adsorption on biochar surface.

**TABLE 4.1**
**Application of Pristine/Surface-Modified Biochar for Resource Recovery**

| S. No. | Biochar Feedstock/ Raw Material | Pyrolysis Conditions Temperature | Time | Surface Modification | Recovered Nutrient | Source of Pollutants | References |
|---|---|---|---|---|---|---|---|
| 1. | Peanut shell<br>Corncob<br>Cotton stalk | 300°C–600°C | 2 hours | – | $NH_4^+$ N | Synthetic solution prepared with ammonia nitrogen | Gao et al. (2015) |
| 2. | Raw sesame straw | 300°C–700°C | 2 hours | – | $NH_4^+$ N<br>$PO_4^{-3}$ P | Synthetic solution prepared with $NH_4Cl$ and $KH_2PO_4$ | Yin et al. (2018) |
| 3. | Pine sawdust<br>Wheat straw | 300°C–550°C | 2 hours | – | $NH_4^+$ N | Synthetic solution prepared with ammonium sulfate $(NH_4)_2SO_4$ | Yang et al. (2018) |
| 4. | Date palm rachis | 300°C–700°C | 4 hours | – | $NO_3^-$ N<br>$BrO_3^-$ | Synthetic solutions spiked with nitrate and bromate | Alsewaileh et al. (2019) |
| 5. | Peanut shell | 700°C | 3 hours | – | $PO_4^{-3}$ P | Synthetic solution prepared with $KH_2PO_4$ | Jung et al. (2015) |
| 6. | *Phragmites australis* | 600°C | 1 hours | Impregnated with $MgCl_2{\cdot}6H_2O$ | $NH_4^+$ N<br>$PO_4^{-3}$ P | Synthetic solution prepared with $NH_4Cl$ and $KH_2PO_4$ | Gong et al. (2017) |
| 7. | Poplar chips | 550°C | 2 hours | Impregnated with $AlCl_3{\cdot}6H_2O$ | $NO_3^-$ N<br>$PO_4^{-3}$ P | Synthetic solution prepared with $NaNO_3$ and $KH_2PO_4$ | Yin et al. (2018) |
| 8. | Wheat straw | 550°C | 1 hours | Impregnated with $MgCl_2$ pretreated with $NH_3{\cdot}H_2O$ | $NH_4^+$ N<br>$PO_4^{-3}$ P | Biogas slurry from pig farm | Xiao et al. (2020) |
| 9. | Palygorskite (PAL) | 400°C | 4 hours | Impregnated with $MgCl_2{\cdot}6H_2O$ | $NH_4^+$ N<br>$PO_4^{-3}$ P | Swine wastewater | Wang et al. (2019) |
| 10. | Poultry litter | 500°C | 2 hours | Impregnated with magnesium chloride ($MgCl_2$) | $PO_4^{-3}$ P | Synthetic solution prepared with $KH_2PO_4$ | Nardis et al. (2021) |
| 12. | Sawdust and sediment | 500°C | 2 hours | Impregnated with $MgCl_2$ and $FeCl_3$ | $PO_4^{-3}$ P | Synthetic solution prepared with $KH_2PO_4$ | Yang et al. (2021) |
| 13. | Korean pine residues | 500°C | 15 minutes | Impregnated with AlC13 and $Al_2(SO_4)_3$ | $PO_4^{-3}$ P | Synthetic solution prepared with $KH_2PO_4$ | Van Truong et al. (2023) |
| 14. | Tomato tissues (leaves) | 600°C | 1 hours | Mg- and Ca-enriched | $PO_4^{-3}$ P | Synthetic solution prepared with $KH_2PO_4$ | Yao et al. (2013) |
| 15. | Wheat straw | 550°C | 30 minutes | Chitosan, quaternary ammonium salt and lanthanum | $PO_4^{-3}$ P | Synthetic solution prepared with $KH_2PO_4$ | Huang et al. (2020) |

with strong CEC (e.g., montmorillonite and bentonite) into biochar may substantially increase the maximum ammonium adsorption capacity (Zhang et al., 2023). In another study, Gong et al. (2017) utilized Mg-modified biochar produced from *Phragmites australis* (MPB) for ammonium and phosphate adsorption. They have reported that the application of MgO on the MPB surface significantly enhances the adsorption capacity for both $NH_4^+$ N and $PO_4^{-3}$ P, achieving maximum capacities above 30 mg/g and 100 mg/g, respectively. Authors ascribe this phenomenon to the $Mg^{2+}$ cation exchange mechanism for $NH_4^+$ N adsorption, rather than to functional groups. Xiao et al. (2020) utilized $MgCl_2$ and ammonium hydroxide ($NH_3.H_2O$) to prepare a novel MgO-biochar for nutrient recovery. They reported that pretreatment with $NH_3.H_2O$ resulted in enhanced CEC and specific surface area, which facilitated better $NH_4^+$ N adsorption. The maximum adsorption capacity for $NH_4^+$ N was 17.5 mg/g.

Significant modifications of biochar surfaces may enhance their nitrate ($NO_3^-$ N) adsorption capability, particularly when such changes eliminate the electrostatic repulsion between $NO_3^-$ N and biochar. Wang et al. (2021) utilized a novel zeolite-based modified biochar for $NO_3^-$ N and $NH_4^+$ N adsorption. They obtained maximum $NO_3^-$ N and $NH_4^+$ N adsorption capacities of 24.45 and 24.63 mg/g, respectively. They reported that the strong adsorption of $NO_3^-$ N was due to the excess accessibility of active sites on the adsorbent surface.

Similar to $NO_3^-$ N and $NH_4^+$ N removal, researchers have aimed to enhance phosphate adsorption by including metal oxides onto the biochar surface. Magnesium is the most extensively researched, as it may substantially enhance the biochar's $PO_4^{-3}$ P adsorption capacity attributed to the strong divalent cation bridging between Mg and P. Shin et al. (2020) prepared Mg-modified biochar from ground coffee waste for $PO_4^{-3}$ P adsorption. They reported a maximum phosphate adsorption capacity of 56 mg P/g for the Mg-modified biochar. XRD and FT-IR analyses indicated that phosphates were attached to the Mg-biochar by both electrostatic adsorption and covalent bonding, resulting in the formation of magnesium phosphate crystals, $Mg_3(PO_4)_2$.

Similarly, Luo et al. (2023a) also reviewed several works that utilized Mg-, Ca-, and Fe-rich biochar for phosphate adsorption from aqueous solutions and explored fertilizer potential of these biochars. Nardis et al. (2021) found that Mg-modified biochar shows better $PO_4^{-3}$ P adsorption than the pristine biochar. They have prepared biochar from three different sources, namely, pig manure, poultry litter, and sewage sludge, and modified it with $MgCl_2$. Maximum adsorption capacity (68.0 mg/g) was observed for modified pig manure biochar. The observed adsorptions can be attributed to electrostatic attraction and ligand exchange reactions involving Mg cations.

## 4.5 BIOCHAR-BASED SLOW-RELEASE FERTILIZERS PREPARATION

The increasing price of synthetic commercial fertilizers and the limited availability of non-renewable P have led to the search for alternative nutrient blends and soil amendments. Biochar has been extensively utilized as a carrier material or adsorbent to produce carbon-based SRFs (Das & Ghosh, 2023). Biochar positively influences several soil physicochemical and biological features, including greater water retention, improved ion exchange capacity, and elevated soil pH. Biochar addition enhances soil microbial activity or can be utilized to sequester hazardous pollutants, including heavy metals (Marcińczyk & Oleszczuk, 2022a). Biochar can reduce nutrient loss in soil and extend fertilizer efficacy for plant absorption. The pristine biochar possesses less nutrients; thus, it is essential to supplement biochar with nutrients to enhance its fertilizing capabilities. Recent studies on biochar-based SRFs have been reported (Das & Ghosh, 2023; Marcińczyk & Oleszczuk, 2022b; Samoraj et al., 2022; Wang et al., 2022). Biochar-based fertilizers can be synthesized using many methods, including co-pyrolysis, combining with fertilizers, impregnation, and nutrient sorption (Figure 4.3).

### 4.5.1 Co-Pyrolysis

Co-pyrolysis involves the mixing of nutrient-rich minerals with biomass, followed by the pyrolysis process. Co-pyrolysis may be performed using a range of feedstocks, nutrient sources, and temperature ranges. Zhang et al. (2023) investigated sludge-derived multifunctional value-added

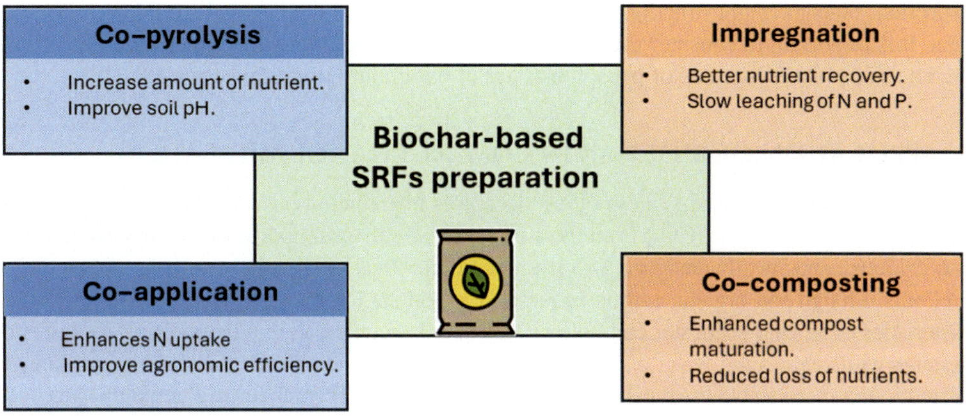


**FIGURE 4.3** Preparation method of biochar-based slow-release fertilizers.

fertilizers using the co-pyrolysis of sewage sludge and *Pistia stratiotes*, enhanced by liquid-phase activation utilizing $MgCl_2$ and $K_2CO_3$, respectively. The desorption study shows remarkably low N (20.34%) and P (33.97%) release rate for the biochar fertilizer during 0–40 minutes of desorption time. In another study, Liu et al. (2021) prepared biochar-based fertilizer via microwave-assisted co-pyrolysis by premixing $K_3PO_4$ and MgO with cotton stalk. They reported that the synthetic Biochar-based fertilizers (BBFs) exhibited a multi-level nutrient structure that was characterized by water-soluble P, precipitated P, and organic P. This structure was the result of the reactions between $K_3PO_4$ and cotton stalk and MgO.

### 4.5.2 Co-Application

The co-application of biochars with synthetic fertilizers in soil may address the issues related to inorganic fertilizers and the nutritional deficiencies inherent in biochars. Numerous studies indicate that the simultaneous application of biochar and inorganic nitrogen enhances N uptake with suboptimal physicochemical characteristics. Bai et al. (2022) reviewed published works on the co-application of biochar with synthetic fertilizers. They showed that individually biochar and synthetic fertilizers resulted in 25.3% and 21.9% increases in crop yield, respectively, while the co-application of these resulted in a 179% increase in crop yield. They reported an interesting fact that the most significant yield improvement was observed in highly acidic soils (pH ≤5); however, the advantages of biochar were unaffected by the application rate and the duration post-application.

### 4.5.3 Impregnation

Another approach for the production of biochar-based fertilizers involves the impregnation and sorption of nutrients into biochar. After sorption, the nutrients are extracted from the solution, and the fertilizer is subsequently dried out and processed (Samoraj et al., 2022). Das and Ghosh (2023) prepared a modified biochar by nutrient impregnation and mixed it with PVA solution to achieve even coating and later utilized as SRFs. They reported that the biochar inhibited the leaching of N and K in a packed soil column. Interestingly, the mobility of P was augmented by SRF in the column, hence increasing its release potential in the soil column. The biochar-based SRFs enhanced soil quality by enhancing available nutrients (5.20%–15.71%), oxidizable carbon (19.01%–37.18%), and reducing soil pH (11.74%–3.73%). Luo et al. (2023b) reviewed the application of biochar for nutrient recovery and its potential applications as SRFs. The possibility of P release has been explained by the theory of bond order. $MgHPO_4$ and $Mg_3(PO_4)_2$, resulting from P adsorption on the biochar surface, are likely to transform into more stable P analogues and MgO. This suggests that following

P adsorption, biochar might function as an SRFs to enhance soil quality and stimulate crop development. It is important to consider the soil type and the planned crop when utilizing biochar-based SRFs. The use and release rates of phosphorus are of paramount importance.

## 4.6 APPLICATION OF BIOCHAR IN CLIMATE CHANGE MITIGATION

Climate change mitigation necessitates not just decreases in greenhouse gas (GHG) emissions but also the removal of carbon dioxide ($CO_2$) from the atmosphere. A surface-modified biochar substance identical to charcoal, specifically designed as a soil conditioner to (1) reduce atmospheric carbon dioxide ($CO_2$) levels through soil organic carbon sequestration and (2) improve soil fertility. The recalcitrant characteristics of biochar sequester carbon from the atmosphere, creating a carbon sink for terrestrial ecosystems while enhancing water and soil quality (Lehmann et al., 2021). The carbon in biomass is converted into a more stable form called biochar, which persists in soils for millennia or longer. The $CO_2$ taken by the plants during photosynthesis is ultimately released into the atmosphere during breakdown or combustion. Transforming biomass into biochar would sequester the $CO_2$ absorbed during photosynthesis, preventing its release and thereby reducing $CO_2$ emissions (Oliveira et al., 2017).

Furthermore, the characteristics of biochar and soil management techniques can diminish $N_2O$ emissions by as much as 80%. The hypothesized processes involve $N_2O$ trapping inside water-saturated soil pores and microbial activity linked to biological denitrification, which would reduce the $N_2O/(N_2O + N_2)$ ratio. Consequently, the use of biochar may restrict nitrogen availability to crops, hence decreasing nitrogen losses via both direct and indirect $N_2O$ emissions, while simultaneously enhancing crop production (Oliveira et al., 2017). Thus, biochar can significantly aid in climate change mitigation through soil organic carbon sequestration at greater soil depths.

## 4.7 CONCLUSIONS AND PERSPECTIVES

This chapter provides an overview of the application of biochar for nutrient recovery and its subsequent use as a soil amendment to promote a CE, particularly in the context of climate change mitigation. Biochar adsorption procedures offer advantages over alternative methods due to their simplicity, cost-effectiveness, and environmentally friendly nature. The porous structure and numerous functional groups on pristine biochar facilitate the effective enrichment of nutrients, while surface-modified biochar enhances selective adsorption of essential macronutrients (N, P, K) from solutions. This nutrient-rich biochar can serve as an SRF, optimizing nutrient utilization by plants and contributing to sustainable agricultural practices. Although the adsorptive properties of biochar are significantly improved through metal modification (e.g., Mg, Al, Fe, Zn) or by combining biochars with minerals, challenges remain regarding the scalability and safety of biochar-based SRFs. The potential release of hazardous compounds under varying environmental conditions necessitates further investigation. Additionally, conducting economic feasibility studies is essential for assessing the cost-effectiveness and viability of large-scale SRF production systems. In the broader context of a CE, developing technologies that reduce our dependency on non-renewable resources is crucial for addressing the pressing issues of climate change and environmental degradation.

## REFERENCES

Agarwal, H., Kashyap, V. H., Mishra, A., Bordoloi, S., Singh, P. K., & Joshi, N. C. (2022). Biochar-based fertilizers and their applications in plant growth promotion and protection. *3 Biotech*, *12*(6). https://doi.org/10.1007/s13205-022-03195-2.

Allohverdi, T., Mohanty, A. K., Roy, P., & Misra, M. (2021). A review on current status of biochar uses in agriculture. *Molecules*, *26*(18). https://doi.org/10.3390/molecules26185584.

Alsewaileh, A. S., Usman, A. R., & Al-Wabel, M. I. (2019). Effects of pyrolysis temperature on nitrate-nitrogen ($NO^{3-}$-N) and bromate ($BrO^{3-}$) adsorption onto date palm biochar. *Journal of Environmental Management*, *237*, 289–296. https://doi.org/10.1016/j.jenvman.2019.02.045.

Bai, S. H., Omidvar, N., Gallart, M., Kämper, W., Tahmasbian, I., Farrar, M. B., Singh, K., Zhou, G., Muqadass, B., Xu, C. Y., Koech, R., Li, Y., Nguyen, T. T. N., & van Zwieten, L. (2022). Combined effects of biochar and fertilizer applications on yield: A review and meta-analysis. *Science of the Total Environment, 808*. https://doi.org/10.1016/j.scitotenv.2021.152073.

Buss, W., Bogush, A., Ignatyev, K., & Mašek, O. (2020). Unlocking the fertilizer potential of waste-derived biochar. *ACS Sustainable Chemistry and Engineering, 8*(32), 12295–12303. https://doi.org/10.1021/acssuschemeng.0c04336.

Das, S. K., & Ghosh, G. K. (2023). Developing biochar-based slow-release N-P-K fertilizer for controlled nutrient release and its impact on soil health and yield. *Biomass Conversion and Biorefinery, 13*(14), 13051–13063. https://doi.org/10.1007/s13399-021-02069-6.

Deng, S., Wang, C., Ngo, H. H., Guo, W., You, N., Tang, H., Yu, H., Tang, L., & Han, J. (2023). Comparative review on microbial electrochemical technologies for resource recovery from wastewater towards circular economy and carbon neutrality. *Bioresource Technology, 376*. https://doi.org/10.1016/j.biortech.2023.128906.

Gao, F., Xue, Y., Deng, P., Cheng, X., & Yang, K. (2015). Removal of aqueous ammonium by biochars derived from agricultural residuals at different pyrolysis temperatures. *Chemical Speciation and Bioavailability, 27*(2), 92–97. https://doi.org/10.1080/09542299.2015.1087162.

Gong, Y. P., Ni, Z. Y., Xiong, Z. Z., Cheng, L. H., & Xu, X. H. (2017). Phosphate and ammonium adsorption of the modified biochar based on Phragmites australis after phytoremediation. *Environmental Science and Pollution Research, 24*(9), 8326–8335. https://doi.org/10.1007/s11356-017-8499-2.

Huang, Y., Lee, X., Grattieri, M., Yuan, M., Cai, R., Macazo, F. C., & Minteer, S. D. (2020). Modified biochar for phosphate adsorption in environmentally relevant conditions. *Chemical Engineering Journal, 380*. https://doi.org/10.1016/j.cej.2019.122375.

Jing, H. P., Li, Y., Wang, X., Zhao, J., & Xia, S. (2019). Simultaneous recovery of phosphate, ammonium and humic acid from wastewater using a biochar supported Mg(OH)2/bentonite composite. *Environmental Science: Water Research and Technology, 5*(5), 931–943. https://doi.org/10.1039/c8ew00952j.

Jones, E. R., Van Vliet, M. T. H., Qadir, M., & Bierkens, M. F. P. (2021). Country-level and gridded estimates of wastewater production, collection, treatment and reuse. *Earth System Science Data, 13*(2), 237–254. https://doi.org/10.5194/essd-13-237-2021.

Jung, K. W., Hwang, M. J., Ahn, K. H., & Ok, Y. S. (2015). Kinetic study on phosphate removal from aqueous solution by biochar derived from peanut shell as renewable adsorptive media. *International Journal of Environmental Science and Technology, 12*(10), 3363–3372. https://doi.org/10.1007/s13762-015-0766-5.

Kumari, K., Kumar, R., Bordoloi, N., Minkina, T., Keswani, C., & Bauddh, K. (2023). Unravelling the recent developments in the production technology and efficient applications of biochar for agro-ecosystems. *Agriculture (Switzerland), 13*(3). https://doi.org/10.3390/agriculture13030512.

Lehmann, J., Cowie, A., Masiello, C. A., Kammann, C., Woolf, D., Amonette, J. E., Cayuela, M. L., Camps-Arbestain, M., & Whitman, T. (2021). Biochar in climate change mitigation. *Nature Geoscience, 14*(12), 883–892. https://doi.org/10.1038/s41561-021-00852-8.

Li, D., Manu, M. K., Varjani, S., & Wong, J. W. C. (2022) Role of tobacco and bamboo biochar on food waste digestate co-composting: Nitrogen conservation, greenhouse gas emissions, and compost quality. *Waste Management, 156*, 44–54. https://doi.org/10.1016/j.wasman.2022.10.022.

Liu, H., Kumar, V., Yadav, V., Guo, S., Sarsaiya, S., Binod, P., Sindhu, R., Xu, P., Zhang, Z., Pandey, A., & Kumar Awasthi, M. (2021). Bioengineered biochar as smart candidate for resource recovery toward circular bio-economy: A review. *Bioengineered, 12*(2), 10269–10301. https://doi.org/10.1080/21655979.2021.1993536.

Luo, D., Wang, L., Nan, H., Cao, Y., Wang, H., Kumar, T. V., & Wang, C. (2023a). Phosphorus adsorption by functionalized biochar: A review. *Environmental Chemistry Letters, 21*(1), 497–524. https://doi.org/10.1007/s10311-022-01519-5.

Luo, L., Wang, J., Lv, J., Liu, Z., Sun, T., Yang, Y., & Zhu, Y. G. (2023b). Carbon sequestration strategies in soil using biochar: Advances, challenges, and opportunities. *Environmental Science and Technology, 57*(31), 11357–11372. https://doi.org/10.1021/acs.est.3c02620.

Marcińczyk, M., & Oleszczuk, P. (2022a). Biochar and engineered biochar as slow- and controlled-release fertilizers. *Journal of Cleaner Production, 339*. https://doi.org/10.1016/j.jclepro.2022.130685.

Marcińczyk, M., Ok, Y. S., & Oleszczuk, P. (2022). From waste to fertilizer: Nutrient recovery from wastewater by pristine and engineered biochars. *Chemosphere, 306*. https://doi.org/10.1016/j.chemosphere.2022.135310.

Maroušek, J., & Gavurová, B. (2022b). Recovering phosphorous from biogas fermentation residues indicates promising economic results. *Chemosphere, 291*. https://doi.org/10.1016/j.chemosphere.2021.133008.

Nardis, B. O., Santana Da Silva Carneiro, J., De Souza, I. M. G., De Barros, R. G., & Azevedo Melo, L. C. (2021). Phosphorus recovery using magnesium-enriched biochar and its potential use as fertilizer. *Archives of Agronomy and Soil Science*, *67*(8), 1017–1033. https://doi.org/10.1080/03650340.2020.1771699.

Nguyen, M. N. (2021). Potential use of silica-rich biochar for the formulation of adaptively controlled release fertilizers: A mini review. *Journal of Cleaner Production*, *307*. https://doi.org/10.1016/j.jclepro.2021.127188.

Oliveira, F. R., Patel, A. K., Jaisi, D. P., Adhikari, S., Lu, H., & Khanal, S. K. (2017). Environmental application of biochar: Current status and perspectives. *Bioresource Technology*, *246*, 110–122. https://doi.org/10.1016/j.biortech.2017.08.122.

Patel, P., Muteen, A., & Mondal, P. (2020). Treatment of greywater using waste biomass derived activated carbons and integrated sand column. *Science of the Total Environment*, *711*. https://doi.org/10.1016/j.scitotenv.2019.134586.

Puyol, D., Batstone, D. J., Hülsen, T., Astals, S., Peces, M., & Krömer, J. O. (2017). Resource recovery from wastewater by biological technologies: Opportunities, challenges, and prospects. *Frontiers in Microbiology*, *7*. https://doi.org/10.3389/fmicb.2016.02106.

Rawat, S., Wang, C. T., Lay, C. H., Hotha, S., & Bhaskar, T. (2023). Sustainable biochar for advanced electrochemical/energy storage applications. *Journal of Energy Storage*, *63*. https://doi.org/10.1016/j.est.2023.107115.

Saheed, I. O., Azeez, S. O., & Suah, F. B. M. (2022). Imidazolium based ionic liquids modified polysaccharides for adsorption and solid-phase extraction applications: A review. *Carbohydrate Polymers*, *298*. https://doi.org/10.1016/j.carbpol.2022.120138.

Samoraj, M., Mironiuk, M., Witek-Krowiak, A., Izydorczyk, G., Skrzypczak, D., Mikula, K., Baśladyńska, S., Moustakas, K., & Chojnacka, K. (2022). Biochar in environmental friendly fertilizers - Prospects of development products and technologies. *Chemosphere*, *296*. https://doi.org/10.1016/j.chemosphere.2022.133975.

Samuel Olugbenga, O., Goodness Adeleye, P., Blessing Oladipupo, S., Timothy Adeleye, A., & Igenepo John, K. (2024). Biomass-derived biochar in wastewater treatment- a circular economy approach. *Waste Management Bulletin*, *1*(4), 1–14. https://doi.org/10.1016/j.wmb.2023.07.007.

Shakoor, M. B., Ye, Z. L., & Chen, S. (2021). Engineered biochars for recovering phosphate and ammonium from wastewater: A review. *Science of the Total Environment*, *779*. https://doi.org/10.1016/j.scitotenv.2021.146240.

Sharma, A. K., Ghodke, P. K., Goyal, N., Bobde, P., Kwon, E. E., Lin, K. Y. A., & Chen, W. H. (2023). A critical review on biochar production from pine wastes, upgradation techniques, environmental sustainability, and challenges. *Bioresource Technology*, *387*. https://doi.org/10.1016/j.biortech.2023.129632.

Shin, H., Tiwari, D., & Kim, D. J. (2020). Phosphate adsorption/desorption kinetics and P bioavailability of Mg-biochar from ground coffee waste. *Journal of Water Process Engineering*, *37*. https://doi.org/10.1016/j.jwpe.2020.101484.

Sun, D., Hale, L., Kar, G., Soolanayakanahally, R., & Adl, S. (2018). Phosphorus recovery and reuse by pyrolysis: Applications for agriculture and environment. *Chemosphere*, *194*, 682–691. https://doi.org/10.1016/j.chemosphere.2017.12.035.

Thengane, S. K., Kung, K. S., Gupta, A., Ateia, M., Sanchez, D. L., Mahajani, S. M., Lim, C. J., Sokhansanj, S., & Ghoniem, A. F. (2020). Oxidative torrefaction for cleaner utilization of biomass for soil amendment. *Cleaner Engineering and Technology*, *1*. https://doi.org/10.1016/j.clet.2020.100033.

Van Truong, T., Kim, Y. J., & Kim, D. J. (2023). Study of biochar impregnated with Al recovered from water sludge for phosphate adsorption/desorption. *Journal of Cleaner Production*, *383*. https://doi.org/10.1016/j.jclepro.2022.135507.

Varjani, S., Yan, W., Priya, A., Xin, F., & Lin, C. S. K. (2023). Management and valorisation strategies for transforming food waste into bio-based products: Roadblocks and the way forward, *Current Opinion in Green and Sustainable Chemistry*, *41*,100806. https://doi.org/10.1016/j.cogsc.2023.100806.

Vo, T. D. H., Pham, M. D. T., Dang, B. T., Tran, C. S., Le, T. S., Nguyen, V. T., Nguyen, T. B., Ling, C., Varjani, S., Dao, T. S., Bui, T. V., Huynh, K. P. H., & Bui, X. T. (2022). Influence of nitrogen species and biomass retention time on nutrient removal and biomass productivity in a microalgae-based bioreactor, *Environmental Technology and Innovation*, *28*, 102880. https://doi.org/10.1016/j.eti.2022.102880.

Wang, C., Luo, D., Zhang, X., Huang, R., Cao, Y., Liu, G., Zhang, Y., & Wang, H. (2022). Biochar-based slow-release of fertilizers for sustainable agriculture: A mini review. *Environmental Science and Ecotechnology*, *10*. https://doi.org/10.1016/j.ese.2022.100167.

Wang, H., Wang, X., & Zhao, J. (2019). Application of MgO-modified palygorskite for nutrient recovery from swine wastewater: Effect of pH, ions, and organic acids. *Environmental Science and Pollution Research*, *26*(19), 19729–19737. https://doi.org/10.1007/s11356-019-05254-3.

Wang, H. S. H., & Yao, Y. (2023). Machine learning for sustainable development and applications of biomass and biomass-derived carbonaceous materials in water and agricultural systems: A review. *Resources, Conservation and Recycling*, *190*. https://doi.org/10.1016/j.resconrec.2022.106847.

Wang, P., Kumar, R., Xue, W., Luo, L., Varjani, S., Wong, J. W. C., & Zhao, J. (2024). Strategies for enhancing catalytic efficiency and stability of MgO-biochar catalysts in glucose isomerization to fructose. *Industrial Crops and Products*, *207*(2), 117769 https://doi.org/10.1016/j.indcrop.2023.117769.

Wang, Y., Song, X., Xu, Z., Cao, X., Song, J., Huang, W., Ge, X., & Wang, H. (2021). Adsorption of nitrate and ammonium from water simultaneously using composite adsorbents constructed with functionalized biochar and modified zeolite. *Water, Air, and Soil Pollution*, *232*(5). https://doi.org/10.1007/s11270-021-05145-9.

Wu, H., & Vaneeckhaute, C. (2022). Nutrient recovery from wastewater: A review on the integrated Physicochemical technologies of ammonia stripping, adsorption and struvite precipitation. *Chemical Engineering Journal*, *433*. https://doi.org/10.1016/j.cej.2021.133664.

Xiao, R., Zhang, H., Tu, Z., Li, R., Li, S., Xu, Z., & Zhang, Z. (2020). Enhanced removal of phosphate and ammonium by MgO-biochar composites with $NH_3 \cdot H_2O$ hydrolysis pretreatment. *Environmental Science and Pollution Research*, *27*(7), 7493–7503. https://doi.org/10.1007/s11356-019-07355-5.

Yang, F., Chen, Y., Nan, H., Pei, L., Huang, Y., Cao, X., Xu, X., & Zhao, L. (2021). Metal chloride-loaded biochar for phosphorus recovery: Noteworthy roles of inherent minerals in precursor. *Chemosphere*, *266*. https://doi.org/10.1016/j.chemosphere.2020.128991.

Yang, H., Ye, S., Zeng, Z., Zeng, G., Tan, X., Xiao, R., Wang, J., Song, B., Du, L., Qin, M., Yang, Y., & Xu, F. (2020). Utilization of biochar for resource recovery from water: A review. *Chemical Engineering Journal*, *397*. https://doi.org/10.1016/j.cej.2020.125502.

Yang, H. I., Lou, K., Rajapaksha, A. U., Ok, Y. S., Anyia, A. O., & Chang, S. X. (2018). Adsorption of ammonium in aqueous solutions by pine sawdust and wheat straw biochars. *Environmental Science and Pollution Research*, *25*(26), 25638–25647. https://doi.org/10.1007/s11356-017-8551-2.

Yao, Y., Gao, B., Chen, J., Zhang, M., Inyang, M., Li, Y., Alva, A., & Yang, L. (2013). Engineered carbon (biochar) prepared by direct pyrolysis of Mg-accumulated tomato tissues: Characterization and phosphate removal potential. *Bioresource Technology*, *138*, 8–13. https://doi.org/10.1016/j.biortech.2013.03.057.

Yin, Q., Zhang, B., Wang, R., & Zhao, Z. (2018). Phosphate and ammonium adsorption of sesame straw biochars produced at different pyrolysis temperatures. *Environmental Science and Pollution Research*, *25*(5), 4320–4329. https://doi.org/10.1007/s11356-017-0778-4

Zhang, H., Liu, W. H., Liu, C. X., Sun, P., Zeng, Y. P., Gao, Y. Y., Wang, H. F., & Zeng, R. J. (2023). Enhancing waste management and nutrient recovery: Preparation of adsorption-type sludge-biochar value-added fertilizer from sewage sludge and Pistia stratiotes. *Journal of Cleaner Production*, *429*. https://doi.org/10.1016/j.jclepro.2023.139642.

# 5 Bioconversion Route of Oil Palm Residues to Produce Holocellulose-Based Products

*Made Tri Ari Penia Kresnowati*[1,2], *Ibnu Maulana Hidayatullah*[3,4], *Heri Hermansyah*[3,4], *and Tjandra Setiadi*[1]

[1]Department of Chemical Engineering, Faculty of Industrial Technology, Institut Teknologi Bandung, Bandung, Indonesia

[2]Food Engineering Program, Faculty of Industrial Technology, Institut Teknologi Bandung, Kampus Jatinangor, Sumedang, Indonesia

[3]Department of Chemical Engineering, Faculty of Engineering, Universitas Indonesia, Depok, Indonesia

[4]Research Centre for Biomass Valorization, Faculty of Engineering, Universitas Indonesia, Depok, Indonesia

## 5.1 INTRODUCTION

Oil palm is the most important plantation commodity with the largest cultivation in Southeast Asia. Indonesia, together with Malaysia, contributes 85% of total oil palm production worldwide (Purnomo et al. 2020). Several industrial sectors need oil palm as a raw material for making their products, such as foods, cosmetics, pharmaceuticals, energy, chemicals, and other industries. The contribution of oil palm products and their derivatives boosts the economy of developing countries, making plantation commodities quite an attractive and promising business.

Oil palm solid waste comprises a number of biomass types, which are quite abundant as a logical consequence of crude palm oil (CPO) production activities. Oil palm solid waste is composed of varying levels of lignocellulosic structures (cellulose, hemicellulose, and lignin) (Ahmad et al. 2019). The utilization of oil palm solid waste is seen to be crucial for increasing the use of biomass waste into semi-finished products and final products with high market value. Holocellulose (cellulose and hemicellulose) is a major polymer component that can be used to produce valuable chemical products.

Throughout its history, waste biomass has been utilized as a limited source of renewable energy and fuels. In general, pyrolysis in a biomass gasification system is used in the production of sustainable energy and fuels (Parthasarathy et al. 2022). The production method evidently necessitates high temperatures and expenditure, and it frequently generates environmental issues due to the usage of environmentally harmful chemical solvents (Kahar et al. 2022). Biological processing can be used for alternative uses of biomass waste. The advantage of biological processing is not only to mitigate the drawbacks of chemical processes but also to produce multiple useful chemical products simultaneously.

Pretreatment, enzymatic hydrolysis, and fermentation are all required steps in the bioconversion of oil palm solid waste into valuable chemical products (Figure 5.1). Pretreatment is an important step in obtaining holocellulose monomer sugar (reducing sugar) as the primary precursor of its derivative products. The appropriate pretreatment strategy impacts the success of the enzymatic hydrolysis process, which is indicated by high reducing sugar production and good


DOI: 10.1201/9781032676890-6

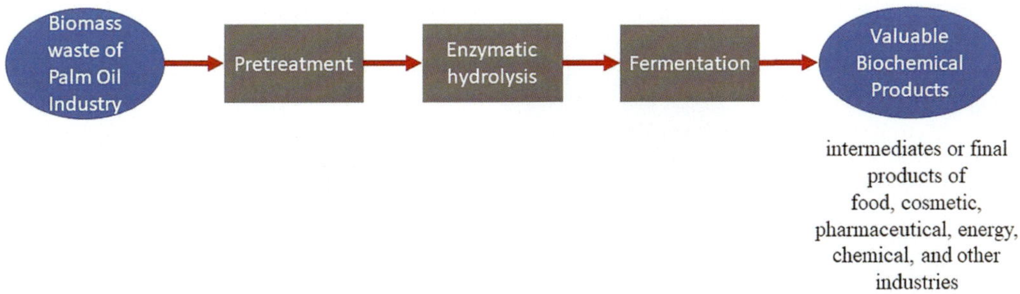


**FIGURE 5.1** Bioconversion of oil palm solid waste into valuable chemical products.

hydrolysing enzyme activity. Another critical aspect is the selection of target product-producing microorganisms. The target product's biosynthetic agents are determined by microbes. Factors such as influencing metabolic systems towards target products, managing cultivation conditions, and genetic alteration are thought to be crucial to undertake in order to maximize product yield.

This book chapter discusses the most recent bioconversion technology to assess the future potential and problems of biosynthesis of valuable chemical compounds from oil palm solid waste.

## 5.2 OIL PALM RESIDUES AS A SOURCE OF HOLOCELLULOSE-BASED CHEMICAL PRODUCTS

Different species of oil palm can vary in both the chemical composition and the structure of the plant. By examining the fruit's structure, it is possible to distinguish between the three most common types of oil palm. The most frequent wild variety is the dura, which has a very thin coating of flesh covering the seed and the thickest shell compared to other varieties. A small percentage of pisifera oil palms produce fruit with a thick coating of flesh and a small kernel with a very thin or absent shell. Tenera, a small proportion of plants in the wild, has rather thick flesh with a medium-sized shell and kernel (Berger 2003). Tenera is a result of a hybrid between the thick-shelled dura (D) parent and the shellless (often female-sterile) pisifera (P) parent, and it has been the dominant commercial planting material to this day (Soh 2012).

Holocellulose is abundant in oil palm solid waste. Oil palm solid wastes include fronds (oil palm fronds [OPFs]), trunks (oil palm trunk [OPT]), empty fruit bunches (EFBs), seed shells (oil palm seed shell [OPS]), and fibres from pressed fruit's mesocarp (PFF) (Figure 5.2). As the oil palm tree grows and produces fruit bunches, it produces biomass such as OPF, EFB, OPS, and PFF. For every 25 metric tonnes of fresh fruit bunches, 16 metric tonnes of EFBs are produced (Sridhar and AdeOluwa 2009). When the seed shells and mesocarp fibre (MF) from the pressed fruit are included, the total residues of the complete fruit bunches will be larger, yielding tonnes of residues each harvest.

Holocellulose consists of the whole polysaccharide component (cellulose and hemicellulose) of extractives-free biomass (Miyazaki 2005). The concentration of holocellulose in a biomass may vary and is inversely proportional to the amount of lignin. Lignin is the encrusting material in which cellulose microfibrils are embedded and dispersed alongside hemicellulose in the voids of intercellulose microfibrils in primary and secondary walls (Donaldson et al. 2016; Lu and Ralph 2010). Hemicelluloses are hetero-polysaccharides with complex structures comprising varying proportions of glucose, xylose, arabinose, galactose, mannose, fucose, galacturonic acid, and glucuronic acid, depending on their source (Ren and Sun 2010). Cellulose is a linear homopolymer consisting of D-glucose; it consists of a cluster of threads that may contain 6,000–16,000 glucose units and serves as the framework of the cell wall (Liu and Sun 2010). The interaction between these three elements determines the structure and rigidity of various plant parts.

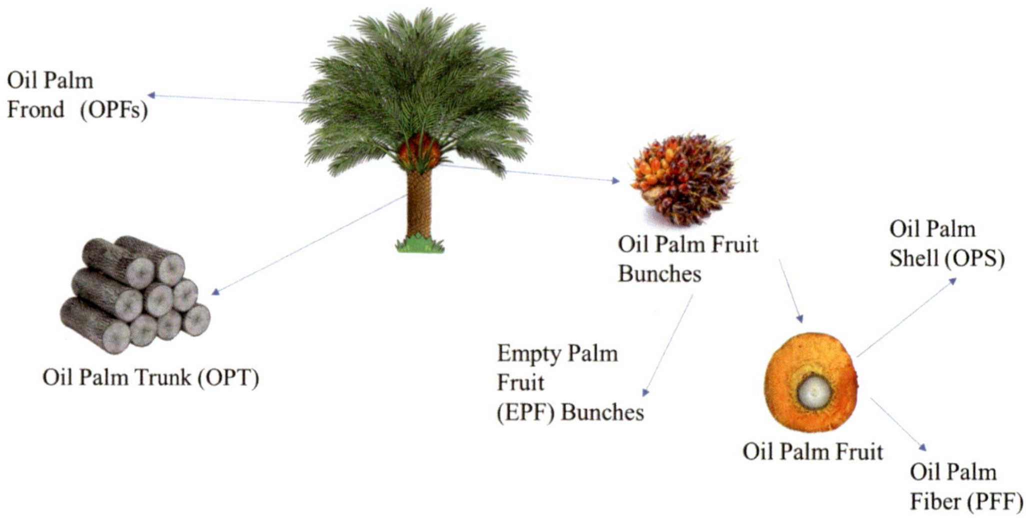


**FIGURE 5.2** Oil palm residues.

**TABLE 5.1**
**Chemical Composition of Oil Palm Biomass Waste**

| Biomass Types | Holocellulose (%) | Cellulose (%) | Hemicellulose (%) | Lignin (%) | Extractive (%) | Ash (%) |
|---|---|---|---|---|---|---|
| EFB[a] | 68–86 | 43–65 | 17–33 | 13–37 | 2–4 | 1–6 |
| OPF[a] | 80–83 | 40–50 | 34–38 | 20–21 | 2–5 | 2–3 |
| OPT[a] | 42–45 | 29–37 | 12–17 | 18–23 | 4–7[a] | 2–3 |
| OPS[a] | 40–47 | 27–35 | 15–19 | 48–55 | 0.9–2[a] | 1–4 |
| PFF[b] | N/A | 21–44 | 23–38.2 | 17.28–33 | N/A | N/A |

*Note:* Data of EFB, OPF, OPT, and OPS are from Dungani et al. (2018)[a] and data of PFF are from Ng et al. (2022)[b].
*EFBs, empty fruit bunches; OPFs, oil palm fronds; OPTs, oil palm trunks; OPSs, oil palm seed shells; PFF, oil palm press fruit fibre; N/A, not available.*

Table 5.1 presents the chemical composition of oil palm biomass. Several factors, such as plant type, age, geographical condition, location, and soil conditions, might influence chemical composition (Dungani et al. 2018). The compositions of cellulose, hemicellulose, and lignin are altered by pretreatment methods for biomass. The content of cellulose, hemicellulose, and lignin varied after pretreatment of palm MF by a range of 37.7%–66%, 14.4%–34.67%, and 4.9%–25%, respectively (Ng et al. 2022). Variations in the content of pretreated biomass may result from the application of different chemicals and processes (Ying et al. 2014; Ng et al. 2022).

## 5.3 DELIGNIFICATION STRATEGY

Delignification is the removal of lignin from plant sources through the disintegration of the lignocellulosic structure into its fibrous constituents (Wool and Sun 2005). The aim is to separate holocellulose from its lignin bond so that it can be utilized in subsequent processing steps. By adjusting its delignification/pretreatment procedure, holocellulose yield can be increased. There are numerous pretreatment processes available, and their combination can increase the holocellulose content. The four primary strategies for delignification are physical, chemical, physicochemical, and biological methods. Some information on the delignification strategy is shown in Figure 5.3.

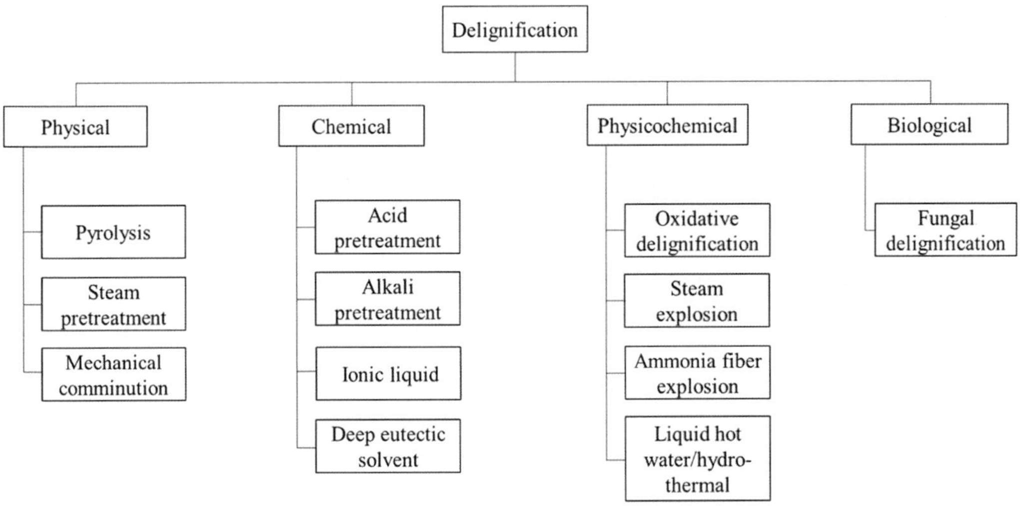


**FIGURE 5.3** Types of oil palm biomass delignification strategy.

## 5.3.1 Physical Delignification

### 5.3.1.1 Mechanical Pretreatment

The simplest delignification involves the mechanical fragmentation of lignocellulosic chains into smaller pieces. The purpose of mechanical pretreatment is to shear and reduce the particle size and crystallinity of a biomass. The increase in specific surface area, the decrease in the degree of polymerization (DP), and the shearing are all factors that increase the total hydrolysis yield of lignocellulose, in most cases by 55%–25% (depending on the type of biomass, the type of milling, and the duration of the milling), while decreasing the technical digestion time by 23%–59% (thus an increase in hydrolysis rate) (Sołowski et al. 2020). Ethanol and methanol can be produced from biomass using mere milling as a pretreatment, which improves yield and hydrolysis rate (Hendriks and Zeeman 2009).

Another study implied that a delignification procedure involving only mechanical pretreatment was not commercially viable (Ramos 2003). Grinding is one of the most prevalent and energy-intensive operations in contemporary technology; it is one of the top five processes that consume the most energy. The energy needed for mechanical comminution is primarily dependent on the properties of the biomass and the required final particle size. There are a variety of milling machines available, including vibratory milling, ball milling, disc milling, and hammer milling. Different milling machines can determine the ultimate particle size of biomass, as well as the amount of energy required for grinding (Singh et al. 2014; Mussato 2016).

Oil palm biomass is inflexible and has a low specific surface area when compared to other biomass such as rice straw. Mechanical pretreatment, or at the very least basic cutting of the biomass, has become prevalent when using oil palm biomass such as OPT, OPF, and EFB due to their low surface area, which minimizes delignification (Hidayatullah et al. 2021a; Indera Luthfi et al. 2016).

### 5.3.1.2 Pyrolysis

Pyrolysis is a type of thermochemical process that occurs in the absence of oxygen at high temperatures. During the conversion of lignocellulosic biomass with pyrolysis technology, typically the three basic constituent components of biomass—cellulose, hemicellulose, and lignin—are thermally decomposed (Lu and Gu 2022). Cellulose has been utilized widely in the pyrolysis process because of its comparatively simple structure. In summary, this process begins with depolymerization at approximately 350°C and continues with the conversion of cellulose into active cellulosic species such as levoglucosan (LGA), levoglucone (LGO), and other monosaccharides via bond breakage and dehydration. Anhydrous sugars are the primary pyrolytic products of cellulose, and at higher temperatures, they can be converted into a small number of light oxygenates (Gu et al. 2013). Extensive usage of pyrolysis of oil palm biomass is applied for bio-oil production. The majority of oil palm biomass pyrolysis is conducted between 450°C and 600°C in order to boost the production of bio-oil and the key compounds such as phenols, LGA, and furans. Bio-oil production from co-pyrolysis and catalytic co-pyrolysis of oil palm biomass has not yet been demonstrated on a large scale; therefore, an ongoing study is necessary to speed its future development (Terry et al. 2021).

#### 5.3.1.3 Steam Pretreatment

Delignification of biomass by increasing temperature and pressure with steam is referred to as steam pretreatment. Under high pressure, hot steam (190°C–240°C) is used to process biomass. At the high steam temperature of 240°C, pyrolysis occurs, and it was determined that it is an unfavourable side reaction that results in low yields of reducing sugars from enzymatic hydrolysis (Saddler et al. 1986). Various studies have demonstrated that steam pretreatment effectively hydrolyses hemicelluloses, modifies lignin, improves surface area, and reduces cellulose crystallinity and polymerization degree (Singh et al. 2014).

A study indicated that the explosion phase of the steam explosion (SE) process contributes minimally or not at all to the accessibility of cellulose in aspen wood (Öhgren et al., 2005). In recent years, SE has remained the treatment of choice, as it requires quick heating and decompression and so reduces treatment time. Few researchers have detected its potential; however, chemical compounds are frequently utilized to improve the process. Using corn stover soaked in $SO_2$ and treated with steam at 200°C for 10 minutes, Öhgren et al. (2005) obtained a maximum overall yield of 89% glucose and 79% xylose. There is currently no oil palm delignification using steam pretreatment; SE is commonly employed to delignify oil palm residue.

### 5.3.2 Chemical Delignification

#### 5.3.2.1 Acid Pretreatment

Acid delignification uses an organic or inorganic acid catalyst to dissolve lignin and improve cellulose accessibility. Various organic acids, including nitric acid, acetic acid, and maleic acid, as well as inorganic acids such as hydrochloric acid and sulphuric acid, have been explored for their potential use. Acid catalysts are often employed for the delignification of biomass due to their low cost, capacity to alter the crystallinity of cellulose, and ability to weaken the hydrogen bonds between glucose units in cellulose, hence decreasing the recalcitrance of lignocellulosic biomass. Numerous studies have demonstrated that acid pretreatment increases sugar yields during enzymatic saccharification (Banu et al. 2022).

Acidic pretreatment of oil palm residual biomass yields encouraging results. Aishah et al. (2019) examined the effects of nitric acid on EFB delignification. The maximum amount of xylan, 0.5 g/L, was achieved following agitation-free pretreatment for 24 hours at 0.01% acid concentration. Temperature affects xylan recovery more than any other factor, according to the findings. Recent research indicates that inorganic acid has a significant delignification potential for oil palm residual biomass. Using 1% sulphuric acid yields the greatest xylose recovery from OPT, whereas 5% oxalic acid yields a comparable result (Bukhari et al. 2020). Unfortunately, the acid's corrosive qualities may limit its applications.

#### 5.3.2.2 Alkali Pretreatment

Even under ambient conditions, alkaline pretreatment is performed at low temperature and pressure. It was found that the effects could take anywhere between hours and days. Temperature increase can improve the efficacy of the treatment. Alkaline reactants include calcium, potassium, sodium, barium, lithium, and ammonium hydroxides ($NH_4OH$). The initial processes of alkaline pretreatment include solvation and saponification, which swell the biomass and make it more accessible to enzymes and microorganisms. Alkaline hydrolysis involves the saponification of intermolecular ester bonds cross-linking xylan hemicelluloses and other components, and the breakdown of the cross-links increases the porosity of the lignocellulosic biomass. Alkaline hydrolysis is more successful for biomass with a low lignin content, such as agricultural leftovers, depending on the substrate and treatment parameters. Sodium hydroxide (NaOH) increases the internal surface of cellulose, reduces polymerization and crystallinity, and disturbs the structure of lignin (Xu and Sun 2016).

The significance of alkaline reactant for delignification of oil palm EFB is demonstrated by a study comparing $NH_4OH$ and NaOH. Pretreatment with NaOH and $NH_4OH$ results in the loss of aromatic functional groups C=O acetyl group of hemicellulose ester or carbonyl ester of p-coumaric lignin, as seen by FTIR spectra. Crystallinity of $NH_4OH$-treated OPEFB is comparable to that of NaOH, indicating that both are theoretically capable of enhancing the efficiency of enzyme hydrolysis. Enzymatic hydrolysis results indicate that NaOH pretreatment produced 351.51 mg/ml of glucose, whereas $NH_4OH$ only produced 168.58 mg/ml (Zulkiple et al. 2016).

#### 5.3.2.3 Ionic Liquid and Deep Eutectic Solvents

Pretreatment of lignocellulosic biomass by dissolving lignin with innovative green solvents has proven to be effective and environmentally friendly. Ionic liquids (ILs) refer to a class of molten salts with a low melting point (100°C) and a combination of cations and anions that imparts many distinct features. The classification of ILs includes aprotic ionic liquids (AILs), solvate ionic liquids (SILs), and protic ionic liquids (PILs). By breaking the hydrogen bond network in lignocellulose, IL acts as a solvent for cellulose, lignin, and hemicellulose, thereby increasing its susceptibility to further transformation (Pang et al. 2016). Other kinds of solvent delignification involve a deep eutectic solvent (DES), which is a mixture of two substances formed by the interaction of hydrogen bonds; the melting point of the eutectic mixture is lower than that of either component. DES can be used to replace IL, which is notoriously harmful and costly.

The application of IL for delignification of oil palm biomass typically involves submerging the biomass in a heated tank. With the addition of an anti-solvent, IL would dissolve lignin and hemicellulose, leaving the majority of cellulose precipitated. Pretreatment of OPF with 1-ethyl-3-methylimidazolium diethyl phosphate decreases the lignin percentage of the generated material from 24.0%–8.5% wt.% (Financie et al. 2016). Another study used oxidative IL pretreatment by delivering oxygen to biomass- and IL-containing reactors. With $O_2$-BmimCl (1-butyl-3-methylimidazolium chloride) pretreatment, the hydrogen bond network can be effectively broken and mass transfer can be improved, hence enhancing lignin depolymerization and extraction (Pang et al. 2016).

The delignification of EFB by DES demonstrates that DES effectively reduces the crystallinity of cellulose. Even though the functional groups of lignin and hemicellulose were still intact, DES produced from choline chloride and urea was able to disclose the structure of cellulose. Unfortunately, the glucose and xylose yields from DES pretreatment following enzymatic hydrolysis were considerably lower than those from NaOH pretreatment (Nor et al. 2016). Thi and Lee (2019) conducted a comprehensive comparison of DES on the pretreatment of EFB and found that choline chloride-lactic acid produced the maximum reducing sugar yield (20.7%) after enzymatic hydrolysis. Comparing the yield of DES pretreatment to that of standard dilute $H_2SO_4$ and NaOH pretreatment, DES pretreatment produced a much lower yield of reducing sugar under the same pretreatment conditions.

### 5.3.3 Physicochemical Delignification

#### 5.3.3.1 Oxidative Delignification

Using an oxidant to delignify lignocellulosic biomass lowers cellulose crystallinity and destroys the bond between carbohydrates and lignin. Methods for delignification include pretreatment with alkaline peroxide, ozonolysis, and wet oxidation. Wet oxidation is accomplished by rapidly treating biomass with water, air, or oxygen at a high temperature. Hemicellulose is extensively solubilized and predominantly recovered as oligosaccharides. Lignin undergoes fragmentation and oxidation to produce phenolic chemicals and aliphatic carboxylic acids.

Hermansyah et al. (2019) conducted a study that combined acidic and oxidative pretreatments. The delignification of EFB was accomplished with peracetic acid and alkaline peroxide, with ultrasound aiding in both pretreatments. The procedure involves peracetic acid pretreatment, filtration, neutralization, washing, and drying, followed by an alkaline peroxide pretreatment, filtration, neutralization, washing, and drying. Compared to untreated EFB, the technique reduces lignin content by 68.37% while increasing cellulose content by 121.82%. It should be highlighted that 89.38% of hemicellulose was lost throughout the entire pretreatment procedure.

In recent years, ozonolysis has garnered considerable interest for the processing of oil palm residue. The ozonolysis pretreatment brings ozone gas to the surface of the wetted biomass to destroy lignin linkages with minimum degradation of cellulose and hemicelluloses. Due to its high oxidative properties, ozone permits simultaneous delignification and oxidation without the formation of hazardous chemicals. Using a new ozone reactor, the pretreatment can degrade lignin by as much as 78% by weight (Shamjuddin et al. 2021). In a lab-scale ozonolysis pretreatment, previous research has demonstrated a high lignin degradation rate of 84.7%, while total reducing sugar recovery reaches 99.9% (Wan and Amin 2016). Another EFB pretreatment study demonstrates that ozonolysis is an efficient approach for delignifying lignin up to 63.86% and increasing cellulose up to 40.9% (Mardawati et al. 2019).

#### 5.3.3.2 Steam Explosion

Using steam, the biomass is swiftly heated to the desired temperature without considerable dilution of the resultant sugars. SE is a hydrothermal pretreatment in which biomass is rapidly heated in a reactor by adding high-pressure saturated steam at temperatures typically between 160°C and 260°C (0.69–4.83 MPa) over a brief length of time that can range from seconds to several minutes. Theoretically, the quick release of pressure assists in defibrillating the cellulose bundles, hence enhancing the accessibility of cellulose for enzymatic hydrolysis and fermentation. While this phenomenon is controversial, there is no doubt that steam improves cellulose accessibility. Autohydrolysis and depressurization are the two steps of an SE. During the autohydrolysis process, a high temperature promotes the creation of acetic acid from acetyl groups attached to hemicellulose. The acetic acid produced further accelerates the hydrolysis of the hemicelluloses. The stage of depressurization reduces the biomass particle size and liberates the biomass's particulate structure, hence increasing the cellulose's enzymatic accessibility (Mosier et al. 2005; Duque et al. 2016). Table 5.2 shows the SE treatment that is applied to various oil palm biomass.

**TABLE 5.2**
**Oil Palm Biomass Delignification Using Steam Explosion**

| | | After Pretreatment | | |
|---|---|---|---|---|
| **Biomass** | **Conditions** | **Cellulose** | **Hemicellulose** | **Lignin** |
| Empty fruit bunch[a] | 195°C, 6 minutes | Increase 34.69% glycan (mostly cellulose) compared to raw biomasses | Reduction of 68.12% hemicellulose compared to raw biomasses | N/A |
| Oil palm mesocarp[b] | 168°C, 10 minutes | N/A | N/A | Increase lignin yield by ±15% in pretreatment water |
| Oil palm frond[c] | 210°C, 4 minutes | 6.84 glucose in soluble fraction, 60.83% cellulose in water insoluble fractions | 23.49% in water-soluble fractions, 6.91% in water insoluble fractions | 4.27 in water-soluble fraction, 21.23 in water insoluble fraction |

*Note:* Data of empty fruit bunch are from Medina et al. (2016)[a], oil palm mesocarp is from Pereira Marques et al. (2021)[b], and oil palm frond is from Sabiha-Hanim et al. (2015)[c]; N/A = not available.

The glycan content of oil palm biomass delignified with SE is greater than that of untreated biomass. The content of acetic acid in liquid hydrolysate increased according to the intensity of treatment, as a result primarily of autocatalytic hydrolysis and deacetylation of hemicellulose and lignin. The severity of pretreatment (time and temperature) has a positive effect on the enzymatic hydrolysis of steam-exploded biomass. However, by intensifying the pretreatment, more cellulose and hemicellulose are removed into the liquid fraction. For hemicellulose and cellulose, the proportion of soluble sugar polysaccharides in the liquid fraction is less than 25%; therefore, if just the solid fraction is utilized, a portion of sugar polysaccharides will be lost. According to studies, the presence of levulinic and formic acid indicates that furfural and hydroxymethylfurfural (HMF) have become hydrated. The aliphatic acids generated function as inhibitors for subsequent processes; solids are often washed to remove inhibitory substances (Medina et al. 2016).

The SE method can be combined with alkaline preparation. Soaking OPF in water before SE resulted in a greater yield of the water-soluble fraction than soaking OPF in potassium hydroxide (KOH) solution prior to SE; however, soaking OPF in a KOH solution prior to SE resulted in a greater fractionation efficiency for hemicellulose extraction (Sabiha-Hanim et al. 2015).

#### 5.3.3.3 Ammonia Fibre Explosion

Pretreatment with ammonia is regarded as superior to acid pretreatment. High lignin removal (30%–80%) can be accomplished while preserving cellulose and minimizing the formation of inhibitory components. A method of maize stover pretreatment by ammonia recycle percolation can solubilize 50% of xylan while retaining greater than 92% of the cellulose content, as well as a high enzyme digestibility of 93% at 10 FPU/g-glucan enzyme loading (Tae and Lee 2005). The effectiveness of ammonia as a pretreatment has been improved by including the SE principle in the pretreatment procedure. Ammonia fibre expansion (AFEX) utilizes a batch reaction in which liquid ammonia is injected into the biomass under regulated pressure (35–50 bar) and temperature (135°C). By evaporating ammonia molecules, the quick fall in pressure in the treatment vessel depolymerizes the lignocellulosic biomass. The expansion of ammonia exerts a shear force on the surrounding structure, causing the EFB fibrils to be mechanically divided further. AFEX processing resulted in the deformation of waxy lignin and the separation of initially linked microfibrils, thereby exposing the inner surface and making the cellulose more accessible. Probably, the deformation of the silicified waxy surface increased the surface area, resulting in enhanced enzyme accessibility and enzymatic saccharification (Abdul et al. 2016; Table 5.3).

#### 5.3.3.4 Hydrothermal/Liquid Hot Water

Instead of chemical reactants or steam, hot water can be used to delignify wood simply and efficiently. The goal of the liquid hot water pretreatment is to dissolve hemicellulose and lignin in order to improve cellulose accessibility. During the pretreatment process, the pH is kept between

**TABLE 5.3**
**Acid Pretreatment for Oil Palm Residue Delignification**

| Biomass | Acid | Acid Concentration (%) | Sugar Concentration |
|---|---|---|---|
| Oil palm frond[a] | Nitric acid | 0.01 | 0.5 g/L xylan |
| Oil palm empty fruit bunches[b] | Maleic acid | 1.5 | 22.32% reducing sugar yield |
| Oil palm trunk[c] | Citric acid | 5 | Xylose recovery 12.9% |
| | Formic acid | 5 | Xylose recovery 20.1% |
| | Oxalic acid | 5 | Xylose recovery 61.2% |
| | Sulphuric acid | 1 | Xylose recovery 64.8% |

*Note:* Data are from Aishah et al. (2019)[a], Fatriasari et al. (2021)[b], and Bukhari et al. (2020)[c].

4 and 7 to minimize the development of monosaccharides, hence reducing the formation of product degradation, which can result in undesirable reactions and inhibitor chemicals (Hendriks and Zeeman 2009).

High-temperature hydrothermal processing of EFB results in a greater monomeric sugar concentration. As solid loading increases, glucose and xylose recovery will decrease, and the addition of reactants may either increase or decrease monomeric sugar. Recent research indicates that hydrothermal glucose extraction at a concentration of 200 mg/g EFB was possible (Yusuf and Kresnowati 2019). The monomeric sugar production is unaffected by the combination of hydrothermal pretreatment with acetic acid and ammonia. It appears that extra reactant may be necessary to boost glucose production, although xylose yield is low in comparison with glucose yield (Hidayatullah et al. 2021b).

### 5.3.4 Biological Delignification

Microorganisms and their enzymes provide an eco-friendly way for delignifying lignocellulosic biomass. Several species of microorganisms are capable of delignifying biomass by generating extracellular enzymes that break down lignin, including manganese peroxidase, lignin peroxidase, and laccase. The prominent microorganisms that produce these types of enzymes are typically white-rot fungi. Enzymes that degrade lignin target lignin bundles, thereby exposing the matrix of hemicellulose and cellulose and increasing their susceptibility to further conversion.

Delignification by *Marasmius* sp. has been shown to break down lignin by 26.67%. Xylose production reaches 166 mg/g EFB, whereas glucose production reaches 283 mg/g OPEFB; however, pretreatment by *Phanerochaete chrysosporium* produces minor results. The combination of fungal delignification and hydrothermal pretreatment increased the yield of monomeric sugars (Hidayatullah et al. 2021a). *Pycnoporus sanguineus*, another type of white-rot fungi, can also delignify biomass by generating laccase. A study demonstrates that *P. sanguineus* can break down 11.67% of total lignin from EFB after 10 days of incubation at 30°C (Hafid et al. 2021).

Delignification is an essential step before polysaccharides can be utilized for the desired products. Determining the best delignification process provides a high yield of sugar monomers. However, several of the previously stated delignification ways are expensive and difficult to carry out. As a result, alternative solutions to this problem are being explored by improving operational variables (temperature, processing duration, solvent selection) in different delignification procedures that are cost-effective. A different approach is to employ a combination of several delignification procedures. Hidayatullah et al. (2021a) and Putri et al. (2023) showed that combining the delignification approach resulted in increased recovery of sugar monomers (glucose and xylose) and fermentation products (succinic acid).

## 5.4 BIOLOGICAL PROCESS TO PRODUCE HOLOCELLULOSE-BASED CHEMICAL PRODUCTS

Various beneficial holocellulose-based chemical compounds can be synthesized via biological processing, that is fermentation, beginning with biofuel, biogas, and alcohols, organic acids such as succinic acid, 5-HMF, and levulinic acid are produced. The majority of these chemicals are derived from glucose, which can be obtained by hydrolysing holocellulose. In lignocellulosic material bioconversion, there are three essential processes: delignification, saccharification, and fermentation (Figure 5.1). Due to the fact that each microbe produces different enzymes, each process can be completed using a distinct microorganism. This section will solely address the saccharification and fermentation processes, as delignification was covered in the preceding section (Figure 5.3).

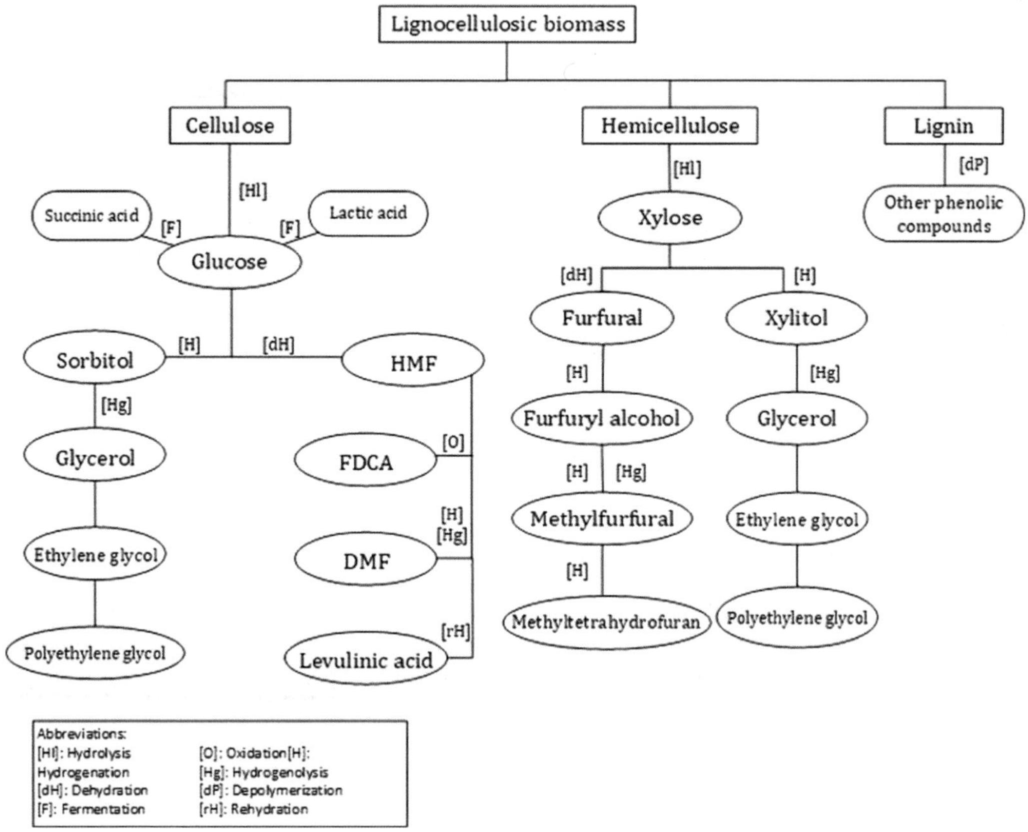


**FIGURE 5.4** Reaction pathways for lignocellulosic materials. (Adapted from Putro et al. 2016.)

### 5.4.1 Saccharification

The majority of microbes that devour cellulose are capable of performing this activity. This stage involves the breakdown of hollocellulose into simpler carbohydrates: C6 sugar (glucose) and C5 sugar (xylose). Cellulases and hemicellulases are engaged in this process as enzymes. The 44 groups of cellulases that disintegrate plant cell walls are categorized as follows: Endoglucanases (EGs) commence the enzymatic hydrolysis of celluloses by splitting the chains in the amorphous areas of cellulose, hence decreasing the DP. Exoglucanases generate cellobiose at both the reducing and non-reducing ends of cellulose chains, whereas glucosidases convert cellobiose to glucose. Hemicellulases are particular for the breakdown of polysaccharides composed of glucose, mannose, galactose, xylose, and various acids, including galacturonic, acetic, and ferulic (Vasić et al. 2021). Cellulase enzymes convert cellulose to glucose under mild circumstances, such as temperatures between 40°C and 50°C and pH between 4.5 and 5. The crystalline structure of cellulose influences the pace of hydrolysis due to the presence of hemicellulose and lignin, which form a bond with cellulose and hence slow down the hydrolysis process. Non-ionic surfactants or polymers containing poly(ethylene glycol) such as PEG are known to improve the efficacy of hydrolysis, as they can alter the surface characteristics of cellulose and thus reduce the enzyme loading (Vasić et al. 2021). Additional research indicated that the addition of Tween 80 greatly increases the glucose yield (up to 602 mg/g EFB), whereas an increase in solid loading decreases the xylose yield and raises the glucose yield (Yusuf and Kresnowati 2019).

### 5.4.2 Fermentation

This step has the most complicated reaction/process and a more specified microbe to produce the wanted result. This section will explain the biological processes to produce some holocellulose-based chemical products. Alcohol, especially bioethanol, is produced through the glycolysis and fermentation processes. The overall reaction of alcohol fermentation is $C_6H_{12}O_6 \rightarrow 2CO_2 + 2C_2H_5OH$. The reaction pathways are summarized by Maicas (2020). Glucose is converted into glucose-6-P by the addition of one molecule of adenosine triphosphate (ATP). This process is called "phosphorylation" because glucose is phosphatized. This process is catalysed by the enzyme hexokinase. The phosphoglucose isomerase enzyme then converts glucose-6-P to fructose-6-P. Using phosphofructokinase with magnesium as a cofactor, a second ATP is added to create fructose 1,6-P by adding a second ATP. The aldolase enzyme then transformed fructose 1,6-P into dihydroxyacetone phosphate (DHAP) and glyceraldehyde-3-P (GAP). These two carbohydrates are isomers. GAP is subsequently oxidized by NAD+ with the assistance of GAP dehydrogenase, yielding 1,3-diphosphoglycerate. The product of this reaction is 3-phosphoglycerate. Phosphoglycerate mutase then converted 3-phosphoglycerate into 2-phosphoglycerate. Enolase then took a water molecule from 2-phosphoglycerate to produce phosphoenolpyruvate acid (PEP). The pyruvate kinase enzyme then transferred phosphate from PEP to ADP, resulting in the production of more ATP and pyruvate. Pyruvate is the ultimate product of glycolysis and is required for the fermentation of either ethanol or lactic acid (Aryal 2022). Pyruvate is non-oxidatively decarboxylated to acetaldehyde and carbon dioxide during ethanol fermentation, catalysed by pyruvate decarboxylase (PDC). The conversion of acetaldehyde to ethanol is catalysed by ADH (Eram and Ma 2013). This process generates one molecule of ethanol and $CO_2$ for every molecule of pyruvate. This process occurs in the absence of oxygen, hence the demand for an anaerobic environment. Figure 5.5 depicts the glucose absorption process utilized by yeast to make ethanol.

*Saccharomyces cerevisiae* is mostly utilized in ethanol production. The procedure requires a lignocellulosic material to be broken down into glucose. Norhazimah et al. (2020) employed the *Saccharomyces cerevisiae* Kyokai No. 7 (ATCC 26622) strain from Japan to produce ethanol with the submerged fermentation (SmF) method. Oil palm tree trunk (OPT) was the substance used. Fermentation was carried out at a constant temperature of 30°C without pH change. Approximately 10% v/v of the inoculum was utilized to inoculate the fermentation medium. The shaking flask was purged with nitrogen to eliminate oxygen and establish an anaerobic state, and after 30 hours of fermentation, the maximum productivity of 1.033 g/L.h was obtained.

Lactic acid is another substance that can be derived from pyruvate, albeit with the assistance of other bacteria and enzymes. Lactic acid is synthesized by reducing pyruvate using lactate dehydrogenase, an enzyme typically derived from *Lactobacillus* sp. This reaction oxidizes NADH molecules into NAD+, hence regenerating NAD+ for use in the subsequent glycolysis step. Two molecules of lactic acid can be produced from one molecule of glucose. In hetero-fermentation, other byproducts such as carbon dioxide and ethanol are produced. Figure 5.6 depicts the processes of lactic acid fermentation.

Lactic acid can be derived from microorganisms, particularly bacteria and mould. There are two routes for lactic acid fermentation: homo- and hetero-fermentation. In homo-fermentation, lignocellulose biomass carbohydrate is transformed by the Embden-Meyerhof-Parnas (EMP) glycolysis route, which creates pyruvate, and lactate dehydrogenase produces lactic acid as the sole product. *Lactobacillus acidophilus*, *L. amylophilus*, *L. bulgaricus*, *L. helveticus*, and *L. salivarius* are all homofermentative bacteria. In hetero-fermentation, in addition to lactic acid, other minor products such as ethanol, diacetyl, formate, acetoin, or acetic acid and carbon dioxide are produced. *L. brevis*, *L. fermentum*, *L. parabuchneri*, and *L. reuteri* are heterofermentative-obligate bacteria. On the contrary, *L. alimentarius*, *L. plantarum*, *L. casei*, *L. rhamnosus*, *L. lactis*, *L.*

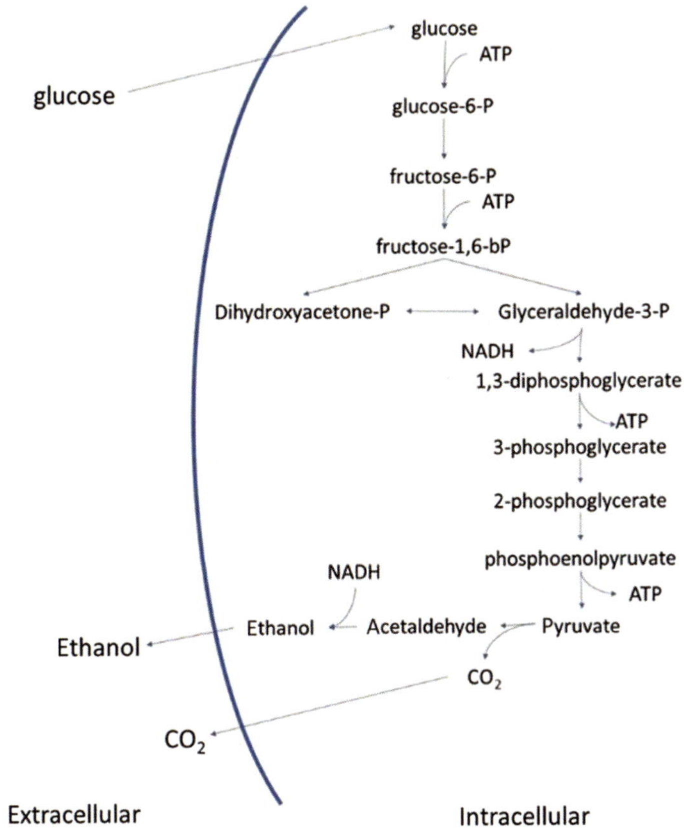


**FIGURE 5.5** Reaction pathways of glucose conversion to ethanol.

*pentosus*, and *L. xylosus* are heterofermentative-facultative bacteria (Putro et al. 2016; Komesu et al. 2017).

Succinic acid is a tricarboxylic acid (TCA) cycle metabolite. Food additives, cosmetics, pigments, toners, detergents, cement additives, soldering fluxes, and pharmaceutical intermediates are classic uses for succinic acid (Nghiem et al. 2017). *Actinobacillus succinogenes*, *Anaerobiospirillum succiniciproducens*, *Mannheimia succiniciproducens*, and recombinant *Escherichia coli* are all bacteria that have been extensively examined for their ability to produce succinic acid (Putro et al. 2016). Succinic acid is created via a series of complex and lengthy processes involving multiple enzymes. This reaction occurred within the human body, particularly within the Krebs cycle.

Succinic acid can also be produced during pyruvate fermentation. This molecule can be produced under aerobic or anaerobic settings. However, the vast majority of succinic acid–producing bacteria are capnophilic and so do not require oxygen. PEP is the substrate for a carboxylase-catalysed anaplerotic reaction during anaerobic fermentation and can be converted to oxaloacetate by PEP carboxylase (*ppc*) or PEP carboxykinase (*pck*). Pyruvate, the end product of glycolysis, can also be combined with $CO_2$ by pyruvate carboxylase (*pyc*) or malic enzyme (*mae*), resulting in the formation of oxaloacetate or malate. The enzymes malate dehydrogenase (*mdh*) and fumarase can convert oxaloacetate and malate into fumarate (*fum*). Fumarate reductase (*frd*) is able to convert fumarate to succinate (Cao et al. 2013).

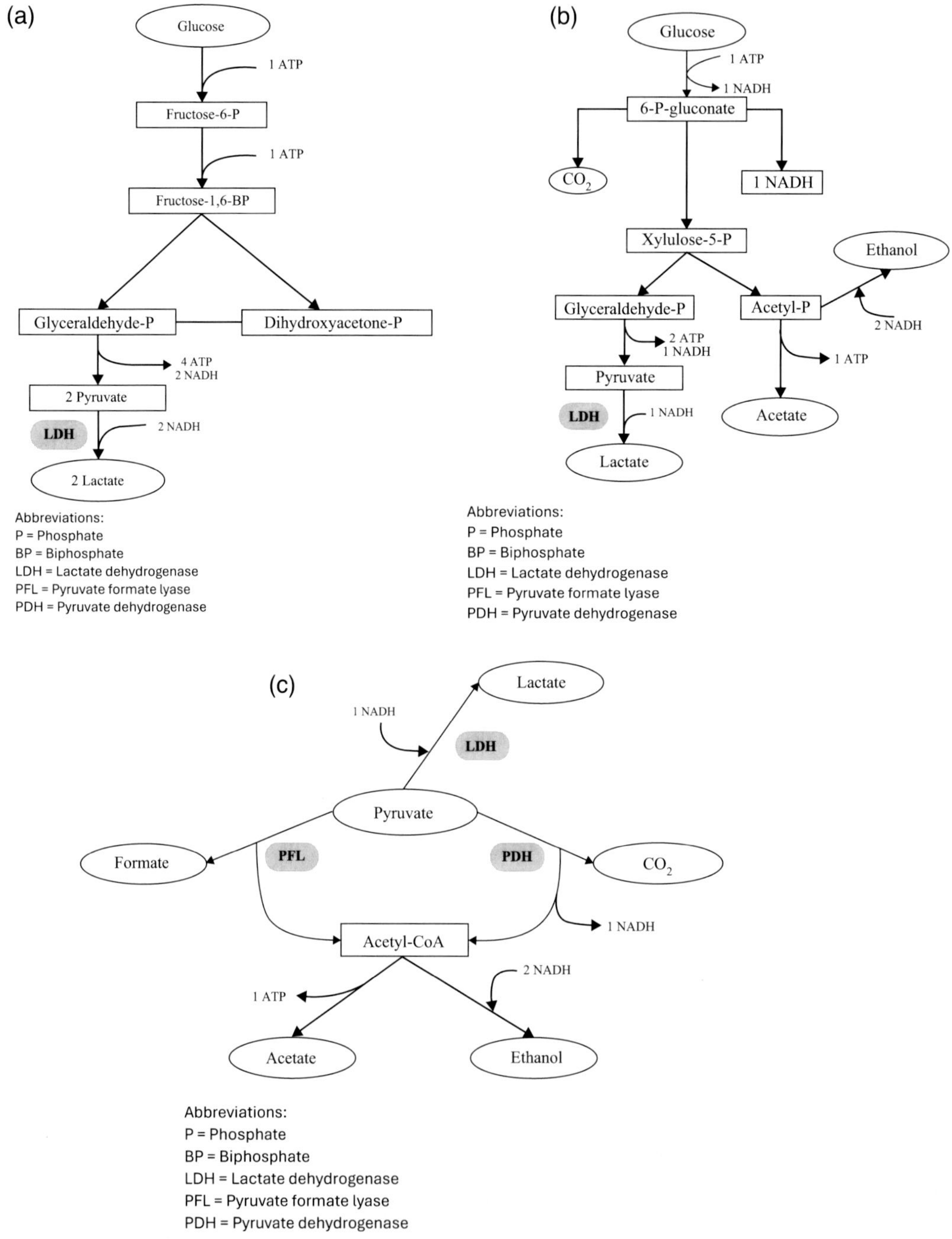


**FIGURE 5.6** Reaction pathways of lactic acid fermentation. (a) Homofermentative reactions. (b) Heterofermentative reactions. (c) Mixed acid fermentation. P = phosphate, BP = biphosphate, LDH = lactate dehydrogenase, PFL = pyruvate formate lynase, and PDH = pyruvate dehydrogenase. (Adapted from Hofvendahl & Hahn-Hägerdal, 2000)

Since 5-hydroxymethyl-2-furan carboxylic acid (HMFCA) and furan dicarboxylic acid (FDCA) are both byproducts of HMF oxidation, the bioconversion of HMF into HMFCA and FDCA is a parallel process. These processes are conducted by the enzyme HMF oxidase. There are two mechanisms for the oxidation of HMF to FDCA. In the first process, the aldehyde group is oxidized to generate HMFCA. The alcohol group of HMFCA is then oxidized to an aldehyde group, producing

2-formyl-5-furancarboxylic acid (FFCA), which is subsequently oxidized to FDCA (Sajid et al. 2018). In the second pathway, the alcohol group of HMF is preferentially oxidized to an aldehyde to produce 2,5-diformylfuran (DFF), and then, the aldehyde groups of DFF are successively oxidized to carboxyl groups to produce FDCA via FFCA (Sajid et al. 2018).

There exists a genetically engineered strain of bacteria *R. ornithinolytica* BF60 that contains HMFO and HmfH (HMF oxidation enzymes). This reaction simultaneously oxidizes aldehyde and alcohol groups. During biotransformation, strain ROBF60-H first converted HMF to HMF alcohol in 24 hours before oxidizing it to FDCA. After 24 hours, HMF alcohol and HMF concentrations decreased rapidly in the cultivated strain (ROBF60-H), which behaved similarly to the wild strain.

HMF is typically produced during the processing of lignocellulosic biomass and is regarded as a microbial inhibitor since it is cytotoxic to many lignocellulose-degrading microorganisms. It is produced via dehydration of hexoses (such as the cellulose-derived glucose). HMF is a chemical building block for several more valuable and useful chemicals (derivatives), including FDCA, BHMF, and hydroxymethyl furoic acid (HMFCA) (Cunha et al. 2022).

FDCA is one of the HMF derivatives that can be incorporated into polymers, particularly as a terephthalate replacement. By oxidizing HMF with nanoparticles as a catalyst (Pt/C, Pt/$ZrO_2$, Pd), FDCA can be produced (Sousa et al. 2015; Ait Rass et al. 2015). FDCA can also be created biologically through the combination of degradation and fermentation through the consolidation of bioprocessing (CBP). Sheng et al. (2020) developed FDCA utilizing *Acinetobacter calcoaceticus* native strain. The experiment determined that 5 g/L of HMF could be converted by 100% in 48 hours using 0.5 g/L of sodium carbonate ($Na_2CO_3$), yielding 0.31 g/L of FDCA. Other bacteria with the potential to make FDCA include *Methylobacterium radiotolerans* (G-2), *Burkholderia cepacia* (H-2), and *R. ornithinolytica* BF60 and *Cupriavidus basilensis* HMF14 (Sajid et al. 2018).

HMFCA, the oxidation product of the aldehyde group in HMF, is a promising precursor for the synthesis of various polyesters. To generate HMFCA, the aldehyde group in HMF must be oxidized selectively, while the alcohol group remains unaltered. Utilizing chemo-selective oxidation procedures, noble metal catalysts are typically used to produce HMFCA from HMF. In a recent experiment, HMF was oxidized preferentially to HMFCA in three hours by an immobilized molybdenum complex in toluene, with an 87% yield (Cang et al. 2019). Few bacteria are capable of oxidizing HMF selectively into FDCA. Nonetheless, Cang et al. (2019) identified *D. wulumuqiensis* R12 is a strain of bacteria that oxidizes HMF to HMFCA preferentially (Figure 5.7).

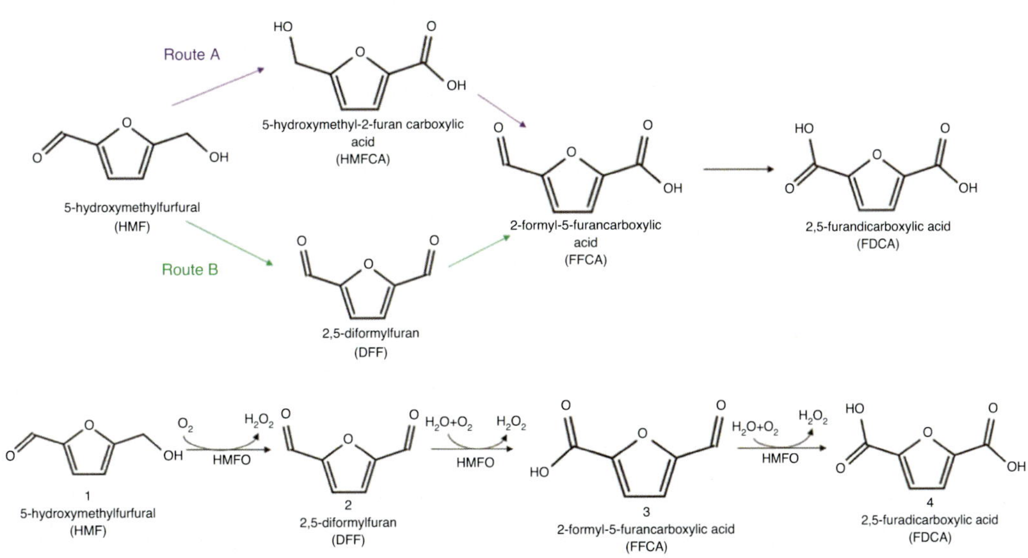


**FIGURE 5.7** Reaction pathways of FDCA formation and HMF oxidation. (Adapted from Sajid et al. 2018.)

With the assistance of the xylose reductase enzyme, D-xylose is directly converted to xylitol, resulting in the formation of xylitol. Other substrates, such as L-arabinose and D-glucose, can also be utilized to produce xylitol. These substrates, however, necessitate additional reaction steps since they follow a distinct catabolic pathway of the bacteria. L-arabinose can be turned into xylitol by using arabitol reductase and NADPH/NADH to reduce it into L-arabitol, which is then oxidized into D-xylulose by arabitol dehydrogenase and finally reduced into D-xylitol by xylulose reductase. The entire biochemical pathways are depicted in Figure 5.8.

Xylitol is a pentitol (five-carbon sugar alcohol) that is commonly used as a sugar substitute and has numerous applications in the pharmaceutical business (Umai et al. 2022). The majority of xylitol is derived from D-xylose, an abundant pentose found in the hemicellulose matrix. Xylitol can be generated by *Candida shehatae*, *Pichia stipites*, and *Pachysolen tannophilus* (Chandel et al. 2011). Xylitol-producing enzymes (xylose reductase) from *Pichia stipites* can be genetically engineered into other microorganisms (Fernandes and Murray 2010).

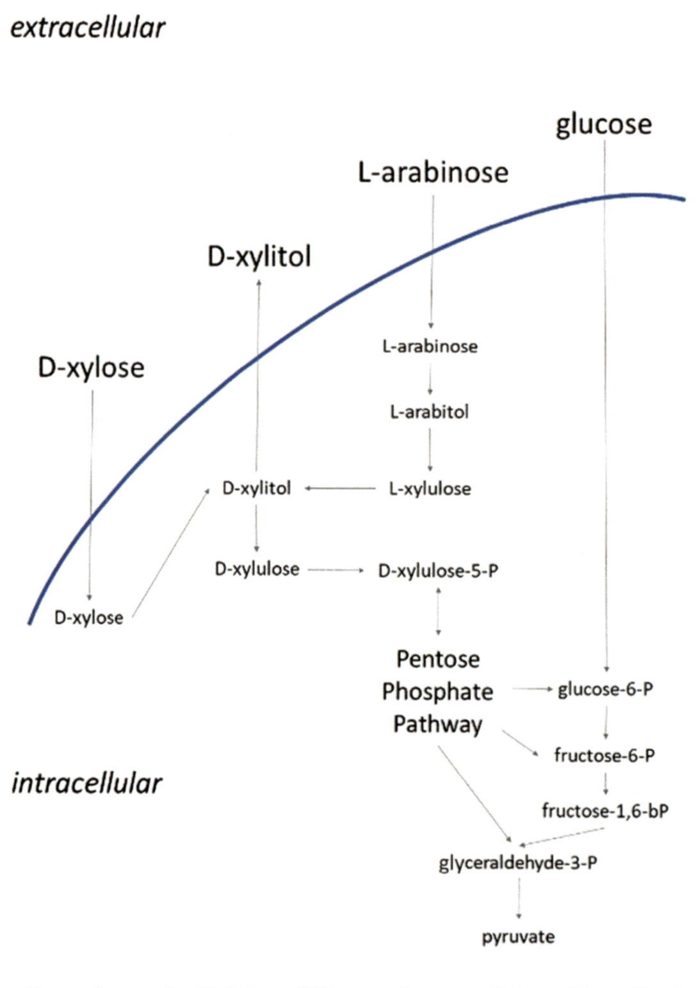


**FIGURE 5.8** Reaction pathway of xylitol from different substrates. (Adapted from Umai et al. 2022.)

## 5.5 MICROBIAL SELECTION'S CRITERIA

Bioconversion of lignocellulosic materials requires the assistance of several microorganisms, including fungi and bacteria. These bacteria utilize enzymes to convert specific substrates into their respective products. Every microbe produces unique enzymes, requires unique cultivation conditions, and yields unique compounds. *Saccharomyces cerevisiae*, also known as yeast, is one of the most well-known microorganisms that transform lignocellulose resources into valuable chemicals. Yeast can convert glucose to ethanol. Identifying and classifying lignocellulosic-digesting microorganisms is the key to successfully creating holocellulose-based chemical compounds. Therefore, this section will explain microbial selection criteria and how to pick microbes for specific items.

The general rules in picking the microbial strain for industrial process are the strain has a high specific growth rate, the strain can utilize low-cost substrate, the strain produces a single product, and at a high product yield, the strain is resistant to contamination and can be easily maintained for a long period. In the case of microbial utilization of lignocellulosic-derived materials from the oil palm industry, the ability to utilize low-cost substrate also means how much the strain can tolerate the presence of toxic substances that are coproduced during the pretreatment or hydrolysis process prior to the fermentation. Examples of microbes that are used to produce holocellulose-based chemicals are listed in Table 5.4.

**TABLE 5.4**
**Examples of Microbes Producing Holocellulose-Based Chemicals**

| Product | Substrate | Microbe Name | Operating Condition | Result |
|---|---|---|---|---|
| Ethanol[a] | Oil palm tree sap (glucose, fructose) | *Saccharomyces cerevisiae* ATCC 26622 | Anaerobic flask; 30°C temperature in stable pH; 72 hours | 1.033 g/L·h ethanol productivity |
| Lactic acid[b,c] | Glucose[b] | *Lactobacillus lactis* BME5-18M | Fed batch; 37°C temperature; 100 rpm | 210 g/L concentration; 97% yield; 2.2 g/L·h productivity |
| | Cellulose[c] | *Bacillus coagulans* 36D1 | Fed batch; 50°C temperature; pH 6.0 | 80 g/L concentration; 80% yield; 0,30 g/L·h productivity |
| Succinic acid[d,e] | Glucose | *Actinobacillus succinogenes* ATCC 55618[d] | Incubated shaker; 38°C temperature; 210 rpm | 33.4g/L concentration; 47% yield |
| | | *Mannheimia succiniciproducens* MBEL55E[e] | Anaerobic condition; pH 6.5; 39°C temperature; 100 mL MH medium | 14 g/L concentration; 70% yield; 1.87 g/L·h productivity |
| FDCA[f,g] | HMF | *Pseudomonas putida* S12[f] | 30°C temperature; pH 7; 150 rpm; mineral medium glycerol | 193 mM concentration; 97% yield; 1.26 mM/h productivity |
| | | *Burkholderia cepacia* H-2[g] | Batch; 28°C; pH 7; 120 rpm; mineral salt media | 8.17 mM concentration; 64% yield; 0.34 mM/h productivity |

*(Continued)*

**TABLE 5.4 (*Continued*)**
**Examples of Microbes Producing Holocellulose-Based Chemicals**

| Product | Substrate | Microbe Name | Operating Condition | Result |
|---|---|---|---|---|
| HMFCA[h,i] | HMF | *Serratia marcescens*[h] | Fed batch; 30°C temperature; pH 8.0; 200 rpm; minimum salt medium | 5.56 mM concentration; 28%; 0.278 mM/h productivity |
| | | *Pseudomonas putida* ATCC 47054[i] | Batch reactor; 35°C temperature; pH 6; 200 rpm; 200 mM phosphate buffer | 73.9 mM concentration; 98%; 73.9 mM/h productivity |
| Xylitol[j,k] | Xylose[j] | *Kluyveromyces marxianus* | 42°C temperature | 1.49 g/L.h productivity from 100 g/L xylose |
| | EFB[k] | *Debaryomyces hansenii* ITBCC R8 | 30°C temperature; 5.0 pH | 0.201 g/g yield |

*Note:* Data of ethanol production is from Norhazimah et al. (2020)[a]; lactic acid from Bai et al. (2003)[b] and Ou et al. (2011)[c]; succinic acid from Akhtar and Idris (2017)[d]; Lee et al. (2002)[e]; FDCA from Koopman et al. (2010)[f] and Yang and Huang (2016)[g]; HMFCA from Muñoz et al. (2020)[h] and Liu et al. (2020)[i]; data of xylitol production is from Zhang et al. (2014)[j] and Mardawati et al. (2020)[k].

*FDCA, furan dicarboxylic acid; HMF, hydroxymethylfurfural; HMFCA, hydroxymethyl furoic acid; EFBs, empty fruit bunches.*

### 5.5.1 Cultivating Conditions

Each microbe has unique requirements. Each requires a unique environment to survive and ferments the necessary substrate to produce the intended output. In bioprocess engineering, there are other variables, such as inoculum media composition, enzyme loading, substrate conversion rate, and enzyme inhibitors. Typically, yeast (*S. cerevisiae*) ferments glucose in an anaerobic environment to make ethanol. In addition to delignification and hydrolysis, bioconversion requires delignification and hydrolysis. Cellulase enzymes convert cellulose to glucose under mild circumstances, such as temperatures between 40°C and 50°C and pH levels between 4.5 and 5. The pretreatment of lignocellulosic biomass contributes significantly to the efficiency of hydrolysis. This pretreatment method involves lignin removal, hemicellulose solubility, hydrolysis time, and enzyme loading (Vasić et al. 2021). The use of a non-ionic surfactant or polymers containing poly(ethylene glycol), such as PEG, is known to improve the efficiency of hydrolysis, as these substances can alter the surface characteristics of cellulose and hence reduce the enzyme loading. Since it is easier to apply optimal conditions in a fed-batch stirred tank reactor, ethanol fermentation is typically conducted there (Vasić et al. 2021).

Bacteria and fungi can generate lactic acid. Most lactic acid–producing bacteria are facultative anaerobes, catalase-negative, non-motile, and spore-free. Typically, they have a high acid tolerance and may live at pH 5.0 or below. They have a competitive advantage over other bacteria due to their strong acid tolerance. Nonetheless, a number of research studies indicate that a pH of 6.5 is ideal for microbial growth (Guilherme 2009). The ideal development temperature ranges from 20°C to 45°C. The manufacture of lactic acid also needs a pH regulator, as the process decreases pH. Typically, calcium carbonate, calcium hydroxide, or NaOH is employed. Lactic acid can also be used for pH regulation by extraction, adsorption, and electrodialysis. This procedure also necessitates supplemental nourishment, specifically nitrogen delivery. Despite its high price, yeast extract is the finest nitrogen source for lactic acid–producing bacteria. Ammonium sulphate is a viable option, mostly due to its reduced price. To substitute commercial peptone and yeast extract, inexpensive nitrogen supply supplements (such as pigeon pea flour, red lentil gram, black gram, Bengal gram, green gram, soya bean, and baker's yeast) have been examined. In general, the addition of additional nutrients (such as peptone and meat extract) to the medium stimulates lactic acid production (Komesu et al. 2017). Generally, lactic acid is manufactured in batch mode; however, continuous and fed-batch techniques are also viable. Compared to continuous fermentation, batch fermentation offers greater conversion and yield, but lower volumetric productivity. In a batch process, all of the substrate is utilized, whereas in a continuous process, residual substrate concentration is constantly present. Due to the high dilution ratio and the ability to keep the process going for an extended amount of time, continuous fermentation can achieve its best possible yield (Komesu et al. 2017).

Microorganisms that produce succinic acid are often anaerobes, grow at a pH close to neutral, are capnophilic, and utilize the reductive TCA cycle. The procedure necessitates a substantial amount of carbon dioxide, which can be supplied by a pH-regulating alkaline or alkaline-earth carbonate solution. Alternatively, carbon dioxide can be sparged directly into the fermentor while adjusting the pH with bases such as NaOH, KOH, and $NH_4OH$ or alkali-earth hydroxides (Nghiem et al. 2017). The use of $MgCO_3$ as a pH regulator led to optimal cell growth, glucose utilization, and succinic acid synthesis. The production of succinic acid requires a nitrogen source, such as yeast extract or corn steep liquor. Simultaneous saccharification and fermentation (SSF), semi-SSF (SSSF), and separate hydrolysis and fermentation (SHF) are all viable fermentation techniques. SSSF is the best operating method since it produced the maximum concentration and yield of succinic acid (Putri et al. 2020).

Bacterial strain *D. wulumuqiensis* R12 generated HMFCA at temperatures between 25°C and 60°C, with a maximum yield of 79% observed at 35°C after four hours of reaction. It actively converted substrate in the pH range of 5.0–10.0, with the ideal pH being 5.0, assimilating a 100 mM phosphate buffer, after a 4-hour reaction (Cang et al. 2019). The strain was pre-cultivated at 30°C and 200 rpm for 24 hours in tryptone glucose yeast (TGY) extract medium containing 0.5% tryptone, 0.1% glucose, and 0.3% yeast extract. Then, 1% of the preculture grown overnight was transferred to fresh TGY media. To harvest cells, the culture was incubated at 30°C and 200 rpm for 48 hours before being centrifuged at 5,000 rpm for 10 minutes. The cell pellet was washed twice with 100 mM phosphate buffer (pH 7.4) and resuspended with a final concentration of 0.12 g/mL in phosphate buffer (cell wet weight). The production was conducted in a fed-batch bioreactor. Five millilitres of phosphate buffer (0.1 M, pH 7.0) containing 0.2 g/mL of bacteria culture and 150 mM HMF substrate was incubated at 35°C with 850 rpm. Five millimoles of HMF was added to the reaction mixture every hour, while the pH was adjusted with NaOH solution between 7.0 and 8.0. The substrate concentration can be increased to 300 mM in order to obtain the best result, which is a 90% yield with a 98% HMFCA selectivity. The process takes a maximum of 36 hours. As a result of HMF inhibition, an increase in substrate concentration would lower the HMFCA yield. In addition, a fed-batch approach produced up to 511 mM of HMFCA in 20 hours, resulting in a productivity of 44 g/L per day (Cang et al. 2019).

The optimal conditions for xylitol production include an acidic pH and a high temperature. Candida fungal strains have been observed to be active at pH values between 5.0 and 7.0, with an optimum pH range of 5.0–5.5. Depending on the species, yeasts create xylitol at temperatures between 30°C and 37°C. To produce a high concentration of xylitol, a high concentration of xylose must be adjusted higher, but not so high as to produce excessive amounts of inhibition products, such as furanic compounds, although this can be avoided by increasing the concentration of microbe inoculum or by employing a detoxification process. Some yeasts also prefer a low oxygen concentration in the broth because it increases the synthesis rate of xylitol (Umai et al. 2022; Table 5.5).

### 5.5.2 Genetic Modification

Multiple strains of a single microorganism can impact its resilience to environmental conditions, product conversion yield, and substrate range. Effective lignocellulosic-digesting microbes should be able to meet all of the aforementioned parameters. This is possible through genetic modification of the microbial strain. Typically, thermophilic bacteria are utilized to treat lignocellulose

**TABLE 5.5**
**Removal of Inhibitor for Detoxification Method**

| Method | Removal of Inhibitor | Note |
|---|---|---|
| Neutralization | Acetic acid, furfural, and HMF | Poor ability to remove toxic compounds |
| Overliming | Furfural and HMF | Sugar loss due to hydroxide-catalysed degradation reactions, no alteration in acetic acid concentration |
| Adsorptions | Furans, phenolic and acetic acid | Good removal of acetic acid and phenolic compounds |
| Ion exchange resin | Furans, phenolic and acetic acid | High removal of furan, total phenolic compounds, and acetic acid |
| Electrodialysis | Acetic acid, furfural, phenolic compounds | Remove 90% of acetic acid, low sugar losses (<5%), environmentally friendly, high instrument cost, better fermentability of the hydrolysate |
| Enzyme detoxification | Phenolic compounds | Excellent selectivity removal of phenolic content |

materials. Several thermostable or thermophilic bacteria that decompose lignocellulosic materials have been researched. These microbes offer a number of benefits, including a shorter hydrolysis time, a strong tolerance to low or high pH, a lower danger of contamination, and a low energy cost (Bhalla et al. 2013).

*S. cerevisiae* is one of the examples that are used in industrial processing. Industrial strains of *S. cerevisiae* are regarded as more resistant to varied stress conditions than laboratory strains. All industrial strains had varying amounts of DNA ranging between 1.3 and 3*n* (*n* being the DNA content of a haploid strain). Incomplete ploidy values showed partial amplifications of some chromosomal areas and the presence of additional chromosome sets. The genomes of industrial yeasts contain several copies of various *SUC* genes, whereas laboratory strains only carry the *SUC2* gene. The *SUC* genes code for the invertase enzyme, which breaks down sucrose, the primary carbon source in molasses and sugarcane juice, into glucose and fructose. This shows the amplification of specific genes to prepare industrial yeasts for biological stress resistance in industrial processes (Le Borgne 2012).

Additionally, genetic engineering enables the modification of specific bacteria to digest substrates that they were previously incapable of. *Escherichia coli* and *Saccharomyces cerevisiae* are commonly employed as media in genetic engineering. Kim et al. (2015) engineered *E. coli* for increased xylitol production by introducing the following modifications to the gene structure: (1) deletion of arabinose transcriptional regulator (*araC*), (2) intergenic mutation in xylose isomerase (*xylA*) and orotate phosphoribosyl transferase (*pyrE*), and (3) missense mutation in arabinose-proton symporter (*araE*) and putative undecaprenyl diphosphate synthase (*ybjG*).

Even in the presence of glucose in the media, the resultant mutant *E. coli* GX20 strain exhibited rapid xylose to xylitol conversion. Pratter et al. (2015) transferred xylose reductase from *Candida tenuis* into *Saccharomyces cerevisiae*, allowing it to produce 1.16 g/L.h xylitol.

## 5.6 TECHNOLOGICAL DEVELOPMENTS

### 5.6.1 Processing Methods

Currently, industrial ethanol production from lignocellulosic biomass is feasible. Even if all phases of the process can be improved, pretreatment, pentose fermentation, solid-state fermentation (SSF), downstream processing, and the reduction in byproducts remain relevant research areas. Despite the tools developed through research, pretreatment remains one of the most expensive process phases (hydrolysis to carbohydrates, fermentation, distillation, and purification). Preprocessing expenses for lignocellulosic ethanol production range from 11% to 27%. This variation will be determined by the method of preprocessing employed. For the pretreatment step to be as efficient as possible, it must ensure high sugar yields required for the subsequent fermentation stage, a very low inhibitor content in the fermentation stages, and the separation of lignin and inadvertently converted hemicellulose (Su et al. 2020).

Currently, there is lack of efficient methods for the pretreatment of lignocellulosic biomass with a predominant composition. The choice of pretreatment technology depends on a variety of aspects, including the type of biomass employed. Nonetheless, SE is recognized as one of the most efficient and cost-effective pretreatment processes for agricultural residues, particularly for large-scale operations (Su et al. 2020).

After the pretreatment and valorization of lignin and hemicellulose, it will also be necessary to develop an effective method for separating biomass elements. Currently, lignin and hemicellulose can be converted into products with additional value. Similarly, fermented broths and solid wastes (including bacterial/yeast cell masses) are nutrient-dense and can be recycled, utilized as animal feed, or added to biogas plants (Su et al. 2020).

Due to the high cost of particular microorganisms, biological pretreatment is the most expensive pretreatment approach. Various researchers have undertaken extensive investigations on the use of microbes for the pretreatment of lignocellulosic material, although the use of microorganisms for lignocellulosic material degradation is still far from industrial application. Prior to the microbial process, a combination of physical or chemical pretreatment is required due to the complicated coupling of lignin–hemicellulose–cellulose, which is the primary obstacle in the use of microbial processes (Zeng et al. 2014).

Several variables, including pH, temperature, and inhibitors, influence the biological pretreatment procedure. The existence of these parameters is problematic for functioning microorganisms, resulting in suboptimal output. Recent research emphasizes the significance of the detoxification process prior to the involvement of microbes. In later stages, widespread inhibitors such as furfural, vanillin, p-hydroxybenzaldehyde (PHB), and syringaldehyde can suppress microorganisms. Therefore, a process called detoxification is required to eradicate the inhibiting factor. Cardona et al. (2010) discussed some detoxification methods that have been developed, such as neutralization, overliming, adsorption, ion exchange, and electrodialysis. In addition, enzymatic and microbial detoxification has also been performed successfully to eliminate inhibitors (Ujor and Okonkwo 2022).

The choice of feedstock, size reduction, hydrolysis, fermentation, product recovery, residue processing, and co-product potential are all significantly influenced by pretreatment. Several pretreatments involving biological, chemical, physical, and thermal approaches have been studied over the years, but only those involving chemicals give the necessary high yields and low prices for commercial success at this time (Yang and Wyman 2008).

Yoricya et al. (2016) conducted research employing a delignification method involving NaOH in ionic and non-IL systems. Choline chloride is chosen as the IL because it is a mild acid that can be used at room temperature, is non-toxic, and is eco-friendly (Lee et al. 2013). Choline chloride (ChCl) delignification is generated by the interaction of fibre free radical binding between Cl- and hydroxyl groups in lignin, which breaks the lignin bond and increases cellulose levels (Yoricya et al. 2016). The use of Ionic ChCl liquid can boost delignification by 61.45% when compared to the absence of ChCl.

The hydrolysis/saccharification process presents another challenge, in particular when bulk-biochemical products such as bioethanol are targeted. In this process, the downstream processing of ethanol from the fermentation broth contributes a significant part of the overall process, and thus, the use of sugar-concentrated hydrolysate solution as the substrate for fermentation will be an advantage. Methods for producing sugar-concentrated hydrolysate solution, for example by strategic fed-batch hydrolysis of pretreated lignocellulosic materials as presented by Sugiharto et al. (2016), need to be developed.

Process intensification may also contribute to the economics of the overall process. Research on the simultaneous hydrolysis and fermentation process that is combining both the hydrolysis and fermentation processes in one reactor has been evaluated. Saha et al. (2011) evaluated the simultaneous hydrolysis and fermentation configuration for the ethanol production from wheat straw. Burhan et al. (2019) and Hidayatullah et al. (2021b) evaluate the simultaneous hydrolysis and fermentation for the xylitol production from oil palm EFB hydrolysate. This process configuration may cut the total processing time from 128 to 77 hours or increase xylitol productivity by 167%.

### 5.6.2 Microorganisms

Utilizing microorganisms is crucial for producing the desired output. Recent research has produced numerous thermostable and thermophilic microbes that are pH resistant at high/low temperatures, as well as inhibitors (Bhalla et al. 2013). These microorganisms, some of which are listed in Table 5.6, are examined for their ability to digest lignocellulose.

Although thermophilic bacteria have been cultivated as described previously, the final selection for an industrial-scale process depends on other factors such as energy consumption, production costs, catalytic efficiency, enzymes, and the environmental impact of the overall lignocellulose conversion process.

### 5.6.3 Products

The fermentation process is currently the most prevalent method since it is non-toxic, environmentally benign, and inexpensive. Table 5.7 displays the results of processing various oil palm biomass residues as substrate using the fermentation technique (Said et al. 2021). So far, the majority of bioproducts derived from oil palm waste (OPW) substrates are enzymes that can be used in industry. Xylanase, for instance, is used in the food, feed, and pulp/pulp bleaching industries. In addition, chemical substances such as ethanol, succinic acid, biohydrogen, methane, and biodiesel sources are produced as products.

As for OPW from the plantation and agricultural industry, its fibres can be made into several useful chemical products, as has been discussed in the previous sections. However, with current industrial and environmental challenges, much further research still needs to be conducted before appropriate technology can be applied (Cheah et al. 2023).

Biorefining (the conversion of solid waste into valuable chemical compounds) is a growing industry across the world, not only for the purpose of reducing dependence on nonrenewable sources but also to limit the availability of solid waste without further processing (Barragán-Ocaña et al. 2023). In addition, biorefining is also projected to increase a country's economy if the production technology used is mature.

**TABLE 5.6**
**The Effect of Thermophilic Bacteria in Lignocellulose Biomass Pretreatment**

| Organism | Note |
|---|---|
| *Clostridium thermocellum* | At 60°C, effectively degrade crystalline cellulose, produce a large multiprotein complex known as a cellulosome, and enhance ethanol tolerance and product yields |
| *Caldicellulosiruptor saccharolyticus* | A biohydrogen production candidate that produces thermostable cellulolytic and xylanolytic enzymes and grows best at 70°C on several lignocellulose biomass types |
| *Caldicellulosiruptor bescii* DSM 6725 | The most thermophilic organism, with an ideal development temperature of 80°C, is capable of digesting large amounts of both untreated switchgrass and crystalline cellulose (up to 200 g/L) |

**TABLE 5.7**
**Biochemical Production from Different Organisms**

| Substrate | Organism | Fermentation | Product(s) |
|---|---|---|---|
| OPT[a,b,c] | *Aspergillus fumigatus* SK1[a] | SSF | Enzymes, CMCase, Fpase, β-glucosidase, xylanase |
| | *Saccharomyces cerevisiae* Kyokai No. 7 (ATCC 26622)[b] | SmF | Ethanol |
| | *Aspergillus niger*[c] | SSF | Enzymes, CMCase, Fpase, β-glucosidase, xylanase |
| EFB[d,e,f] | *Botryosphaeria* sp.[d] | SSF | Enzymes, CMCase, Fpase, β-glucosidase |
| | *Aspergillus niger* EFB1[e] | SmF | Enzymes, CMCase, Fpase, β-glucosidase |
| | Yeast[f] | SmF | Ethanol |
| OPF[g,h,i] | *Aspergillus niger* EFB1[g] | SSF | Enzymes, endoglucanase, β-glucosidase, exoglucanase, xylanase |
| | *Trichoderma asperellum* UC1[h] | SSF | Enzymes, CMCase, Fpase, β-glucosidase, xylanase |
| | *Actinobacillus succinogenes* | SmF | Succinic acid |
| POME[j,k,l] | *Clostridium beijerinckii* ATCC 8260[j] | SmF batch | Biohydrogen |
| | Mixed culture SmF[k] | SmF continuous | Methane |
| | Mixed microalgae[l] | SmF batch | Biodiesel |

*Note:* Data of OPT-based chemicals are obtained from Ang et al. (2013)[a]; Norhazimah et al. (2020)[b]; Ang et al. (2015)[c]; EFB-based chemicals from Ang et al. (2013)[d]; Noratiqah et al. (2013)[e]; Fitriasari et al. (2018)[f]; OPF-based chemicals from Ikubar et al. (2018)[g]; Ezeilo et al. (2019)[h]; Tan et al. (2015)[i]; POME-based chemicals from Rosa et al. (2020)[j]; Mamimin et al. (2015)[k]; Kamyab et al. (2018)[l].

OPTs, oil palm trunks; EFBs, empty fruit bunches; OPFs, oil palm fronds; POME, palm oil mill effluent.

Researchers are beginning to consider applying more complex technological ideas, such as the application of genetic modification to enzyme-producing microbes, cell metabolism engineering efforts, integration of delignification processes, consolidated bioprocess technologies, and integrated feeding strategies for hydrolysis-fermentation processes both simultaneously and separately. These technical innovations are undoubtedly future problems that are currently being investigated in order to accomplish the long-term goal of a sustainable and optimum biorefining process.

## 5.7 CHALLENGES AND OPPORTUNITIES

Indonesia is currently the top producer of oil palm in the world. In 2021, CPO production reached 46.888 million tonnes or 0.31% less than 2020's 47.034 million tonnes. The drop is attributed to the recovery year from COVID-19, the weather, and minimal fertilization, according to estimates. However, the Indonesian Oil Palm Association (IPOA) projects that CPO production will increase by 4.87% to 48 million tonnes in 2022 (IPOA 2022). According to data compiled by the Badan Pusat Statistik (BPS) on Indonesia's oil palm production, oil palm production in Indonesia in the form of CPO has increased by an average of 11.13% every year. Figure 5.9 depicts the production graph for oil palms. However, a rapidly expanding sector will also generate a substantial increase in environmental waste. Consequently, the potential for employing OPW remains substantial.

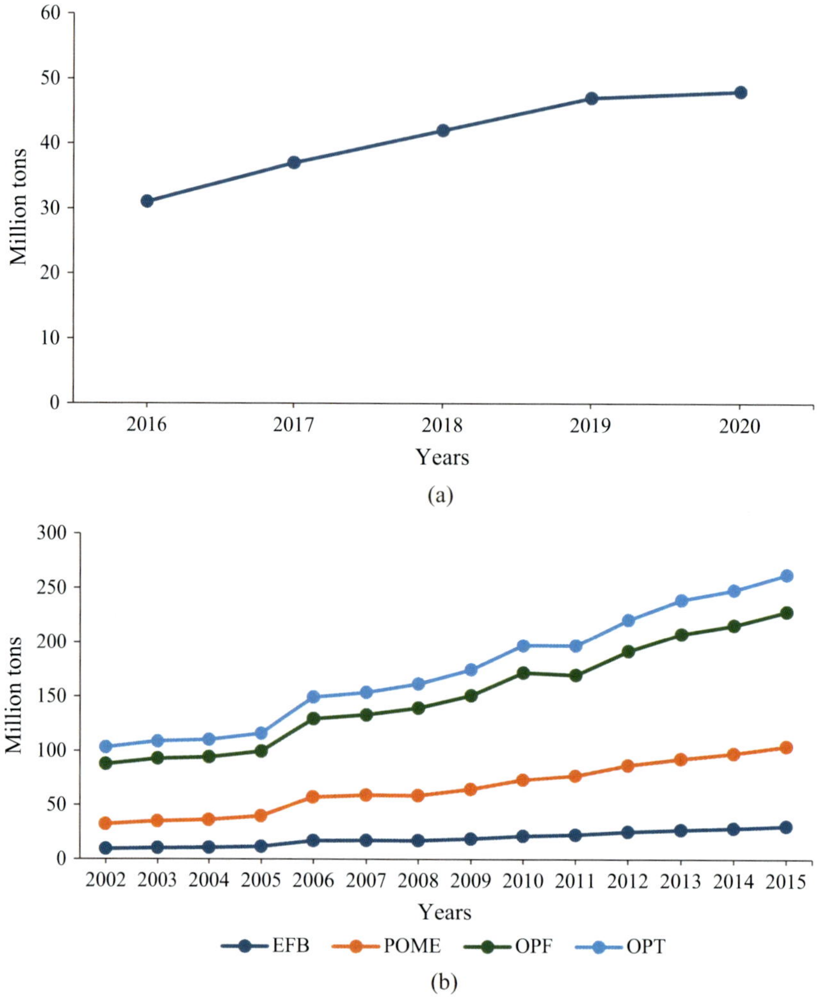


**FIGURE 5.9** (a) CPO production in Indonesia (BPS, 2021), and (b) rise in oil palm waste from 2002 to 2015.

Palm biomass is one of the most significant sources of lignocellulosic material, particularly in Asia, with four times as much biomass produced as oil palm. Unfortunately, the bulk of waste is disposed of in landfills, resulting in grave environmental issues such as global warming and the greenhouse effect (Said et al. 2021). From 2002 to 2015, OPW production increased considerably. In Indonesia, Hambali and Rivai (2017) examined the volume of OPW. Using graphical and cross-tabulation techniques, as well as statistical tools for data processing, the analysis provided an explanation and a thorough projection. This method employed trend analysis, which generated a mathematical equation line for making future projections. In 2030, it is anticipated that 54 million tonnes of OPEFB, 130 million tonnes of palm oil mill effluent (POME), 115 million tonnes of OPF, and 59.7 million tonnes of OPT will be produced.

It is known that this waste contains large quantities of cellulose and hemicellulose (Said et al. 2021). Consequently, OPW has the potential to become a future feedstock for the low-cost production of a number of chemical goods. The potential generation of solid and liquid wastes by the oil palm sector in the future is enormous. The liquid waste produced by the processing of POME for biogas/biomethane is still nutrient-rich and can be used as fertilizer, land application, and nutrient-rich medium for microalgae development. This procedure can also be used to cultivate microalgae by producing $CO_2$. OPF and OPT can be used as mulching material to preserve soil moisture and limit weed growth, as a raw material for composting, as a material for covering plant vegetation, and as a source of fibre for various composite goods (Hambali and Rivai 2017).

Some companies and governments are interested in the use of biomass. For example, the Japanese company eREX Singapore Pte Ltd., a subsidiary of eREX, bought a palm kernel shell (PKS), which is palm waste in Indonesia from 2020 for the following 15 years to be used as electrical energy in Japan (detikFinance, 2020). This proves that there is a potential for biomass power plants that are economically attractive and have high sustainability.

OPW has a high economic potential due to its abundance and its chemical derivatives' value. Parthasarathy et al. (2022) have done an economic calculation regarding OPW, which includes POME, EFBs, MF, PKS, OPF, and OPT. These materials can be pyrolysed to obtain fuel such as bio-oil, biochar, and syngas. The study took each material into calculation and then determined the most economical outcome, such as payback period, investment return, and annual profit. The economic evaluation indicated the CAPEX was in the range of US dollars (USD) 75–80 M, and OPEX was between USD 17 and 20 M/year. PKS showed the highest CAPEX USD 79.88 M, whereas the EFB demonstrated a lower CAPEX of USD 75.24 M. OPF produced the largest profit of USD 17 M/year, with a 22% return on investment (ROI), among the feedstocks (Parthasarathy et al. 2022). All OPW feedstocks indicated a reasonable payback period of 4–6 years in terms of investment capital. Overall, OPW bioconversion is a promising prospect since it is sustainable and environmentally friendly.

Mariyana et al. (2021) conducted a study on the monetization of oil palm EFB to bio-based chemicals, in which the lignocellulosic biomass was converted into lignin, xylose, and glucose. A biorefinery of 60 kilotons/year EFB would give an IRR of 12.4% and a payback period of 7 years. Zamri et al. (2022) projected that 5,000 MW of electricity could be generated using palm oil biomass of PKS, EFB, and MF under 40% of operation efficiency. However, there is no current information on the specifics of the economic potential resulting from the utilization of solid waste from the plantation industry and the processing of palm oil into products that are useful for human life, so a special study on the economic feasibility of biorefining palm oil solid waste is required.

Production of chemical products derived from holocellulose, like bioethanol, necessitates a multistep procedure that greatly raises the price of biofuels. Finding quick, inexpensive, and effective production procedures is the greatest problem (Carrillo-Nieves et al. 2019). Typically, the collected biomass contains pollutants, such as stone and soil, which might hinder sugar recovery by blocking the pretreatment reaction and clogging reactor lines. Moreover, these factors have a significant impact on the operating and capital costs of the total biomass conversion process (Antunes et al. 2019).

Delignification, enzymatic hydrolysis, and fermentation are all steps of the process that begin with the first usage of whole biomass into chemical derivative products. When compared to the previous processes, the delignification stage produces the most gas emissions. In principle, every prospective biomass delignification process emits distinct gases. The ensuing gas emissions are dependent on the rise in temperature and pressure provided prior to delignification. No one has before discussed the quantity of gas emissions produced during the processing of bio-based cellulose and hemicellulose into important chemical derivative products (solid-liquid based) as defined in the text. This is achievable because both processes (enzymatic hydrolysis and fermentation) employ operating conditions that are moderate (do not require high temperatures or pressures). Another example is lignin. Many researchers have examined the comparison of gas emissions produced by different lignin-based biomass. Bhatt et al. (2022) investigated the possibility for gas emissions from operational units in an effort to convert lignin into fuel and pellets. According to the study, CO, $NO_x$, $SO_2$, volatile organic compound (VOC), and hazardous air pollution (HAP) can occur. This gas emission is caused by the core operating unit, the supporting operating unit, equipment leakage, and biomass-transport truck traffic (on an industrial scale). Using the same source (Bhatt et al. 2022), enzymatic hydrolysis produces no gas emissions, whereas fermentation produces $NO_x$ and CO, albeit in extremely tiny levels.

In addition to altering the structure of biomass, pretreatment methods aim to remove hemicellulose and lignin to make the carbohydrate fraction amenable to enzymatic hydrolysis.

Each pretreatment technique has its own set of benefits and obstacles to overcome. Biological pretreatment techniques are eco-friendly and environmentally beneficial, but they are intrinsically sluggish and challenging to scale up (Antunes et al. 2019).

Although modern technologies have substantially improved production, there are still obstacles that require additional exploration. The exceptional focus is essential for developing a comprehensive understanding of lignocellulose, the primary structural material in plants, from cellulose synthesis and fibril formation to mature plant cell walls, thereby laying the groundwork for significant advances in sustainable energy and materials. For the development of economically viable enzymatic treatment-based processes, it is necessary to characterize, comprehend, and overcome impediments to the enzymatic hydrolysis of various raw materials (Cardona et al. 2010).

Due to the absence of an integrated process configuration, it is frequently difficult to quantify precisely the progress made in the cost-effective conversion of lignocellulose biomass. Regarding the commercial performance of second-generation biofuels, many environmental and sustainability variables continue to provide significant uncertainty despite current expectations. Outstanding R&D studies enhance the process of conversion, but the availability of raw materials and the stability of income remain unclear and subject to political threats. Realistically, it can be inferred that some components must be joined and optimized before an effective industrial configuration can be achieved (Menon and Rao 2012).

The conversion of lignocellulose biomass to natural product sources has significant promise. OPW still offers a great deal of potential for processing into products with commercial value. In the business world, the configuration of uncertain processes is a significant challenge. To produce an effective and efficient system setup, further research and development is required.

## 5.8 CONCLUSIONS AND PERSPECTIVES

Solid waste from oil palm may be effectively handled by converting it into useful substances. This biomass waste source possesses considerable economic potential owing to its elevated holocellulose concentration. Lignin must be removed before processing holocellulose derivatives. Delignification is a process that requires evaluation of its value. The selection of the delignification procedure, the enzyme cocktail ratio, microbial cultivation conditions, and genetic modifications are intricate and challenging factors in obtaining the desired output. To optimize product production and enhance the fermentation process, these initiatives remain in progress. The abundant availability of oil palm solid biomass waste presents significant prospects for the development of an oil palm biomass-based chemical industry.

In recent years, researchers have investigated many techniques to improve the efficiency of biomass conversion processes. Advancements in enzymatic treatments, microbial strains, and fermentation technologies may substantially enhance the output of desired compounds. Utilizing customized enzyme cocktails can enhance the degradation of lignocellulosic materials, hence enabling more efficient access to holocellulose. Moreover, choosing resilient microbial strains that can flourish in adverse situations and refining fermentation parameters might result in increased product yields and reduced processing durations. Furthermore, the valorization of oil palm biomass is consistent with the tenets of a circular economy, wherein waste is reduced, and resources are repurposed. The oil palm business may diminish its environmental impact and support sustainable development objectives by transforming waste into profitable goods. This method not only alleviates waste disposal problems but also improves resource efficiency in agricultural techniques.

The building of a biomass-based chemical business can generate economic possibilities in rural regions, creating employment and promoting community development. This industry may operate as a stimulus for local economies, fostering investments in infrastructure and technology that facilitate sustainable practices. The effective incorporation of oil palm biomass waste into the chemical industry necessitates cooperation among many stakeholders, including researchers, industry participants, and policymakers. Such collaborations can enhance the exchange of information and resources, fostering innovation and establishing a conducive climate for the development of this nascent industry. Leveraging the potential of oil palm solid waste through efficient biomass valorization processes offers a promising avenue for a sustainable and economically viable future for the oil palm industry and beyond. Moreover, the establishment of a biomass-based chemical industry can create economic opportunities in rural areas, providing jobs and fostering community development. This sector can serve as a catalyst for local economies, promoting investments in infrastructure and technology that support sustainable practices.

The successful integration of oil palm biomass waste into the chemical industry requires collaboration between various stakeholders, including researchers, industry players, and policymakers. Such partnerships can facilitate the sharing of knowledge and resources, driving innovation and creating a supportive environment for the growth of this emerging industry.

Ultimately, harnessing the potential of oil palm solid waste through effective biomass valorization strategies presents a promising pathway towards a sustainable, economically viable future for the oil palm sector and beyond.

## REFERENCES


Abdul, P. M., Jahim, J. M., Harun, S., Markom, M., Lutpi, N. A., Hassan, O., Balan, V., Dale, B. E., & Mohd Nor, M. T. (2016). Effects of changes in chemical and structural characteristic of ammonia fibre expansion (AFEX) pretreated oil palm empty fruit bunch fibre on enzymatic saccharification and fermentability for biohydrogen. *Bioresource Technology*, *211*, 200–208. https://doi.org/10.1016/j.biortech.2016.02.135.

Ahmad, F. B., Zhang, Z., Doherty, W. O. S., & O'Hara, I. M. (2019). The outlook of the production of advanced fuels and chemicals from integrated oil palm biomass biorefinery. *Renewable and Sustainable Energy Reviews*, *109(April)*, 386–411. https://doi.org/10.1016/j.rser.2019.04.009.

Aishah, N., Abdul, K., & Diyana, N. (2019). Factorial analysis on nitric acid pretreatment of oil palm frond bagasse for xylan recovery. *Materials Today: Proceedings*, *19*, 1189–1198. https://doi.org/10.1016/j.matpr.2019.11.122.

Ait Rass, H., Essayem, N., & Besson, M. (2015). Selective aerobic oxidation of 5-HMF into 2,5-furandicarboxylic acid with Pt catalysts supported on $TiO_2$- and $ZrO_2$-based supports. *ChemSusChem*, *8*(7), 1206–1217. https://doi.org/10.1002/cssc.201403390.

Ang. S. K., S. E.M., A. Y., S. A.A, and M. M.S. (2013). "Production of cellulases and xylanase by *Aspergillus fumigatus* SK1 using untreated oil palm trunk through solid state fermentation," Process Biochem., vol. 48, no. 9, pp. 1293–1302, doi: 10.1016/j.procbio.2013. 06.019.

Ang. S, Yahya. A, Aziz. A, and Salleh. M, (2015) "Isolation, screening, and identification of potential cellulolytic and xylanolytic producers for biodegradation of untreated oil palm trunk and its application in saccharification of lemongrass leaves," Prep. Biochem. Biotechnol., vol. 45, no. 3, pp. 279–305, 2015, doi: 10.1080/10826068.2014.923443.

Antunes, F. A. F., Chandel, A. K., Terán-Hilares, R., Ingle, A. P., Rai, M., dos Santos Milessi, T. S., da Silva, S. S., & dos Santos, J. C. (2019). Overcoming challenges in lignocellulosic biomass pretreatment for second-generation (2G) sugar production: Emerging role of nano, biotechnological and promising approaches. *3 Biotech*, *9*(6), 230. https://doi.org/10.1007/s13205-019-1761-1.

Aryal, S. (2022). *Glycolysis Explained in 10 Easy Steps*. https://microbiologyinfo.com/glycolysis-10-steps-explained-steps-by-steps-with-diagram/.

Bai, Dong-Mei, Qiang Wei, Zhi-Hui Yan, Xue-Ming Zhao, Xin-Gang Li, and Shi-Min Xu. 2003. "Fed-Batch Fermentation of Lactobacillus Lactis for Hyper-Production of L-Lactic Acid." *Biotechnology Letters*. Vol. 25.

Banu, S., Kurade, M. B., Basak, B., Geun, C., Keun, K., Jeon, B., & Hyun, T. (2022). A review on physico-chemical delignification as a pretreatment of lignocellulosic biomass for enhanced bioconversion. *Bioresource Technology*, *346*. https://doi.org/10.1016/j.biortech.2021.126591.

Barragán-Ocaña, A., Merritt, H., Sánchez-Estrada, O. E., Méndez-Becerril, J. L., & del Pilar Longar-Blanco, M. (2023). Biorefinery and sustainability for the production of biofuels and value-added products: A trends analysis based on network and patent analysis. *PLoS ONE*, 18(1), 1–18. https://doi.org/10.1371/journal.pone.0279659.

Berger, K. G. (2003). Palm oil. *AOCS Lipid Library*, 4325. https://lipidlibrary.aocs.org/edible-oil-processing/palm-oil.

Bhalla, A., Bansal, N., Kumar, S., Bischoff, K. M., & Sani, R. K. (2013). Improved lignocellulose conversion to biofuels with thermophilic bacteria and thermostable enzymes. *Bioresource Technology*, *128*, 751–759. https://doi.org/10.1016/j.biortech.2012.10.145.

Bhatt, A. H., Zhang, Y., Davis, R., Heath, G., & Ravi, V. (2022). Biorefinery upgrading of herbaceous biomass to renewable hydrocarbon fuels, Part 2: Air pollutant emissions and permitting implications. *Journal of Cleaner Production*, *362*, 132409. https://doi.org/10.1016/j.jclepro.2022.132409.

Bukhari, N. A., Jahim, J., Kheang, S., Abu, L., Nasrin, B., & Harun, S. (2020). Organic acid pretreatment of oil palm trunk biomass for succinic acid production. *Waste and Biomass Valorization, 11*(10), 5549–5559. https://doi.org/10.1007/s12649-020-00953-2.

Burhan, K. H., Kresnowati, M. T. A. P., & Setiadi, T. (2019). Evaluation of simultaneous saccharification and fermentation of oil palm empty fruit bunches for xylitol production. *Bulletin of Chemical Reaction Engineering & Catalysis, 14*(3), 559–567.

Cang, R., Shen, L. Q., Yang, G., Zhang, Z. D., Huang, H., & Zhang, Z. G. (2019). Highly selective oxidation of 5-hydroxymethylfurfural to 5-hydroxymethyl-2-furancarboxylic acid by a robust whole-cell biocatalyst. *Catalysts, 9*(6). https://doi.org/10.3390/catal9060526.

Cao, Y., Zhang, R., Sun, C., Cheng, T., Liu, Y., & Xian, M. (2013). Fermentative succinate production: An emerging technology to replace the traditional petrochemical processes. *BioMed Research International*, 723412. https://doi.org/10.1155/2013/723412.

Cardona, C. A., Quintero, J. A., & Paz, I. C. (2010). Production of bioethanol from sugarcane bagasse: Status and perspectives. *Bioresource Technology, 101*(13), 4754–4766. https://doi.org/10.1016/j.biortech.2009.10.097.

Carrillo-Nieves, D., Rostro Alanís, M. J., de la Cruz Quiroz, R., Ruiz, H. A., Iqbal, H. M. N., & Parra-Saldívar, R. (2019). Current status and future trends of bioethanol production from agro-industrial wastes in Mexico. *Renewable and Sustainable Energy Reviews, 102*, 63–74. https://doi.org/10.1016/j.rser.2018.11.031.

Chandel, A. K., Chandrasekhar, G., Radhika, K., Ravinder, R., & Ravindra, P. (2011). Bioconversion of pentose sugars into ethanol: A review and future directions. *Biotechnology and Molecular Biology Reviews, 6*(1), 8–20.

Cheah, W. Y., Siti-Dina, R. P., Leng, S. T. K., Er, A. C., & Show, P. L. (2023). Circular bioeconomy in palm oil industry: Current practices and future perspectives. *Environmental Technology & Innovation, 30*, 103050. https://doi.org/10.1016/j.eti.2023.103050.

Cunha, J.T.; Romaní, A.; Domingues, L. Whole Cell Biocatalysis of 5-Hydroxymethylfurfural for Sustainable Biorefineries. Catalysts 2022, 12, 202. https://doi.org/10.3390/catal12020202

de Guilherme, A. (2009). Optimization of Lactic Acid Production by Lactobacillus casei NRRL B-442 in Cashew Clarified Juice. Federal University of Ceará. https://repositorio.ufc.br/handle/riufc/15751.

Donaldson, L., Nanayakkara, B., & Harrington, J. (2016). Wood growth and development. In B. Thomas, B. G. Murray, & D. J. Murphy (Eds.), Encyclopedia of Applied Plant Sciences (2nd edn., Vol. *1*, pp. 203–210). Elsevier. https://doi.org/10.1016/B978-0-12-394807-6.00114-3.

Dungani, R., Aditiawati, P., Aprilia, S., Yuniarti, K., Karliati, T., Suwandhi, I., & Sumardi, I. (2018). Biomaterial from Oil Palm Waste: Properties, Characterization and Applications. *IntechOpen*. https://doi.org/10.5772/intechopen.76412.

Duque, A., Manzanares, P., Ballesteros, I., & Ballesteros, M. (2016). Steam explosion as lignocellulosic biomass pretreatment. In S. I. B. Mussatto (Ed.), *Biomass Fractionation Technologies for a Lignocellulosic Feedstock Based Biorefinery* (pp. 349–368). Elsevier. https://doi.org/10.1016/B978-0-12-802323-5.00015-3.

Eram, M. S., & Ma, K. (2013). Decarboxylation of pyruvate to acetaldehyde for ethanol production by hyperthermophiles. *Biomolecules, 3*(3), 578–596. https://doi.org/10.3390/biom3030578.

Ezeilo. U, Lee. C, Huyop. F, Zakaria. I, and Wahab. R, "Raw oil palm frond leaves as cost-effective substrate for cellulase and xylanase productions by Trichoderma asperellum UC1 under solid-state fermentation," *J. Environ. Manage.*, vol. 243, pp. 206–217, 2019, doi: 10.1016/j.jenvman.2019.04.113.

Fatriasari. W, Raniya. R, Oktaviani. M, and Hermiati. E, (2018) "The improvement of sugar and bioethanol production of oil palm empty fruit bunches (Elaeis guineensis Jacq) through microwave-assisted maleic acid pretreatment," BioResources, vol. 13, no. 2, pp. 4378–4403, 2018, doi: 10.15376/ biores.13.2.4378–4403.

Fatriasari, W., Ulwan, W., Aminingsih, T., Sari, F. P., Fitria, Suryanegara, L., Iswanto, A. H., Ghozali, M., Kholida, L. N., Hussin, M. H., Fudholi, A., & Hermiati, E. (2021). Optimization of maleic acid pretreatment of oil palm empty fruit bunches (OPEFB) using response surface methodology to produce reducing sugars. *Industrial Crops and Products, 171*, 113971. https://doi.org/10.1016/j.indcrop.2021.113971

Fernandes, S., & Murray, P. G. (2010). Metabolic engineering for improved microbial pentose fermentation. *Bioengineered Bugs, 1*(6), 424–428. https://doi.org/10.4161/bbug.1.6.12724.

Financie, R., Moniruzzaman, M., & Uemura, Y. (2016). Enhanced enzymatic delignification of oil palm biomass with ionic liquid pretreatment. *Biochemical Engineering Journal, 110*, 1–7. https://doi.org/10.1016/j.bej.2016.02.008.

Francisco Pereira Marques, Amanda Kelly Lima Soares, Diego Lomonaco, Lorena Mara Alexandre e Silva, Sandra Tédde Santaella, Morsyleide de Freitas Rosa, Renato Carrhá Leitão, Steam explosion pretreatment improves acetic acid organosolv delignification of oil palm mesocarp fibers and sugarcane bagasse, International Journal of Biological Macromolecules, Volume 175, 2021, Pages 304–312, ISSN 0141-8130, https://doi.org/10.1016/j.ijbiomac.2021.01.174.

Hafid, H. S., Baharuddin, A. S., Mokhtar, M. N., Omar, F. N., Mohammed, M. A. P., & Wakisaka, M. (2021). Enhanced laccase production for oil palm biomass delignification using biological pretreatment and its estimation at biorefinary scale. *Biomass and Bioenergy*, 144, 105904. https://doi.org/10.1016/j.biombioe.2020.105904.

Hambali, E., & Rivai, M. (2017). The potential of palm oil waste biomass in Indonesia in 2020 and 2030. IOP Conference Series: Earth and Environmental Science, 65(1). https://doi.org/10.1088/1755-1315/65/1/012050.

Hendriks, A. T. W. M., & Zeeman, G. (2009). Pretreatments to enhance the digestibility of lignocellulosic biomass. *Bioresource Technology*, *100*(1), 10–18. https://doi.org/10.1016/j.biortech.2008.05.027.

Hermansyah, H., Putri, D. N., Prasetyano, A., Chairuddin, Z. B., Perdani, M. S., Sahlan, M., & Yohda, M. (2019). Delignification of oil palm empty fruit bunch using peracetic acid and alkaline peroxide combined with the ultrasound. *International Journal of Technology*, *10*(8), 1523–1532.

Hidayatullah, I. M., Al Husna, M. D., Radiyan, H., Kresnowati, M. T. A. P., Suhardi, S. H., Setiadi, T., & Boopathy, R. (2021a). Combining biodelignification and hydrothermal pretreatment of oil palm empty fruit bunches (OPEFB) for monomeric sugar production. *Bioresource Technology Reports*, *15*, 100808. https://doi.org/10.1016/j.biteb.2021.100808.

Hidayatullah, I. M., Makertihartha, I. G. B. N., Setiadi, T., & Kresnowati, M. T. A. P. (2021b). Modelling based analysis and optimization of simultaneous saccharification and fermentation for the production of lignocellulosic-based xylitol. *Bulletin of Chemical Reaction Engineering & Catalysis*, *16*(4). https://doi.org/10.9767/BCREC.16.4.11807.857-868.

Hofvendahl, K., & Hahn-Hägerdal, B. (2000). Factors affecting the fermentative lactic acid production from renewable resources. Enzyme and Microbial Technology, 26, 87–107.

Ikubar. M, Manan. A, Salleh. M, and Yahya. A, (2018). "Solid-state fermentation of oil palm frond petiole for lignin peroxidase and xylanase-rich cocktail production," 3 Biotech, vol. 8, no. 5, p. 0, 2018, doi:10.1007/s13205-018-1268-1.

Indera Luthfi, A. A., Jahim, J. M., Harun, S., Tan, J. P., & Mohammad, A. W. (2016). Biorefinery approach towards greener succinic acid production from oil palm frond bagasse. *Process Biochemistry*, *51*(10), 1527–1537. https://doi.org/10.1016/j.procbio.2016.08.011.

IPOA, 2022. Press Release Indonesia Palm Oil Associate, Palm oil performance in 2021 and Prospect in 2022. Online: https://gapki.id/en/news/21136/palm-oil-performance-in-2021-and-prospect-in-2022. January 2022.

Kahar, P., Rachmadona, N., Pangestu, R., Palar, R., Triyono Nugroho Adi, D., Betha Juanssilfero, A., Yopi, Manurung, I., Hama, S., & Ogino, C. (2022). An integrated biorefinery strategy for the utilization of palm-oil wastes. *Bioresource Technology*, *344*(*PB*), 126266. https://doi.org/10.1016/j.biortech.2021.126266.

Kamyab, H., Chelliapan, S., Din, M. F. M., Lee, C. T., Rezania, S., Khademi, T., & Bong, C. P. C. (2018). "Isolate new microalgal strain for biodiesel production and using FTIR spectroscopy for assessment of pollutant removal from Palm Oil Mill Effluent (POME)," Chem. Eng. Trans., vol. 63, no. May, pp. 91–96, 2018, doi: 10.3303/CET1863016.

Kim, S. M., Choi, B. Y., Ryu, Y. S., Jung, S. H., Park, J. M., Kim, G.-H., & Lee, S. K. (2015). Simultaneous utilization of glucose and xylose via novel mechanisms in engineered Escherichia coli. *Metabolic Engineering*, *30*, 141–148. https://doi.org/10.1016/j.ymben.2015.05.002.

Komesu, A., de Oliveira, J. A. R., da Silva Martins, L. H., Maciel, M. R. W., & Filho, R. M. (2017). Lactic acid production to purification: *A review. BioResources*, *12*(2), 4364–4383. https://doi.org/10.15376/BIORES.12.2.KOMESU.

Koopman, Frank, Nick Wierckx, Johannes H. de Winde, and Harald J. Ruijssenaars. 2010. "Efficient Whole-Cell Biotransformation of 5-(Hydroxymethyl)Furfural into FDCA, 2,5-Furandicarboxylic Acid." *Bioresource Technology* 101 (16): 6291–96. https://doi.org/10.1016/j.biortech.2010.03.050.

Le Borgne, S. (2012). Genetic engineering of industrial strains of Saccharomyces cerevisiae. In A. Lorence (Ed.), Recombinant Gene Expression (pp. 451–465). Humana Press. https://doi.org/10.1007/978-1-61779-433-9_24.

Lee, H. J., Sanyoto, B., Choi, J. W., Ha, J. M., Suh, D. J., & Lee, K. Y. (2013). Effects of lignin on the ionic-liquid assisted catalytic hydrolysis of cellulose: Chemical inhibition by lignin. *Cellulose*, *20*(5), 2349–2358. https://doi.org/10.1007/s10570-013-9989-0.

Lee, P., Lee, S., Hong, S., & Chang, H. (2002). Isolation and characterization of a new succinic acid-producing bacterium, Mannheimia succiniciproducens MBEL55E, from bovine rumen. *Applied Microbiology and Biotechnology*, *58*(5), 663–668. https://doi.org/10.1007/s00253-002-0935-6.

Liu, C. F., & Sun, R. C. (2010). Cellulose. In R. C. Sun (Ed.), Cereal Straw as a Resource for Sustainable Biomaterials and Biofuels: Chemistry, Extractives, Lignins, Hemicelluloses and Cellulose (1st edn., pp. 25–46). Elsevier. https://doi.org/10.1016/B978-0-444-53234-3.00005-5.

Lu, F., & Ralph, J. (2010). Lignin. In R. C. Sun (Ed.), Cereal Straw as a Resource for Sustainable Biomaterials and Biofuels: Chemistry, Extractives, Lignins, Hemicelluloses and Cellulose (1st edn., pp. 47–70). Elsevier. https://doi.org/10.1016/B978-0-444-53234-3.00006-7.

Lu, X., Gu, X. A review on lignin pyrolysis: pyrolytic behavior, mechanism, and relevant upgrading for improving process efficiency. Biotechnol Biofuels 15, 106 (2022). https://doi.org/10.1186/s13068-022-02203-0

Lucky Risanto, Deddy Triyono Nugroho Adi, Triyani Fajriutami, Hiroshi Teramura, Widya Fatriasari, Euis Hermiati, Prihardi Kahar, Akihiko Kondo, Chiaki Ogino, Pretreatment with dilute maleic acid enhances the enzymatic digestibility of sugarcane bagasse and oil palm empty fruit bunch fiber, Bioresource Technology, Volume 369, 2023, 128382, ISSN 0960-8524, https://doi.org/10.1016/j.biortech.2022.128382.

Maicas, S. (2020). The role of yeasts in fermentation processes. *Microorganisms*, *8*(8), 1–8. https://doi.org/10.3390/microorganisms8081142.

Mamimin C., Chaikitkaew S., Niyasom C., Kongjan P., O-Thong S, (2015). "Two-stage thermophilic fermentation and mesophilic methanogen process for biohythane production from palm oil mill effluent," Int. J. Hydrogen Energy, vol. 40, no. 19, pp. 6319–6328, doi: 10.1016/j.ijhydene.2015.03.068.

Mariyana, R., Choliq, A. A., Rahmanto, K. A., Nughroho, Y., Purwadi, R., Kresnowati, M. T. A. P., Septiningrum, K., & Dabukke, F. B. M. (2021). Integrated biorefinery technology: Monetization of oil palm empty fruit bunch to biofuel & bio-based chemicals, and beyond. *IOP Conference Series: Materials Science and Engineering, 1143,* 012053. doi:10.1088/1757-899X/1143/1/012053.

Mardawati, E., Herliansah, H., Suryadi, E., Hanidah, I. I., Siti Setiasih, I., Andoyo, R., Sukarminah, E., Djali, M., Rialita, T., & Cahyana, Y. (2019). Optimization of particle size, moisture content and reaction time of oil palm empty fruit bunch through ozonolysis pretreatment. *Journal of the Japan Institute of Energy*, *98*(6), 132–138. https://doi.org/10.3775/jie.98.132.

Mardawati, Efri, Nadya Maharani, Dwi Wahyudha Wira, Budi Mandra Harahap, Tri Yuliana, and Een Sukarminah. 2020. "Xylitol Production from Oil Palm Empty Fruit Bunches (OPEFB) Via Simultaneous Enzymatic Hydrolysis and Fermentation Process." *Journal of Industrial and Information Technology in Agriculture* 2 (1). https://doi.org/10.24198/jiita.v2i1.25064.

Medina, J. D. C., Woiciechowski, A., Filho, A. Z., Nigam, P. S., Ramos, L. P., & Soccol, C. R. (2016). Steam explosion pretreatment of oil palm empty fruit bunches (EFB) using autocatalytic hydrolysis: A biorefinery approach. *Bioresource Technology*, *199*, 173–180. https://doi.org/10.1016/j.biortech.2015.08.126.

Menon, V., & Rao, M. (2012). Trends in bioconversion of lignocellulose: Biofuels, platform chemicals & biorefinery concept. *Progress in Energy and Combustion Science*, *38*(4), 522–550. https://doi.org/10.1016/j.pecs.2012.02.002.

Miyazaki, K. (2005). A hyperthermophilic laccase from Thermus thermophilus HB27. *Extremophiles*, *9*(6), 415–425. https://doi.org/10.1007/s00792-005-0458-z.

Mosier, N., Wyman, C., Dale, B., Elander, R., Lee, Y. Y., Holtzapple, M., & Ladisch, M. (2005). Features of promising technologies for pretreatment of lignocellulosic biomass. *Bioresource Technology*, *96*(6), 673–686. https://doi.org/10.1016/j.biortech.2004.06.025.

Mussato, S. I. (2016). Technologies for a Lignocellulosic Feedstock Based Biorefinery. Elsevier.

Ng, S. W., Chong, W. T., Soo, Y. T., Tang, T. K., Karim, N. A. A., Phuah, E. T., & Lee, Y. Y. (2022). Pickering emulsion stabilized by palm-pressed fiber cellulose nanocrystal extracted by acid hydrolysis-assisted high pressure homogenization. *PLoS ONE*, 17(8 August), 1–23. https://doi.org/10.1371/journal.pone.0271512.

Nghiem, N. P., Kleff, S., Schwegmann, S. (2017). Succinic acid: Technology development and commercialization. *Fermentation*, *3*, 26. https://doi.org/10.3390/fermentation3020026.

Nor, N. A. M., Mustapha, W. A. W., & Hassan, O. (2016). Deep eutectic solvent (DES) as a pretreatment for oil palm empty fruit bunch (OPEFB) in sugar production. *Procedia Chemistry*, 18, 147–154. https://doi.org/10.1016/j.proche.2016.01.023.

Noratiqah. K, Madihah. M, Aisyah. B, Eva. M, Suraini. A, and Kamarulzaman. K, "Statistical optimization of enzymatic degradation process for oil palm empty fruit bunch (OPEFB) in rotary drum bioreactor using crude cellulase produced from Aspergillus niger EFB1," Biochem. Eng. J., vol. 75, pp. 8–20, Jun. 2013, doi:10.1016/j.bej.2013.03.007.

Norhazimah, A. H., Siti, F. M. N., Aida, M., Dilaeleyana, A. B., & Nur Shahirah, M. A. (2020). Direct fermentation of oil palm (Elaeis guineensis) trunk sap to bioethanol by Saccharomyces cerevisiae. *IOP Conference Series: Materials Science and Engineering*, *943*(1). https://doi.org/10.1088/1757-899X/943/1/012012.

Öhgren, K., Galbe, M., & Zacchi, G. (2005). Optimization of steam pretreatment of $SO_2$-impregnated corn stover for fuel ethanol production. Applied Biochemistry and Biotechnology, 121–124, 1055–1067. https://doi.org/10.1385/ABAB:124:1-3:1055

Ou, Mark S., Lonnie O. Ingram, and K. T. Shanmugam. 2011. "L(+)-Lactic Acid Production from Non-Food Carbohydrates by Thermotolerant Bacillus Coagulans." *Journal of Industrial Microbiology and Biotechnology* 38 (5): 599–605. https://doi.org/10.1007/s10295-010-0796-4.

Pang, Z., Lyu, W., Dong, C., Li, H., & Yang, G. (2016). High selective delignification using oxidative ionic liquid pretreatment at mild conditions for efficient enzymatic hydrolysis of lignocellulose. *Bioresource Technology*, *214*, 96–101. https://doi.org/10.1016/j.biortech.2016.04.095.

Parthasarathy, P., Alherbawi, M., Shahbaz, M., Mackey, H. R., McKay, G., & Al-Ansari, T. (2022). Conversion of oil palm waste into value-added products through pyrolysis: A sensitivity and techno-economic investigation. *Biomass Conversion and Biorefinery*, *14*, 9667–9687. https://doi.org/10.1007/s13399-022-03144-2.

Ramos, L. P. (2003). The chemistry involved in the steam treatment of lignocellulosic materials. *Quimica Nova*, *26*(6), 863–871. https://doi.org/10.1590/s0100-40422003000600015.

Rosa, D., Medeiros, A.B.P., Martinez-Burgos, W.J., do Nascimento, J.R., de Carvalho, J.C., Sydney, E.B., Soccol, C.R. (2020), "Biological hydrogen production from palm oil mill effluent (POME) by anaerobic consortia and Clostridium beijerinckii," J. Biotechnol., vol. 323, no. June, pp. 17–23, 2020, doi: 10.1016/j.jbiotec.2020.06.015.

Pratter, S. M., Eixelsberger, T., & Nidetzky, B. (2015). Systematic strain construction and process development: Xylitol production by Saccharomyces cerevisiae expressing Candida tenuis xylose reductase in wild-type or mutant form. *Bioresource Technology*, *198*, 732–738. https://doi.org/10.1016/j.biortech.2015.09.046.

Purnomo, H., Okarda, B., Dermawan, A., Ilham, Q. P., Pacheco, P., Nurfatriani, F., & Suhendang, E. (2020). Reconciling oil palm economic development and environmental conservation in Indonesia: A value chain dynamic approach. *Forest Policy and Economics*, 111, 102089. https://doi.org/10.1016/j.forpol.2020.102089.

Putri, D. N., Pratiwi, S. F., Perdani, M. S., Rosarina, D., Utami, T. S., Sahlan, M., & Hermansyah, H. (2023). Utilizing rice straw and sugarcane bagasse as low-cost feedstocks towards sustainable production of succinic acid. *Science of the Total Environment*, *862*, 160719. https://doi.org/10.1016/j.scitotenv.2022.160719.

Putri, D. N., Sahlan, M., Montastruc, L., Meyer, M., Negny, S., & Hermansyah, H. (2020). Progress of fermentation methods for bio-succinic acid production using agro-industrial waste by Actinobacillus succinogenes. *Energy Reports*, *6*, 234–239. https://doi.org/10.1016/j.egyr.2019.08.050.

Putro, J. N., Soetaredjo, F. E., Lin, S.-Y., Ju, Y.-H., & Ismadji, S. (2016). Pretreatment and conversion of lignocellulose biomass into valuable chemicals. *RSC Advances*, *6*(52), 46834–46852. https://doi.org/10.1039/C6RA09851G.

Ren, J. L., & Sun, R. C. (2010). Hemicelluloses. In R. C. Sun (Ed.), Cereal Straw as a Resource for Sustainable Biomaterials and Biofuels: Chemistry, Extractives, Lignins, Hemicelluloses and Cellulose (1st edn., pp. 71–92). Elsevier. https://doi.org/10.1016/B978-0-444-53234-3.00004-3.

Sabiha-Hanim, S., Mohd Noor, M. A., & Rosma, A. (2015). Fractionation of oil palm frond hemicelluloses by water or alkaline impregnation and steam explosion. *Carbohydrate Polymers*, *115*, 533–539. https://doi.org/10.1016/j.carbpol.2014.08.087.

Saddler, J., Brownell, H. H., & Saddler, J. N. (1986). Steam pretreatment of lignocellulosic material for enhanced enzymatic hydrolysis. *Biotechnology and Bioengineering*, *28*(2), 214–219.

Saha, B. C., Nichols, N. N., Qureshi, N., & Cotta, M. A. (2011). Comparison of separate hydrolysis and fermentation and simultaneous saccharification and fermentation processes for ethanol production from wheat straw by recombinant Escherichia coli strain FBR5. *Applied Microbiology and Biotechnology*, *92*(4), 865–874. https://doi.org/10.1007/s00253-011-3600-0.

Said, F. M., Hamid, N. F., Razali, M. A.-A., & Daud, N. F. S. (2021). Lignocellulosic of oil palm biomass to chemical product via fermentation. In H. Kamyab (Ed.), *Oil Palm Biomass for Bioenergy Production*. IntechOpen. https://doi.org/10.5772/intechopen.99312.

Sajid, M., Zhao, X., & Liu, D. (2018). Production of 2{,}5-furandicarboxylic acid (FDCA) from 5-hydroxymethylfurfural (HMF): Recent progress focusing on the chemical-catalytic routes. *Green Chemistry*, *20*(24), 5427–5453. https://doi.org/10.1039/C8GC02680G.

Shamjuddin, A., Ab Rasid, N. S., Michele Raissa, M. M., Abu Zarin, M. A., Wan Omar, W. N. N., Syahrom, A., Mohd Szali Januddi, M. A. F., & Saidina Amin, N. A. (2021). Kinetic and dynamic analysis of ozonolysis pre-treatment of empty fruit bunch in a well-mixed reactor for sugar production. *Energy Conversion and Management*, *244*, 114526. https://doi.org/10.1016/j.enconman.2021.114526.

Sheng, Y., Tan, X., Zhou, X., & Xu, Y. (2020). Bioconversion of 5-Hydroxymethylfurfural (HMF) to 2,5-Furandicarboxylic Acid (FDCA) by a Native Obligate Aerobic Bacterium, Acinetobacter calcoaceticus NL14. *Applied Biochemistry and Biotechnology*, *192*(2), 455–465. https://doi.org/10.1007/s12010-020-03325-7.

Singh, R., Shukla, A., Tiwari, S., & Srivastava, M. (2014). A review on delignification of lignocellulosic biomass for enhancement of ethanol production potential. *Renewable and Sustainable Energy Reviews*, *32*, 713–728. https://doi.org/10.1016/j.rser.2014.01.051.

Sołowski, G., Konkol, I., & Cenian, A. (2020). Production of hydrogen and methane from lignocellulose waste by fermentation. A review of chemical pretreatment for enhancing the efficiency of the digestion process. *Journal of Cleaner Production*, *267*(May), 0–21. https://doi.org/10.1016/j.jclepro.2020.121721.

Soh, A. C. (2012). Breeding and genetics of the oil palm. In O.-M. Lai, C.-C. Akoh, & G. Choo (Eds.), *Palm Oil: Production, Processing, Characterization, and Uses* (pp. 31–58). AOCS Press. https://doi.org/10.1016/B978-0-9818936-9-3.50005-8.

Sousa, A. F., Vilela, C., Fonseca, A. C., Matos, M., Freire, C. S. R., Gruter, G.-J. M., Coelho, J. F. J., & Silvestre, A. J. D. (2015). Biobased polyesters and other polymers from 2{,}5-furandicarboxylic acid: A tribute to furan excellency. *Polymer Chemistry*, *6*(33), 5961–5983. https://doi.org/10.1039/C5PY00686D.

Sridhar, M. K. C., & AdeOluwa, O. O. (2009). Palm oil industry residues. In P. Singh nee' Nigam & A. Pandey (Eds.), *Biotechnology for Agro-Industrial Residues Utilisation* (pp. 341–355). Springer. https://doi.org/10.1007/978-1-4020-9942-7_18.

Su, T., Zhao, D., Khodadadi, M., & Len, C. (2020). Lignocellulosic biomass for bioethanol: Recent advances, technology trends, and barriers to industrial development. *Current Opinion in Green and Sustainable Chemistry*, *24*, 56–60. https://doi.org/10.1016/j.cogsc.2020.04.005.

Sugiharto, Y. E. C., Harimawan, A., Kresnowati, M. T. A. P., Purwadi, R., Mariyana, R., Andry, Fitriana, H. N., & Hosen, H. F. (2016). Enzyme feeding strategies for better fed-batch enzymatic hydrolysis of empty fruit bunch. *Bioresource Technology*, *207*, 175–179. https://doi.org/10.1016/j.biortech.2016.01.113.

Tae, H. K., & Lee, Y. Y. (2005). Pretreatment and fractionation of corn stover by ammonia recycle percolation process. *Bioresource Technology*, *96*(18), 2007–2013. https://doi.org/10.1016/j.biortech.2005.01.015.

Tan. J, Jahim. J, Harun. S, Wu. T, and Mumtaz. T, (2016). "Utilization of oil palm fronds as a sustainable carbon source in biorefineries," Int. J. Hydrogen Energy, vol. 41, no. 8, pp. 4896–4906, 2016, doi: 10.1016/j.ijhydene.2015.08.034.

Terry, L. M., Li, C., Chew, J. J., Aqsha, A., How, B. S., Loy, A. C. M., Chin, B. L. F., Khaerudini, D. S., Hameed, N., Guan, G., & Sunarso, J. (2021). Bio-oil production from pyrolysis of oil palm biomass and the upgrading technologies: A review. *Carbon Resources Conversion*, *4(October)*, 239–250. https://doi.org/10.1016/j.crcon.2021.10.002.

Thi, S., & Lee, K. M. (2019). Comparison of deep eutectic solvents (DES) on pretreatment of oil palm empty fruit bunch (OPEFB): Cellulose digestibility, structural and morphology changes. *Bioresource Technology*, *282(January)*, 525–529. https://doi.org/10.1016/j.biortech.2019.03.065.

Ujor, V.C & Okonkwo, C. C. (2022). Microbial detoxification of lignocellulosic biomass hydrolysates: Biochemical and molecular aspects, challenges, exploits and future perspectives. *Frontiers in Bioengineering and Biotechnology, 10*, 1061667. https://doi.org/10.3389/fbioe.2022.1061667.

Umai, D., Kayalvizhi, R., Kumar, V., & Jacob, S. (2022). Xylitol: Bioproduction and applications: A review. *Frontiers in Sustainability*, *3(February)*, 1–16. https://doi.org/10.3389/frsus.2022.826190.

Vasić, K., Knez, Z., & Leitgeb, M. (2021). Bioethanol production by enzymatic hydrolysis from different. *Molecules*, *26*(753), 1–23.

Wan Omar, W. N. N., & Amin, N. A. S. (2016). Multi response optimization of oil palm frond pretreatment by ozonolysis. *Industrial Crops and Products*, *85*, 389–402. https://doi.org/10.1016/j.indcrop.2016.01.027.

Wool, R. & Sun, X. S. (2005). Lignin polymers and composites. In R. P. Wool & X. S. Sun (Eds.), *Bio-Based Polymers and Composites* (pp. 551–598). Elsevier.

Xu, J.-K., & Sun, R.-C. (2016). Recent advances in alkaline pretreatment of lignocellulosic biomass. In S. I. B. Mussatto (Ed.), *Biomass Fractionation Technologies for a Lignocellulosic Feedstock-based Biorefinery* (pp. 431–459). Elsevier. https://doi.org/10.1016/B978-0-12-802323-5.00019-0.

Yang, B., & Wyman, C. E. (2008). Pretreatment: The key to unlocking low-cost cellulosic ethanol. *Biofuels, Bioproducts and Biorefining*, *2*(1), 26–40. https://doi.org/10.1002/bbb.49.

Ying, T. Y., Teong, L. K., Abdullah, W. N. W., & Peng, L. C. (2014). The effect of various pretreatment methods on oil palm empty fruit bunch (EFB) and kenaf core fibers for sugar production. *Procedia Environmental Sciences*, *20*, 328–335. https://doi.org/10.1016/j.proenv.2014.03.041.

Yoricya, G., Dalimunthe, S. A. P., Manurung, R., & Bangun, N. (2016). Hidrolisis hasil delignifikasi tandan kosong kelapa sawit dalam sistem cairan ionik choline chloride. *Jurnal Teknik Kimia USU*, *5*(1), 27–33.

Yusuf, A. A. I. S., & Kresnowati, M. T. A. P. (2019). Evaluation of surfactant addition to enzymatic hydrolysis of oil palm empty fruit bunch (OPEFB). AIP Conference Proceedings, *2197*, 020010.

Zamri, M. F. M. A., Milano, J., Shamsuddin, A. H., Roslan, M. E. M., Salleh, S. F., Rahman, A. A., Bahru, R., Fattah, I. M. R., & Mahlia, T. M. I. (2022). An overview of palm oil biomass for power generation sector decarbonization in Malaysia: Progress, challenges, and prospects. *WIREs Energy and Environment*, *11*(4), e437. https://doi.org/10.1002/wene.437.

Zeng, Y., Zhao, S., Yang, S., & Ding, S.-Y. (2014). Lignin plays a negative role in the biochemical process for producing lignocellulosic biofuels. *Current Opinion in Biotechnology*, *27*, 38–45. https://doi.org/10.1016/j.copbio.2013.09.008.

Zulkiple, N., Maskat, M. Y., & Hassan, O. (2016). Pretreatment of oil palm empty fruit fiber (OPEFB) with aqueous ammonia for high production of sugar. *Procedia Chemistry*, *18(Mcls 2015)*, 155–161. https://doi.org/10.1016/j.proche.2016.01.024.

# *Section II*

## *Lignocellulose Waste Management and Climate Change*

# 6 Decoding the Influence of Climate Change and Emerging Contaminants on the Agro-Residue Composting Process and Quality

*Lemnaro Jamir, Shivangani, N. Jaya Lakshmi, and P. Hariprasad*
Environmental Biotechnology Lab, Centre for Rural Development and Technology, Indian Institute of Technology Delhi, New Delhi, India

## 6.1 INTRODUCTION

Climate change (CC) is a natural phenomenon observed during the Earth's evolution. Cycles of ice ages and warmer periods have repeatedly occurred, as observed for the last 800,000 years (https://climate.nasa.gov/). However, several research studies have evidence that the recent accelerated CC phenomenon is linked to anthropogenic activities (Arias et al. 2021). The effects of CC on the atmosphere, land, cryosphere, biosphere, and ocean systems are very well evidenced. It is expected to pose a severe threat to the existence of living organisms in the near future. CC is experiencing increased $CO_2$ (419 ppm as of August 2022; noaa.gov) and other greenhouse gases (GHGs).

Further, the global surface temperature has increased by 1.1°C and is projected to reach 2.6°C–5°C by 2100 (Stocker et al. 2014). The consequences of global warming were seen as shrinking glaciers, rising sea levels, changes in weather patterns like droughts and flooding, and much more. In parallel with CC, emerging contaminants (EC) in the environment are threatening living organisms. EC are chemicals (natural or synthetic), including microorganisms that exhibit a harmful effect on the environment when present above a threshold concentration. They contain industrial chemicals, heavy metals (HMs), pharmaceuticals, hormones, fertilizers, pesticides/herbicides, paints, surfactants, nanomaterials, and personal care products. The origin of these contaminants is linked to a wide range of anthropogenic activities. These EC are present in almost all ecosystem components, like air, water, soil, and biological systems.

As per FAO estimates, based on population growth and trends in food consumption, agricultural production will be required to expand by at least 70% by the year 2050 in order to meet the demand (FAO 2009). In addition, human dependency on the plant for fiber and fuel is increasing. In this CC scenario, getting enough food to feed the world's continuously expanding population is a significant concern threatening global food security. The impact of CC on crop plants is further intensified due to factors like soil erosion/degradation, loss of agro-biodiversity, and EC.

So, to deal with current and future problems sustainably, agricultural practices around the world need to change in a big way. One of the most important factors for sustainable agricultural and livestock production is recycling of organic agricultural waste (Westerman and Bicudo 2005). Some of the old ways of farming that people worldwide still follow, like agroecology, agroforestry, natural farming, conservation agriculture, organic farming, zero-till farming, etc., are more aligned toward

DOI: 10.1201/9781032676890-8

sustainability and environmental protection. However, scientifically upgrading these methods and globalization is the need of the hour. In order to do this, it is essential to know how the different parts of agricultural systems behave and how they work together in different environments. Among the several components of agriculture, nutrient management is a vital process. Sustainable agriculture is based on the idea of recycling organic matter and reducing the usage of synthetic chemical fertilizers. Here, the plant gets most of the nutrients it needs from raw or processed organic matter alone or in combination with chemical fertilizers. Composting is one of the most environmentally friendly ways to use the residues from agriculture to improve soil health and crop production and meet global needs for food security. Compost offers plant nutrients and boosts soil structure, water-holding capacity, soil organic carbon, microbial load, etc. (Insam and de Bertoldi 2007). However, several attempts have been made to evaluate the performance of plants and soil under altered climatic conditions (Baligar et al. 2001). An attempt to comprehend the impact of CC and EC on composting is given little importance.

This chapter discusses the development of the impact of CC and EC on the (1) quality of feedstock, (2) compost process, and (3) quality of the compost. Our primary focus is on agro-residues as they are the most abundant feedstock available and aerobic composting.

## 6.2 COMPOSTING

Composting is a natural process domesticated by humans. Composting started in the Neolithic era when people lived in small groups and made waste pits with organic matter and rotting plants. These pits helped crops grow in the early stages of farming and city life. Composting is a well-known way to turn agricultural waste into bio-fertilizers using microorganisms and exothermic oxidation in an aerobic environment. Composting breaks down complex organic substrates into simpler forms that plants can use as soil amendments. Different types of composting methods generally followed across the world is compiled in Table 6.1.

During the COVID-19 pandemic, the global compost market declined, but there has been a big comeback. According to Lucintel (2020), by 2025, it was expected to reach $8.9 billion, with a 2% compound annual growth rate (CAGR) from 2020 to 2025. In a recent study, Zhao et al. (2022b) revealed the potential of a precise composting strategy (PCS) with an estimated annual increase of 96.3 Tg in the yield of major cereal crops (4% of current production). Moreover, PCS can fix 19.5 Pg of carbon in the top soils of agricultural lands, which is estimated to be equivalent to 26.5% of organic carbon in topsoil. Even though it could have been used as compost, 92 Mt of India's 350 Mt of agro-residue is being burned yearly (NPMCR; United Nations, 1997), turning it into a source of pollution instead. Aside from manure and harvested agro-waste, composting can also be obtained from waste from farms, poultry, slaughterhouses, and municipal solid wastes. However, these feedstocks need to be handled and used judiciously to prevent their detrimental effects on the composting process and quality, affecting human, animal, plant, and soil health. Microbes in such feedstock may be pathogenic carriers of antibiotic resistance (Shrivas et al. 2021). Several EC were reported to enter the food chain through composting. Further, inappropriate composting methods may contribute to the emission of nitrous oxide ($N_2O$) (GHG) and lead to uncertainty in their performance under field conditions. All the possible effects/interactions of CC and EC on feedstock, composting process, and quality is compiled in Figure 6.1.

## 6.3 IMPACT OF CLIMATE CHANGE ON FEEDSTOCK QUALITY

The stability and quality of composts produced from various organic sources (feedstock) vary, and this is further influenced by the makeup of the raw materials used in the compost production process (Azim et al. 2018). Farm waste, animal waste, agro-industrial waste, household trash, and municipal solid waste are the most common raw materials utilized in composting (Füleky and Benedek 2010). Regarding the amount of organic matter in the raw materials, there is a wide range depending on

**TABLE 6.1**
**Most Common Methods of Composting Followed Across the World**

| S. No. | Composting Method | Feedstock | Process Parameters | Duration (Months) | Nutrient Profile | Reference |
|---|---|---|---|---|---|---|
| 1. | Indore composting | Garden waste, animal dung, wood ashes, sawdust, bagasse, and night soil | • Pit/heap or trench method<br>• Feedstock C:N ratio 1:3<br>• Moisture level 60%–80%<br>• Periodic turning | 4–6 | 0.5%–0.8% N<br>0.3%–0.5% $P_2O_5$<br>1.0%–1.5% $K_2O$ | Bhattacharjya et al. (2019)<br>Mawar and Mathur (2022) |
| 2. | Windrow composting | Leaves, bagasse, straw, cotton wastes, poultry manure, wood chips, deoiled cakes, coffee pulp | • Organic waste piled into long rows known as windrows<br>• Moisture level 50%–65%<br>• Active composting phase: 8–16 weeks<br>• Periodic turning: 10–15 days | 4–5 | 0.5%–2.75% N<br>0.5%–2% $P_2O_5$<br>3.0% $K_2O$ | Kuhlman (1990)<br>Michel et al. (2022)<br>Sellami et al. (2008)<br>Abu Qdais and Al-Widyan (2016)<br>Rolz et al. (2010) |
| 3. | Aerated static pile composting | Cattle waste, paddy straw, green residues, husk, leaves, bulking agents | • Feedstock C:N ratio 30:1<br>• Moisture level 50%–60%<br>• Requires continuous aeration | 3–6 | 1.38%–1.94% N<br>0.47%–0.69% $P_2O_5$<br>2.5%–4.7% $K_2O$ | Alkoaik (2019)<br>Cayuela et al. (2006) |
| 4. | Vermicomposting | Sugarcane straw, dry tree leaves, kitchen wastes and cow manure, rice bran, chaff, husk, spent mushroom substrate | • Employs earthworms majorly *Eisenia foetida* and *Eudrilus eugeniae* species<br>• Relative moisture 50%–60%<br>• Requires pH 6.5–7.5 in pile<br>• No turning of the compost heap is required | 2–3 | 1.5%–2% N<br>0.5%–1.5% $P_2O_5$<br>0.5%–1% $K_2O$<br>0.4%–0.8% CaO<br>0.3%–0.6% MgO | Sharma and Garg (2019)<br>Shak et al. (2014)<br>Garg et al. (2006)<br>CRIDA (2003) |
| 5. | NADEP composting | Weeds, farm wastes, straw, leaves, stems, pruning, husk, cattle dung | • Composting carried out in pits or brick tanks<br>• Moisture level 15%–20%<br>• Bio-inoculated with *Azotobacter*, *Trichoderma*, and *Rhizobium* and phosphate-solubilizing microorganisms | 3–5 | 0.5%–1.5% N<br>0.5%–0.9% $P_2O_5$<br>1.2%–1.4% $K_2O$ | Kumar et al. (2012)<br>Directorate of Agriculture (2003)<br>Verma et al. (2014) |

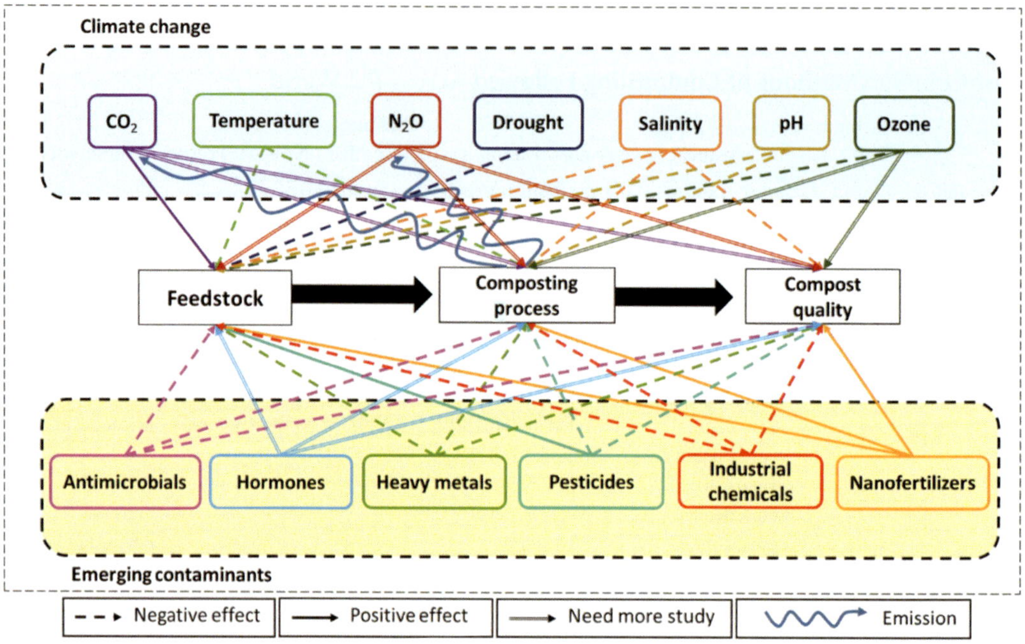


**FIGURE 6.1** Impact of climate change and emerging pollutants on feedstock, compost processing, and quality. Greenhouse gas is primarily represented as $CO_2$ and $N_2O$, and the most common observations reported in the literature are incorporated in this figure.

the raw material's constituents. The lignocellulosic biomass found in agricultural wastes is abundant and comprises cellulose, hemicellulose, lignin, pectin, and protein (Harindintwali et al. 2020). The cell wall composition varies between and within plant species, tissue types, growth stages, etc. Further, other biotic (microbial infection and insect infestation) and abiotic (a nutrient deficiency, drought, salinity, light intensity, ozone, HMs, antibiotics (AB), etc.) factors to which plants are exposed bring variations in the composition of the cell wall (Bhatia et al. 2020).

In lignocellulosic biomass, lignin envelops cellulose and hemicellulose to form a structure that is recalcitrant for bacteria to degrade. Cellulose is a linear polymer of hexose sugars linked by β-1,4-linkage, and hemicellulose is a class of cross-linked heteropolymers made up of hexose (glucose, galactose, mannose, and 4-O-methyl-d-glucuronic acid) and pentose sugars (xylose and arabinose). In comparison, lignin is a three-dimensional, intricate, complex aromatic heteropolymer made up of propyl phenol units (coniferyl alcohol, sinapyl alcohol, and p-coumaryl alcohol) cross-linked together through carbon–carbon, ester, and ether bondings (Saha et al. 2018). Lignin is difficult to depolymerize by enzymatic hydrolysis due to its complex chemical structure, which further inhibits the breakdown and utilization of cellulose and hemicellulose (Chauhan 2020). Hence, lignin concentration in cell walls is negatively correlated with the biodegradability of biomass (Haug 2018).

The environment is negatively affected by CC in various ways, including increased frequency of simultaneous extreme weather events, high temperatures, and changing GHG concentrations like rising atmospheric $CO_2$ along with an increase in soil salinity, drought, etc. CC consequently has several detrimental implications on crop productivity. One favorable response of CC on plants is an increase in the rates of photosynthesis due to an increased $CO_2$ concentration in the atmosphere (Zhang et al. 2019). As observed by Robertson and Leech (1995), 7-day-old wheat seedlings grown at 650 μL/L (elevated) $CO_2$ level recorded accelerated expansion of cell and chloroplast and reduced starch accumulation over the seedlings grown at 350 μL/L (ambient) $CO_2$ level. In these experiments, a linear relationship was also observed between the number or size and chloroplast and mesophyll cell size. A similar finding in arabidopsis was reported by Narendra et al. (2006), where an increase in chloroplast number, starch grains, size, and number, and a decrease in the number of grana thylakoid membranes was observed at a higher $CO_2$ (700 μM/M) level. Moreover, they reported an alteration in the cell wall composition, where cellulose content in leaves was found to increase by 22.3%. Their observations were further supported by Fourier transform infrared (FTIR) (peak corresponding to cellulose at 1,043 and 1,157 $cm^{-1}$) and transmission electron microscopy (TEM) analysis of the cell wall.

The process of photosynthesis also depends on the rate at which $CO_2$ reaches the carboxylation site in chloroplast and cytosol in C3 and C4 plants, respectively. Among the several barriers, the thickness of the cell wall of the mesophyll cells determines the rate of $CO_2$ diffusion into the cell (Flexas et al. 2008). Since the principal sink for carbohydrates on Earth is the cell wall, a high carbon intake also alters the cell wall composition (Delmer et al. 2002; Harrison et al. 2009). This will likely affect plants' primary and secondary metabolisms (Davey et al. 2004). In terms of secondary metabolism, it is well known that high $CO_2$ levels obstruct lignin production. The pace of cell wall elongation and loosening as well as the rate of water and solute absorption by the expanding cell serve as co-regulators of the vast expansion.

As Zhu et al. (2012) reported, different plants differentially respond to elevated $CO_2$ levels. They observed that, under elevated $CO_2$ levels, the rice flag leaf showed a 23.3% increase in cell wall thickness. Under the same conditions, wheat flag leaf cells recorded only a 2.55% increase in cell wall thickness. Similarly, young *Quercus ilex* trees recorded an increase in cellulose and lignin content when exposed to 700 μL/L $CO_2$. Conversely, Brich (*Betula pendula*) plants exposed to 651–729 ppm of $CO_2$ recorded reduced α-cellulose and acid-soluble lignin in mesophyll cells. At the same time, the hemicellulose content was found to increase in young mesophyll cells and to gradually decrease with the age of the leaf (Oksanen et al. 2005). Coffee plants subjected to 37°C experienced alterations in the cell wall composition of their leaves, with pectin levels falling by 50% and hemicellulose levels rising by 40% (Lima et al. 2013).

Other climatic factors, such as temperature, drought, and salinity, can negate the positive effect of elevated levels of atmospheric $CO_2$ on photosynthesis and crop productivity (Ruiz-Vera et al. 2013; Gray et al. 2016). High temperatures cause biomacromolecules like proteins and nucleic acids to break down, accelerate membrane lipid peroxidation, and disrupt the normal metabolism of plants (Liu et al. 2018). Numerous photosynthetic activities in chloroplasts, such as electron transport, photochemical reactions, chlorophyll production, and $CO_2$ assimilation, are adversely impacted by high temperatures. Plastid-based protein translation is impaired by heat stress, and at high temperatures, both newly synthesized proteins and already existing proteins may undergo denaturation (Ellis 1990; Bita and Gerats 2013). Lobell et al. (2011) developed a model correlating crop productivity and increased temperature from 1980 to 2008, revealing that wheat and maize production was reduced by 3.8% and 5.5%, respectively, in response to global warming. However, in the case of rice and soy, the reduction in productivity was insignificant. A similar kind of response in wheat to increased temperature in northwest India was reported by Kalra et al. (2008).

Similarly, ion toxicity and osmotic stress are brought on by excessive sodium buildup in plants, and these effects significantly impact their morphological, physiological, biochemical, and metabolic condition. Plant cell walls serve as the first line of defense against external threats. Therefore, resistance to plant diseases and pests, along with tolerance toward drought, salinity, HMs, extreme temperatures, and other stresses, are all partly correlated with lignin metabolism (Schutzendubel and Polle 2002; Moura et al. 2010). When plants are exposed to a protracted water stress deficit, lignification develops prematurely to prevent cell wall damage (Moura et al. 2010). The other most common adaptation at the cell wall level in plants exposed to drought stress is pectin modification or accumulation as a hydrated form, protecting the cell damage (Gall et al. 2015). Moreover, salt treatment changed many of the lignin production pathway's enzymes (Gall et al. 2015). Feng et al. (2018) reported that high salinity weakens the cell wall by disrupting pectin cross-linkages in arabidopsis, which impedes cell recovery when turgor is restored. *Panicum maximum* seedlings exposed to drought stress recorded reduced cellulose (17.42%) and hemicellulose (15.91%) and increased lignin content (5.03%) compared with control seedlings (cellulose: 29.01%, hemicellulose: 16.96%, and lignin: 3.73%) in leaf tissue (Tiwari et al. 2018). A similar trend was also reported in seedlings exposed to salt stress (Neves et al. 2010; Tiwari et al. 2018). On the contrary, Alvarez et al. (2008) observed that maize plants exposed to drought conditions recorded reduced lignin biosynthesis as evidenced by increased free lignin precursor in xylem sap and reduced anionic peroxidase activity.

It is critical to consider the C/N ratio of the compost feedstock. Carbon is essential for energy production and the formation of microbial cells, whereas nitrogen is necessary for microbial growth and replication due to its role in protein synthesis (Sweeten and Auvermann 2006). Bishop (1983) determined that the optimal C/N ratio for decomposition is between 25 and 35, while bacteria need 30 parts C/N. A high C/N ratio retards decomposition because there is an abundance of degradable substrate for bacteria with low N. When the C/N ratio of a compost pile is low, there is an excess of N per unit of degradable C, and excess inorganic N can be lost as ammonia (volatilized or leached) (Bernal et al. 2009; Kirchmann and Witter 1989; Tiquia et al. 2002). In general, the ideal carbon-to-nitrogen ratio varies greatly depending on the feedstock composition. Shah et al. (2014) provided a list of the C/N ratios of several agro-residues, which aids in developing a combined feedstock suitable for composting. The mean C/N ratio for paddy straw and corn stalks is 65 and 121 with a low moisture content of 10–11% on average while feedstocks such as grass shavings are high in moisture (69%–87%) and low in C/N ratio (19). Yang et al. (2021) investigated the impact of various C/N ratios on lignocellulose degradation during aerobic composting and quantified the metabolic activities of relevant enzymes. The ideal C/N ratio of 25 stimulated the activity of enzymes involved in ligninolytic activities, such as manganese peroxidase, lignin peroxidase, polyphenol oxidase, and peroxidase, which in turn promoted lignocellulose degradation.

The accumulation of biomass and its biochemical characteristics are the product of a complex system of metabolic processes that react to environmental perturbations (Habermann et al. 2021). The impact of CC on plant cell functioning is compiled in Figure 6.2. From the above discussion, it is evident that the factors linked to CC have a detrimental effect on both the quantity and quality of feedstock used for composting. In general, agro-residues are composted with animal excreta/waste

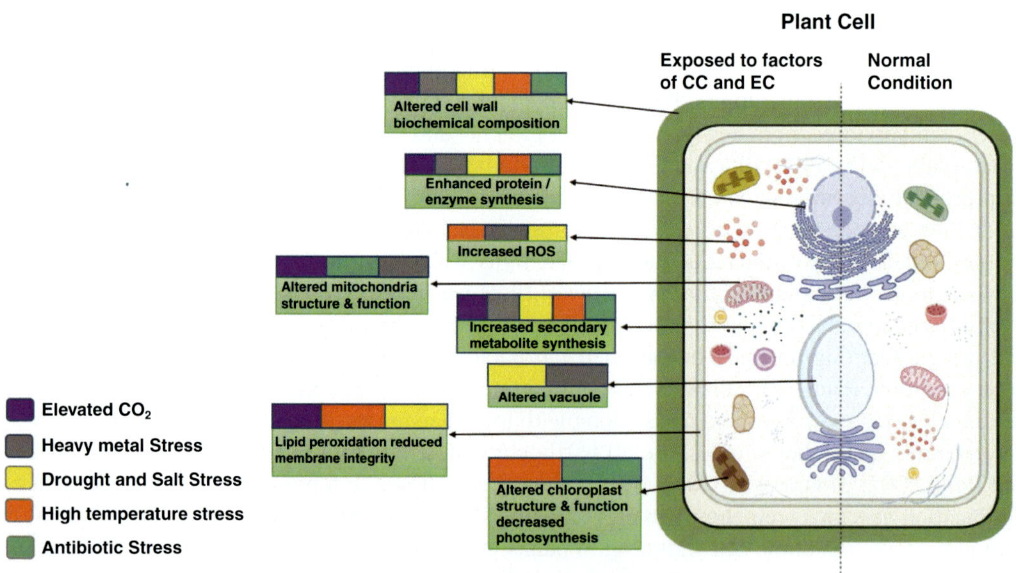


**FIGURE 6.2** Effect of climate change and emerging contaminants on plant cell structure and function. (*Note*: The most common observations reported in the literature are incorporated in this figure.)

to increase the bulk and maintain the C/N ratio. The suitability of animal excreta/waste, which was obtained by feeding crops grown under the altered CC conditions and the impact of CC on animal health are other major research areas that need to be studied. In addition, a detailed study of the impact of CC on the composting process and quality needs to be conducted.

## 6.4 IMPACT OF EMERGING CONTAMINANTS ON FEEDSTOCK QUALITY

During evolution, organisms like microbes and plants responded to EC and developed an adaptive mechanism to overcome their adverse effect. The mechanisms are broadly categorized as (i) escape – the organisms avoid contact with EC either by physically moving or developing some physical barrier (biofilm-produced bacteria and efflux pump activation) or accumulating/immobilizing the EC in the body in compartments like vacuole and (ii) detoxify – the organisms degrade or modify the EC reducing their toxicity to various degrees (Gehlot and Hariprasad 2022; Karmakar et al. 2019; Donlan and Costerton 2002).

HMs are one of the major ECs that negatively impact soil health and plant growth. Though HMs are natural in origin and distributed across the globe, imbalance in their concentration is mediated by anthropogenic activities such as mining and smelting operations, industrial emission/production, and household and agronomic use of metals and metallic compounds (Khan et al. 2008a,b; Zhang et al. 2010; Tchounwou et al. 2012). In a living organism, essential HMs act as cofactors and catalysts for enzyme functioning and oxidation–reduction reactions, respectively. However, their concentrations above the threshold limit may harm the vital functioning of living organisms.

We tried to summarize the complex interaction and effect of the consequences of different HMs on the crops and their consequence in Figure 6.3. In plants, HMs have been reported to affect cell walls, cell membranes, cell organelles (mitochondria, chloroplast, lysosome, endoplasmic reticulum),

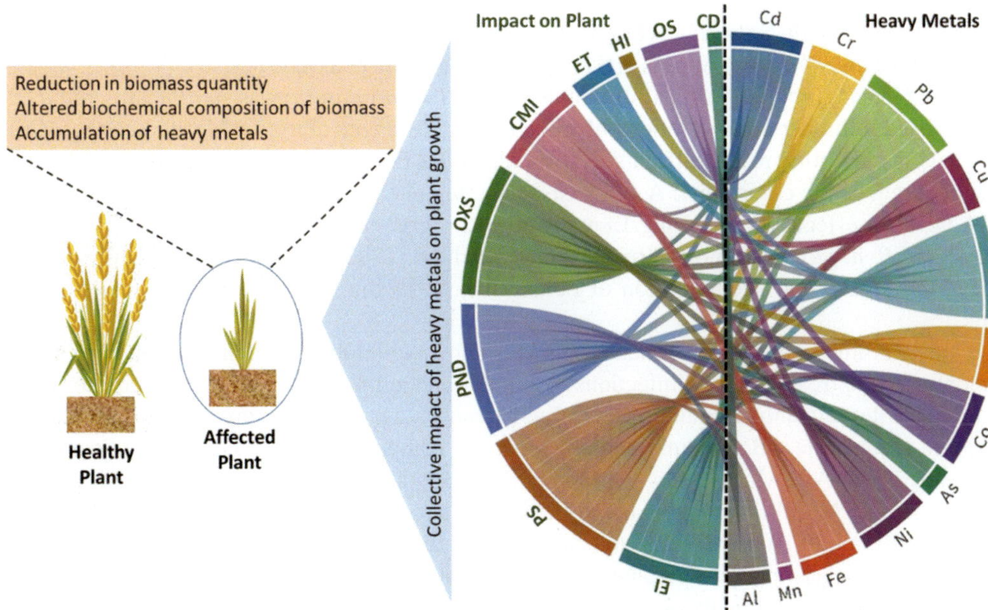


**FIGURE 6.3** Chord graph representing complex interactions of heavy metals with plant physiology and biochemical functioning leads to overall plant growth and health reduction. (Cd, Cadmium; Cr, chromium; Pb, lead; Cu, copper; Hg, mercury; Zn, zinc; Co, cobalt; As, arsenic; Ni, nickel; Fe, iron; Mn, manganese; Al, aluminum; CD, cell division; OS, osmotic stress; HI, hormonal imbalance; ET, electron transport; CMI, cell membrane integrity; OXS, oxidative stress; PND, plant nutrient deficiency; PS, photosynthesis; EI, enzyme inhibition; *Note*: The most common observations reported in the literature are incorporated in this figure.)

structural proteins, and enzymes. The formation of reactive oxygen species by HMs also produces oxidative stress, damaging macromolecules. The overall output of HMs and plant interactions are reduced quality and quantity of biomass and bioaccumulation of HMs. When such deteriorated feedstock is used for compositing, it may affect the microbial composition and functioning, affecting compost quality.

In acidic tropical soils, aluminum (Al) is known to negatively affect plant growth (Panda et al. 2009). The suppression of root development was observed as a common sign of Al toxicity. Cell walls of several plant species like *Zea mays* accumulate more lignin when too much Al is present in the growing medium (Budíková 1999). Sasaki et al. (1996) showed that inhibition of plant growth is significantly connected with the amount of lignin deposited in the cell wall, regardless of the Al resistance in wheat seedlings. Lignin accumulation is even higher in plants vulnerable to Al, significantly decreasing plant growth. Similar observations were reported by Tahara et al. (2005) in plants belonging to the family Myrtaceae when exposed to Al. Mao et al. (2004) have shown that exposure of rice seedlings to Al upregulates the expression of the genes responsible for lignin biosynthesis pathways such as 4-coumarate CoA ligase (4CL), phenylalanine ammonia-lyase (PAL), cinnamyl alcohol dehydrogenase (CAD), and *p*-coumarate 3-hydroxylase (C3H), indicating the influence of Al on the cell wall composition of rice plants. In a similar study, wheat plants exposed to Al recorded higher expression of PAL, which increases lignin formation (Snowden et al. 1993). In contrast, *Camellia sinensis*, tolerant to Al, recorded a decrease in cell wall lignin content when exposed to higher concentrations of Al, which were supported by decreased PAL and peroxidase (POD) activity (Ghanati et al. 2005).

Cadmium (Cd), ranging from 0.2 to 1 mM, inhibited soybean roots' development and increased the amount of lignin in the cell wall. This was accompanied by higher laccase activity, elevated POD (cationic and anionic) activity, and elevated expression of POD-related genes. As observed by the authors, biosynthesis of lignin was initially mediated by laccase, and in the later phase, the association of POD is critical (Bhuiyan et al. 2007). Also, Cd exposure increases the synthesis of lignin in *Phragmites australis* roots (Ederli et al. 2004) and is additionally noticed in callus tissue cultured from *Camellia sinensis* roots and branches (Zagoskina et al. 2007).

The dry biomass production capability of marigold (*Tagetes patula*) was significantly reduced when exposed to higher concentrations of iron (Fe) (3 mM), zinc (Zn) (5 mM), and copper (Cu) (3 mM) along with other micronutrients (Choi et al. 1996). Exposure of marigolds to Cu was also reported to reduce the chlorophyll A synthesis, affecting photosynthesis and biomass accumulation. A negative effect of Zn (6.5 mg/L) and Cu (0.32 mg/L) on the growth of *Brassica juncea*, *B. rapa,* and *B. napus* was reported by Ebbs and Kochian (1997). Further, Zn and Cu exposure impedes the uptake and accumulation of Fe and Mn, reducing the bioavailability of essential metals.

AB are naturally produced or synthetic chemicals known to inhibit or kill microorganisms. Most commonly used to treat microbial infections in humans and animals. The antibiotic can enter agricultural fields through animal manure and water. Moreover, agricultural crops are also reported to uptake and accumulate AB, becoming an AB source to compost. Agricultural fields are regularly contaminated with AB, which enters the food chain when plants absorb them (Alengebawy et al. 2021). Khan et al. (2021) investigated the effects of oxytetracycline (OTC), tetracycline (TC), and norfloxacin (NF) on *Brassica rapa* growth, cell ultrastructure, and metabolite pattern. All antibiotic treatments had a negative impact on plant growth, chlorophyll fluorescence, and antioxidant activities. TEM's ultrastructural analysis of the mesophyll in leaves revealed that antibiotic stress was responsible for alterations in thylakoid orientation, plastoglobule number, and starch grains. The FTIR spectral analysis used to identify functional groups revealed that the carboxyl group ($COO^-$), carbonate ions ($HCO_3^-$), and ammonium ions ($NH_4^+$) are engaged in antimicrobial drug absorption. The production of tryptophan, tyrosine, and glutamine, as well as the metabolism of alanine, aspartate, glutamate, and phenylalanine, was altered by all antibiotic treatments, according to the metabolome analysis of *B. rapa* leaves.

One of the often-used AB that raises environmental concerns is amoxicillin. Amoxicillin has a toxic effect on the photosynthesis efficiency of *Synechocystis* sp. When exposed to amoxicillin, electron transport is slowed on both the acceptor and donor sides, which results in a buildup of $P680^{+}$ (Pan et al. 2008). Similarly, by evaluating the growth rate, chloroplast pigment content, and sulfamethoxazole exposure, researchers examined the effects on the photosynthesis of freshwater algae, *Selenastrum capricornutum* Printz, by three AB (ciprofloxacin, erythromycin, and sulfamethoxazole). The AB considerably reduced the growth rate, chlorophyll content, and photosynthetic rate (Liu et al. 2011). Migliore et al. (2010) reported a hormetic response in growth parameters of *Lythrum salicaria* against sulfadimethoxine concentrations ranging between 0 and 50 mg/L. At 5 mg/L, the dry biomass production was significantly reduced. Similar observations were reported in *Brassica napus*, *Casella bursa-pastoris*, *Triticum aestivum*, and *Apera spica-venti* when exposed to varying concentrations of penicillin G, sulfadiazine, and TC (Minden et al. 2017). Moreover, AB may indirectly affect plant growth and health by altering the structure and function of soil and endophytic microflora (Liu et al. 2009; Karmakar et al. 2019).

In several research studies, the capabilities of different plants to uptake and accumulate the different AB and HMs were studied (Tables 6.2 and 6.3). These AB and HMs may be metabolized/chemically modified or stored in vacuoles and enter into compost when such biomass is subjected to composting. The AB and HMs may also enter the composting environment from the feedstock, animal excreta/waste, water, sludge, etc.

Pesticides can be synthetic or natural in origin, and their purpose is to reduce plant damage caused by pests such as bacteria, insects, nematodes, and rodents. Microbial infections and insect pests damage plants at various developmental phases, reducing biomass quality and quantity. Plants exhibit many physical, physiological, and biochemical modifications in response to pathogen infection, which are energy-dependent processes that ultimately compromise plant growth. Plant diseases always target the plant cell wall to gain access to cellulose as a C source and intracellular components. Previously, it was known that plant infections induce resistance in the host plant, as demonstrated by increased lignin and callose accumulation in the cell wall (Eisa et al. 2013; Kaur et al. 2022). Enzymes such as cinnamyl alcohol dehydrogenase, phenylalanine ammonia-lyase, and peroxidase were discovered to be increased after pathogen infection, suggesting this activity of lignin production (Scott-Craig et al. 1995). The severity of plant diseases was reduced depending on the host resistance and degree of lignification (Purwantisari et al. 2020). In addition, plants' cellulose production pathway is also impaired to varying degrees when plants under stress (Hernandez-Blanco et al. 2007). Furthermore, the process of plant defense against pathogens results in the accumulation of several antimicrobial compounds known as "phytoalexins" in host plants, including scopoletin, camalexin, sakuranetin, momilactone B, and glucosinolates (Gonzale-Lamothe et al. 2009). When such biomass is used to make compost, it is assumed that the increased lignin content and buildup of antimicrobial chemicals will alter the structure and function of compost microorganisms, influencing compost maturation time and final quality. However, few research studies have been conducted to better understand this phenomenon in composting.

Pesticides have played a crucial role in agriculture by minimizing losses due to microbiological diseases, insect pests, and weeds, which can significantly limit the amount of harvestable biomass. The application of pesticides can mitigate the previously mentioned detrimental effects of pathogens on biomass quality in relation to compost production. On the other side, host plants may absorb, biotransform, and accumulate pesticides. Particularly when fungicides and bactericides are accumulated, and the resulting biomass is used for compost production, the native compost microorganisms may be influenced by these chemicals, hence affecting the quality of the compost. On the other hand, several studies indicate that the composting process is more effective at absorbing and degrading pesticides due to the greater carbon sources available for microorganisms to metabolize and eliminate pesticides (Wu and Ma 2002; Tsui and Roy 2007; Fogarty and Tuovinen 1991).

Polycyclic aromatic hydrocarbons (PAHs) are the most dangerous and long-lasting of the various environmental contaminants (Burritt 2008). PAHs are often found in water, air, soil, food, and living

**TABLE 6.2**
**Heavy Metal Bioaccumulation Reported in Different Plant Species**

| Heavy Metal | Plant | Maximum Concentration (PPM) | References |
|---|---|---|---|
| Copper | *Athyrium esculentum* | 18.5 | Sulaiman and Hamzah (2018) |
| | *Chromolaena odorata* | 23.35 | Sulaiman and Hamzah (2018) |
| | *Lantana camara* | 26.85 | Sulaiman and Hamzah (2018) |
| | *Artemisia vulgaris* | 9.65 | Khan et al. (2008a) |
| | *Withania somnifera* | 8.67 | Khan et al. (2008a) |
| | *Stevia rebaudiana* | 7.71 | Khan et al. (2008a) |
| | *Mucuna pruriens* | 8.43 | Khan et al. (2008a) |
| | *Galium aparine* | 6.14 | Khan et al. (2008a) |
| | *Asparagus adscendens* | 5.08 | Khan et al. (2008a) |
| | *Medicago sativa* | 103 | Peralta-Videa et al. (2002) |
| | *Oryza sativa* | 38.12 | Hasan et al. (2022) |
| | *Triticum aestivum* | 2.5 | Chaplygin et al. (2021) |
| | *Saccharum officinarum* | 22.38 | Pandey et al. (2016) |
| | *Pinus brutia* | 10.1 | Başlar et al. (2003) |
| | *Hordeum vulgare* | 5.6 | Wu et al. (2002) |
| Iron | *Athyrium esculentum* | 322.2 | Sulaiman and Hamzah (2018) |
| | *Chromolaena odorata* | 749 | Sulaiman and Hamzah (2018) |
| | *Lantana camara* | 841.6 | Sulaiman and Hamzah (2018) |
| | *Artemisia vulgaris* | 81.39 | Khan et al. (2008a) |
| | *Withania somnifera* | 206.69 | Khan et al. (2008a) |
| | *Mucuna pruriens* | 33.21 | Khan et al. (2008a) |
| | *Galium aparine* | 180.91 | Khan et al. (2008a) |
| | *Asparagus adscendens* | 85.27 | Khan et al. (2008a) |
| | *Stevia rebaudiana* | 201.38 | Khan et al. (2008a) |
| | *Oryza sativa* | 14.89 | Hasan et al. (2022) |
| | *Saccharum officinarum* | 291.37 | Pandey et al. (2016) |
| | *Pinus brutia* | 595.5 | Başlar et al. (2003) |
| | *Hordeum vulgare* | 93.56 | Wu et al. (2002) |
| | *Withania somnifera* | 19.13 | Kulhari et al. (2013) |
| | *Terminalia chebula* | 16.19 | Kulhari et al. (2013) |
| | *Acacia nilotica* | 25.3 | Kulhari et al. (2013) |
| Cadmium | *Athyrium esculentum* | 0.08 | Sulaiman and Hamzah (2018) |
| | *Chromolaena odorata* | 0.06 | Sulaiman and Hamzah (2018) |
| | *Lantana camara* | 0.11 | Sulaiman and Hamzah (2018) |
| | *Medicago sativa* | 202 | Peralta-Videa et al. (2002) |
| | *Oryza sativa* | 1.61 | Hasan et al. (2022) |
| | *Triticum aestivum* | 4.83 | Liu et al. (2009) |
| | *Saccharum officinarum* | 0.15 | Pandey et al. (2016) |
| | *Oryza sativa* | 0.04–1.75 | Sriprachote et al. (2012) |
| | *Glycine max* | 0.07–0.80 | Sriprachote et al. (2012) |
| | *Spinacia oleracea* | 2.88 | Manwani et al. (2022) |
| | *Raphanus sativus* | 1.88 | Manwani et al. (2022) |
| | *Solanum lycopersicum* | 0.266 | Pandey et al. (2016) |
| | *Pinus brutia* | 0.54 | Başlar et al. (2003) |
| | *Hordeum vulgare* | 0.059 | Wu et al. (2002) |
| | *Terminalia chebula* | 0.03 | Kulhari et al. (2013) |
| | *Stevia rebaudiana* | 0.604 | Hajar et al. (2014) |
| | *Athyrium esculentum* | 3.79 | Sulaiman and Hamzah (2018) |

(*Continued*)

**TABLE 6.2 (*Continued*)**
**Heavy Metal Bioaccumulation Reported in Different Plant Species**

| Heavy Metal | Plant | Maximum Concentration (PPM) | References |
|---|---|---|---|
| Lead | *Chromolaena odorata* | 2.64 | Sulaiman and Hamzah (2018) |
| | *Lantana Camara* | 3.05 | Sulaiman and Hamzah (2018) |
| | *Oryza sativa* | 7.98 | Liu et al. (2007) |
| | *Triticum aestivum* | 30.83 | Liu et al. (2009) |
| | *Saccharum* spp. | 7.9 | Segura et al. (2006) |
| | *Spinacia oleracea* | 1.05 | Manwani et al. (2022) |
| | *Raphanus sativus* | 0.072 | Manwani et al. (2022) |
| | *Solanum lycopersicum* | 0.021 | Pandey et al. (2016) |
| | *Zea mays* | 5.12 | Lu et al. (2015) |
| | *Withania somnifera* | 2.64 | Kulhari et al. (2013) |
| | *Terminalia chebula* | 0.62 | Kulhari et al. (2013) |
| | *Acacia nilotica* | 0.25 | Kulhari et al. (2013) |
| | *Stevia rebaudiana* | 1.735 | Hajar et al. (2014) |
| | *Artemisia vulgaris* | 38.14 | Khan et al. (2008a) |
| Zinc | *Withania somnifera* | 43.01 | Khan et al. (2008a) |
| | *Stevia rebaudiana* | 47.18 | Khan et al. (2008a) |
| | *Asparagus adscendens* | 32.87 | Khan et al. (2008a) |
| | *Mucuna pruriens* | 32.48 | Khan et al. (2008a) |
| | *Galium aparine* | 45 | Khan et al. (2008a) |
| | *Medicago sativa* | 160 | Peralta-Videa et al. (2002) |
| | *Oryza sativa* | 121.76 | Hasan et al. (2022) |
| | *Triticum aestivum* | 4.67 | Chaplygin et al. (2021) |
| | *Saccharum* spp. | 177.4 | Segura et al. (2006) |
| | *Hordeum vulgare* | 61.92 | Wu et al. (2002) |
| | *Zea mays* | 6.92 | Lu et al. (2015) |
| | *Acacia nilotica* | 2.42 | Kulhari et al. (2013) |
| | *Terminalia chebula* | 6.32 | Kulhari et al. (2013) |
| | *Artemisia vulgaris* | 4.35 | Khan et al. (2008a) |
| Nickel | *Withania somnifera* | 2.64 | Khan et al. (2008a) |
| | *Stevia rebaudiana* | 2.36 | Khan et al. (2008a) |
| | *Mucuna pruriens* | 3.18 | Khan et al. (2008a) |
| | *Asparagus adscendens* | 0.94 | Khan et al. (2008a) |
| | *Galium aparine* | 4.16 | Khan et al. (2008a) |
| | *Medicago sativa* | 437 | Peralta-Videa et al. (2002) |
| | *Oryza sativa* | 0.18 | Hasan et al. (2022) |
| | *Triticum aestivum* | 1.78 | Chaplygin et al. 2021 |
| | *Saccharum officinarum* | 4.43 | Pandey et al. (2016) |
| | *Pinus brutia* | 10.4 | Başlar et al. (2003) |
| | *Zea mays* | 6.56 | Lu et al. (2015) |
| | *Acacia nilotica* | 0.45 | Kulhari et al. (2013) |
| | *Terminalia chebula* | 0.64 | Kulhari et al. (2013) |
| | *Artemisia vulgaris* | 4.65 | Khan et al. (2008a) |

(*Continued*)

**TABLE 6.2 (*Continued*)**
**Heavy Metal Bioaccumulation Reported in Different Plant Species**

| Heavy Metal | Plant | Maximum Concentration (PPM) | References |
|---|---|---|---|
| Chromium | *Withania somnifera* | 1.13 | Khan et al. (2008a) |
| | *Stevia rebaudiana* | 1.45 | Khan et al. (2008a) |
| | *Mucuna pruriens* | 0.27 | Khan et al. (2008a) |
| | *Galium aparine* | 5.89 | Khan et al. (2008a) |
| | *Asparagus adscendens* | 1.13 | Khan et al. (2008a) |
| | *Saccharum officinarum* | 60.22 | Pandey et al. (2016) |
| | *Oryza sativa* | 23.67 | Hasan et al. (2022) |
| | *Triticum aestivum* | 8.07 | Liu et al. (2009) |
| | *Medicago sativa* | 40.0 | Peralta et al. (2002) |
| | *Pinus brutia* | 12.12 | Başlar et al. (2003) |
| | *Hordeum vulgare* | 1.453 | Wu et al. (2002) |
| | *Zea mays* | 105.21 | Lu et al. (2015) |
| | *Terminalia chebula* | 8.73 | Kulhari et al. (2013) |
| | *Acacia nilotica* | 2.92 | Kulhari et al. (2013) |
| | *Artemisia vulgaris* | 52.94 | Khan et al. (2008a) |
| Manganese | *Withania somnifera* | 23.77 | Khan et al. (2008a) |
| | *Stevia rebaudiana* | 32.87 | Khan et al. (2008a) |
| | *Mucuna pruriens* | 14.7 | Khan et al. (2008a) |
| | *Galium aparine* | 43.16 | Khan et al. (2008a) |
| | *Asparagus adscendens* | 9.7 | Khan et al. (2008a) |
| | *Triticum aestivum* | 1.44 | Chaplygin et al. (2021) |
| | *Saccharum* spp. | 561.6 | Segura et al. (2006) |
| | *Hordeum vulgare* | 26.06 | Wu and Ma (2002) |
| | *Terminalia chebula* | 5.67 | Kulhari et al. (2013) |
| | *Acacia nilotica* | 5.22 | Kulhari et al. (2013) |
| | *Triticum aestivum* | 5.95 | Liu et al. (2009) |
| Arsenic | *Oryza sativa* | 30.42 | Liu et al. (2009) |
| | *Stevia rebaudiana* | 3.909 | Hajar et al. (2014) |
| | *Medicago sativa* | 5.66 | Ibrahim et al. (2016) |
| | *Saccharum* spp. | 0.04 | Segura et al. (2006) |
| Mercury | *Triticum aestivum* | 0.19 | Liu et al. (2009) |
| | *Oryza sativa* | 0.19 | Liu et al. (2007) |
| | *Hordeum vulgare* | 3 | Sierra et al. (2011) |
| | *Zea mays* | 0.29 | Alvarez et al. (2006) |
| | *Withania somnifera* | 0.14 | Rao et al. (2011) |
| | *Terminalia chebula* | 0.19 | Rao et al. (2011) |
| | *Withania somnifera* | 0.14 | Kulhari et al. (2013) |
| Cobalt | *Terminalia chebula* | 0.2 | Kulhari et al. (2013) |
| | *Acacia nilotica* | 0.19 | Kulhari et al. (2013) |
| | *Brassica napus* | 21.87 | Jahani et al. (2020) |
| | *Triticum aestivum* | 300 | Sarma et al. (2014) |
| | *Hordeum vulgare* | 0.05 | Rezvani et al. (2015) |
| | *Stevia rebaudiana* | 0.0842 | Hajar et al. (2014) |

*Note:* The values only represent the highest capability of a plant, irrespective of plant age and tissue type.

**TABLE 6.3**
**Antibiotic Bioaccumulation Reported in Different Plant Species**

| Antibiotics | Plants | Concentration | Reference |
|---|---|---|---|
| Tetracycline | *Oryza sativa* | 269.79 ng/g | Yan et al. (2021) |
| | *Zea mays* | 2,477 ng/g | Geng et al. (2022) |
| | *Raphanus sativus* | 1.4 µg/kg | Hu et al. (2010) |
| | *Coriandrum sativum* | 1.9–5.6 µg/kg | Hu et al. (2010) |
| | *Lactuca sativa* | 4.4–28.3 ng/g | Azanu et al. (2016) |
| | *Daucus carota* | 12.0–36.8 ng/g | Azanu et al. (2016) |
| | *Arachis hypogaea* | 20.06 ng/g | Zhao et al. (2019) |
| Oxytetracycline | *Pisum sativum* | 31–146 µg/g | Ziółkowska et al. (2015) |
| | *Perilla frutescens* | 23.6 µg/kg | Kim et al. (2022) |
| | *Raphanus sativus* | 17–57 µg/kg | Hu et al. (2010) |
| | *Coriandrum sativum* | 78–330 µg/kg | Hu et al. (2010) |
| | *Phragmites australis* | 6,901 ng/g | Liu et al. (2013) |
| | *Arachis hypogaea* | 22.73 ng/g | Zhao et al. (2019) |
| Sulfamethazine | *Zea mays* | 600 ng/g | Geng et al. (2022) |
| | *Zea mays* | 0.1–1.2 mg/kg | Dolliver et al. (2007) |
| | *Lactuca sativa* | 0.1–1.2 mg/kg | Dolliver et al. (2007) |
| | *Lactuca sativa* | 1,700 ng/g | Geng et al. (2022) |
| | *Solanum tuberosum* | 0.1–1.2 mg/kg | Dolliver et al. (2007) |
| | *Raphanus sativus* | 0.9–2.7 µg/kg | Hu et al. (2010) |
| | *Phragmites australis* | 2,047 ng/g | Liu et al. (2013) |
| | *Arachis hypogaea* | 5.26 ng/g | Zhao et al. (2019) |
| Sulfamethoxazole | *Oryza sativa* | 26.45 ng/g | Yan et al. (2021) |
| | *Perilla frutescens* | 4.85 µg/kg | Kim et al. (2022) |
| | *Lactuca sativa* | 471 ng/g | Geng et al. (2022) |
| | *Arachis hypogaea* | 6 ng/g | Zhao et al. (2019) |
| Tylosin | *Arachis hypogaea* | 0.68 ng/g | Zhao et al. (2019) |
| Sulfadiazine | *Salix fragilis* | 5,464 mg/kg | Michelini et al. (2012) |
| | *Zea mays* | 5,331 mg/kg | Michelini et al. (2012) |
| | *Zea mays* | 500 ng/g | Geng et al. (2022) |
| | *Lactuca sativa* | 2,900 ng/g | Geng et al. (2022) |
| Ofloxacin | *Raphanus sativus* | 0.5 µg/kg | Hu et al. (2010) |
| | *Oryza sativa* | 41.89 ng/g | Yan et al. (2021) |
| | *Coriandrum sativum* | 0.7–3.5 µg/kg | Hu et al. (2010) |
| | *Arachis hypogaea* | 6.98 ng/g | Zhao et al. (2019) |
| Ciprofloxacin | *Phragmites australis* | 13,834 ng/g | Liu et al. (2013) |
| | *Arachis hypogaea* | 3.35 ng /g | Zhao et al. (2019) |
| Enrofloxacin | *Hordeum vulgare* | 15.7 | Abou Elezz et al. (2019) |
| | *Arachis hypogaea* | 4.26 ng /g | Zhao et al. (2019) |
| Epitetracycline | *Zea mays* | 226 | Geng et al. (2022) |
| Amoxicillin | *Lactuca sativa* | 13.7–33.6 ng/g | Azanu et al. (2016) |
| | *Daucus carota* | 14.3–45.2 ng/g | Azanu et al. (2016) |
| Chlortetracycline | *Raphanus sativus* | 6–18 µg/kg | Hu et al. (2010) |
| | *Coriandrum sativum* | 54–532 µg/kg | Hu et al. (2010) |
| | *Allium cepa* | 14.4 ng/g | Kumar et al. (2005) |
| | *Brassica oleracea* | 11.4 ng/g | Kumar et al. (2005) |
| | *Arachis hypogaea* | 21.84 ng/g | Zhao et al. (2019) |
| Chloramphenicol | *Raphanus sativus* | 8–30 µg/kg | Hu et al. (2010) |

*Note:* The values only represent the highest capability of a plant, irrespective of plant age and tissue type.

beings, which makes this issue extremely important. PAHs are also shown to interfere with a number of physiological processes in plants, hence reducing crop yields (Kummerová and Vanova 2007). PAHs impair the membrane permeability, enzymatic dysfunction, electron transport, and photosynthesis of chloroplast, consequently diminishing the photosynthetic efficiency of plants (Duxbury et al. 1997; Saxe 1996). Other reported negative effects of PHA on plants include decreased seed germination, decreased shoot and root length, damage to cellular membranes, altered cell expansion, decreased cell size and mitotic activity, and altered auxin biosynthesis and action (Maila and Cloete 2002; Oguntimehin et al. 2010; Achuba 2006; Bona et al. 2011; Wei et al. 2014; Jajoo et al. 2014). In this area, the least studied is the effect of PAHs on the composting process and compost quality. However, the majority of research focuses on the degradation of PHA during composting with various substrates. According to Aemig et al. (2019), benzo[*a*]pyrene, benzo[*b*]fluoranthene, and nonylphenols were eliminated by 31%–52% when sewage sludge and garden waste were utilized in a 1:4 ratio for composting. In a similar effort, Mattei et al. (2016) observed a 26%–57% decrease in PHAs when sediment and yard waste were mixed in proportions ranging from 3:1 to 1:1, respectively. Košnář et al. (2019) reported the elimination of 16 PAHs from fly ash mixed with organic waste. According to their observation, composting can remove up to 84.5% of PHA, which is greater than vermicomposting (61.6%). Löser et al. (1999) discovered that pine wood combined with liquid hog manure and mineral medium as a nitrogen source effectively eliminated spiked PAHs (1,000 ppm) to the amount of 26 ppm of phenanthrene and 83 ppm of pyrene after 61 days of incubation.

Similarly, a number of ECs, such as microplastics, 1,4-dioxane, food additives, and natural and synthetic hormones, must be studied in depth to determine their effect on plant growth and compost quality.

## 6.5 IMPACT OF CLIMATE CHANGE AND EMERGING CONTAMINANTS ON THE COMPOSTING

The main factors controlling the composting process and quality are feedstock composition, environmental conditions, organism population, and function. The impact of CC factors and EC on microbial abundance and structural and functional diversity are of significant concern in composting. It is vital to understand how the microbiome will change when these factors are altered. In this regard, minimal research effort has been made in composting. Hence, we also discussed the research papers revealing the effect of CC and EC on soil microbiomes and tried to correlate them with compost. Different microbes, enzymes, and reactions involved in the composting process are presented in Figure 6.4.

Previous research studies have yielded mixed opinions regarding the impact of elevated $CO_2$ on the soil microbial community. Microbial abundance and activity increased in several plant species' rhizosphere soils exposed to elevated $CO_2$ (Rogers et al. 1992; Runion et al. 1994; Williams et al. 2000). However, an increase in microbial abundance and activity in response to elevated $CO_2$ levels has not been observed in bulk soil (Rogers et al. 1992; Schortemeyer et al. 1996; Schortemeyer et al. 2000). Montealegre et al. (2022) discovered that increased $CO_2$ levels increased the bacterial population of bulk soil in white clover fields. However, in the perennial ryegrass field, the $CO_2$ level did not affect the total and metabolically active bacteria. As per our knowledge and literature survey, we did not find any such correlations between elevated $CO_2$ levels and the microbial community of composting.

The different feedstock used in composting provides different organic carbon and other nutrients and determines the final compost quality. The physiochemical and biological properties of compost are controlled by the compost feedstock.

The most common combinations of feedstock used are agro-residues in combination with animal manure. Waksman et al. (1939) reported that lignin degradation during horse manure and straw

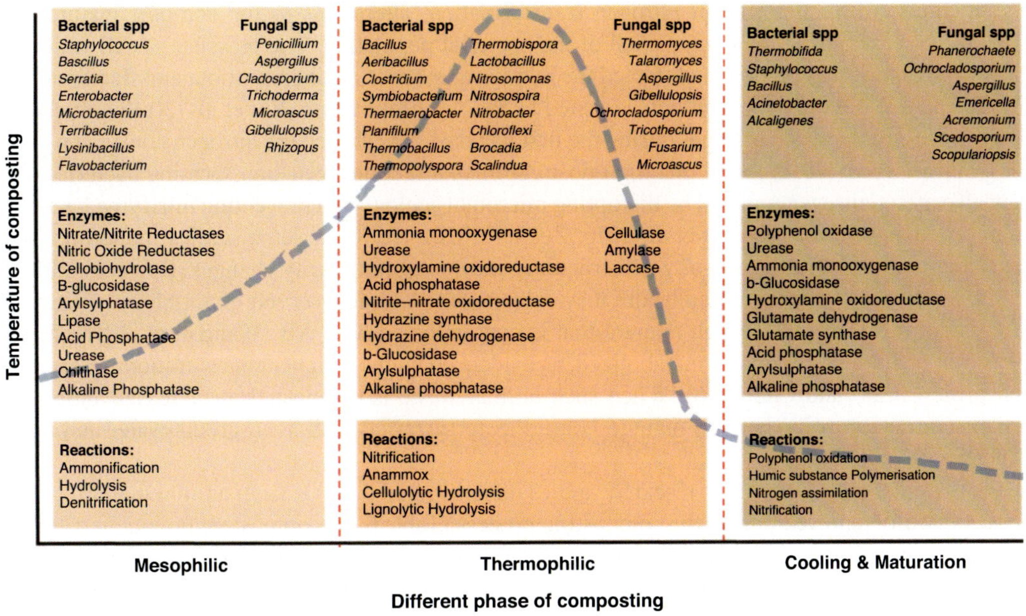


**FIGURE 6.4** Common microbes, enzymes, and reactions involved in different phases of composting.

composting is temperature-dependent. They observed that the lignin degradation significantly increased as the temperature increased from 28°C to 50°C. Further temperature increases (65°C and 75°C) reduced the lignin-degrading efficacy. Supporting this, the experiments of Horwath and Elliott (1996) with ryegrass composting also recorded a significant increase in lignin degradation, from 7% to 27%, when the composting process was carried out at 25°C and 50°C, respectively. In a similar study, Tomati et al. (1995) reported a maximum of 70% lignin degradation when composting was carried out at 50°C, which was negligible during the rest of the composting process. These earlier reports suggested that a more extended thermophilic phase might be needed to break down the lignin-rich feedstock. In this regard, global warming has enhanced the cell wall lignin in field-grown plants and may help degrade lignin during composting. However, higher temperatures may have a negative effect on the mesophilic and maturation phase. By considering the complex biochemical process of composting determined by various environmental factors, more studies are required to understand the impact of global warming on the compost process and quality. Gliniak et al. (2018) revealed that ozone exposure at 10 and 20 mg $O_3/dm^3$ reduced the organic matter (30–60%), moisture (20%), and composting maturation time (50%) in the grass, sawdust, and food waste composting.

AB, being antimicrobial, above specific concentrations, may inhibit bacterial/archaeal and fungal growth, thereby affecting the composting process. With prolonged exposure to a sub-lethal dose of AB, compost microbes may develop resistance and act as carriers of antibiotic-resistant genes (ARGs), posing severe threats to humans and animals. On the other hand, it is unclear how the composting process is affected by different AB and how it works. Wang et al. (2022) investigated how the composting process was affected by combining different AB on broiler manures by studying the fungal and bacterial community architectures' succession in different stages. Rice husk was mixed with broiler dung with varying doses of doxycycline and gatifloxacin. Authors observed a delayed rise in temperature and pH, as well as a gradual increase in total nitrogen when a large number of AB were present at the beginning of the composting process. Bacteria, which may proliferate on readily available resources, are primarily responsible for the rise in pile temperature (Huhe et al. 2017). Inhibition of the growth and functioning of such bacteria by AB may be responsible for the altered composting process. In a study by Liu et al. (2019b), the thermophilic period was postponed due to amoxicillin's impacts on nitrogen transformation, alterations to the compost maturation kinetics, and changes to the composition of the bacterial population. The presence of AB residues during the process of composting affected the relative number of Actinobacteriota and Bacteroidota till the end, even though these effects lessened as the AB broke down. Liu et al. (2019a) observed that the occurrence of AB residues during the first five days also slowed the nitrogen conversion process. Selvam et al. (2012) found that high-dose AB may affect nitrogen-transforming bacteria. AB may also affect nitrogen removal by disrupting microbial ecology and preventing nitrifying bacteria from producing nitrate (Katsou et al. 2016; Zekker 2020). However, some studies found that AB caused minor microbial community perturbations and compost maturity changes (Ray et al. 2017).

Excessive use of AB in animal health may contribute to a large reservoir of antibiotic resistance in agricultural farmland through manure that is contaminated with AB (Wang et al. 2017). Most AB given to animals are poorly metabolized, as evidenced by their higher concentrations in excreta and urine (Tasho and Cho 2016). When such animal waste is composted, it increases antibiotic resistance traits in soil when the manure is used as fertilizer, as well as in groundwater and food (Ghosh and LaPara 2007).

On the other hand, numerous research studies demonstrated that AB from animal manure could be successfully removed during the composting process (Table 6.4). Composting organic fertilizers reduced doxycycline and gatifloxacin residues, but not completely (Wang et al. 2022). Composting removed 70%–99% of AB from TCs, sulfonamides, and macrolides (Wang et al. 2017; Chen et al. 2021). Despite reports of a high fluoroquinolone family removal rate, NF, ciprofloxacin, and ofloxacin were found in compost (Liu et al. 2019a; Khadra et al. 2019).

Also, possibilities of managing AB resistance during composting by using various bulking agents (to increase C/N ratio and lignin) such as sawdust, green waste, and mushroom (Zhang et al. 2019; Chen et al. 2019; Li et al. 2019). According to Kong et al. (2019), microorganisms have the ability to degrade AB. Enzymes from resistant microorganisms can detoxify or use AB as a carbon or nitrogen source (Iranzo et al. 2018). The application of compost containing inefficiently removed toltrazuril along with its metabolites in the soil led to reduced urease activity, altered alkaline phosphatase activity of soil microbes, and increased antioxidant enzyme catalase activity. Overall, undesirable effects in lettuce and radish seed germination and root elongation length were inhibited by 8.33% and 26.74%, respectively (Huo et al. 2022).

Contrarily, the concentrations of HMs were found to be increasing during the composting process (Table 6.4). The presence of HMs, which can present in the form of molecules or ions in organic waste, has a significant impact on the prevalence of ARGs during composting (Yin et al. 2016; Zhang et al. 2017). Metals like As, Cd, Cu, Cr, Pb, As, Mn, and Zn cannot be eliminated or decomposed by composting (Deng et al. 2020); they can concentrate during the process due to the loss in the overall compost mass. HMs potentially act as a long-term selection pressure for composting microbes (Yang et al. 2017). Several earlier research works evidenced the HM-driven co-selection of AB resistance in microbes (Seiler and Berendonk 2012). During the composting of chicken manure, the selection pressure brought on by Cu encouraged the growth expression of *copA*, *sul1*, *sul2*, *tetC*, and *cusA*. Further, Cu is also known to alter the functioning of metallo-enzymes during composting (Hobman and Crossman 2015; Pal et al. 2015; Bao et al. 2020). A similar observation was made between Cu and ARGs (*sul1*, *ermX*, *ermB*, *dfrA7*, and aac(6′)-ib-cr), and Zn and ARGs (*tetQ*, *tetG*, *ermB*, *dfrA7*, and *intI1*) in earlier studies (Zhang et al. 2018; Zhang et al. 2016).

## 6.6 NITROGEN DYNAMICS IN COMPOSTING

Nitrogen is a vital component of all forms of life. It is present in almost all structural and functional components of living organisms, including nucleic acids, amino acids (peptides, proteins, and enzymes), chlorophyll, vitamins, and secondary metabolites. Though the air contains 78% of $N_2$, its availability to living organisms is limited. On the other hand, the cocktail of reactive nitrogen released into the environment due to anthropogenic activities poses a severe threat to climate

**TABLE 6.4**
**Change in the Concentrations of Antibiotics and Heavy Metals in Composting Process**

| Contaminant | Feedstock | Concentration (Initial–Final) | References |
|---|---|---|---|
| Antibiotics | | | |
| Oxytetracycline | Manure (beef calves) | 115–6 µg/g | Arikan et al. (2007) |
| | Poultry manure | 13.3–7.2 µg/kg | Xie et al. (2016) |
| | Swine manure | 19.4–10.5 µg/kg | Xie et al. (2016) |
| Chlortetracycline | Bamboo charcoal + pig manure composting | 0.39–0.13 mg/kg | Wang et al. (2016) |
| Ofloxacin | Poultry manure | 135.3–870.8 µg/kg | Xie et al. (2016) |
| | Swine manure | 20.4–936.3 µg/kg | Xie et al. (2016) |
| Ciprofloxacin | Cattle manure | 107.1–24.2 µg/kg | Xie et al. (2016) |
| | Poultry manure | 34.9–729.3 µg/kg | Xie et al. (2016) |
| | Swine manure | 202.2–41.0 µg/kg | Xie et al. (2016) |
| | Bamboo charcoal + pig manure composting | 1.85–0.02 mg/kg | Wang et al. (2016) |
| Norfloxacin | Bamboo charcoal + pig manure composting | 0.11–0.07 mg/kg | Wang et al. (2016) |
| Salinomycin | Manure (poultry) | 22 mg/g–2 × $10^{-5}$ µg/kg | Ramaswamy et al. (2010) |
| Sulfadiazine | Broiler manure | 30.64–<BDL | Ho et al. (2013) |
| | Cattle manure | 791.4–18.7 µg/kg | Xie et al. (2016) |
| | Poultry manure | 30.0–3.6 µg/kg | Xie et al. (2016) |
| | Swine manure | 12.9–3.3 µg/kg | Xie et al. (2016) |
| Sulfamethazine | Pig manure | 10–0.2 mg/kg | Kim et al. (2012) |
| Sulfamethoxazole | Cattle manure | 15.7–3.6 µg/kg | Xie et al. (2016) |
| Sulfamethoxydiazine | Cattle manure | 6.7–6.8 µg/kg | Xie et al. (2016) |
| | Swine manure | 12.9–5.7 µg/kg | Xie et al. (2016) |
| | Poultry manure | 10.3–4.3 µg/kg | Xie et al. (2016) |
| | Swine manure | 4.6–3.7 µg/kg | Xie et al. (2016) |
| Sulfaclozine | Poultry manure | 29.0–80.4 µg/kg | Xie et al. (2016) |
| | Swine manure | 5.4–3.7 µg/kg | Xie et al. (2016) |
| Tetracycline | Cattle manure | 80.1–28.2 µg/kg | Xie et al. (2016) |
| | Poultry manure | 33.3–32.7 µg/kg | Xie et al. (2016) |
| | Swine manure | 95.0–25.6 µg/kg | Xie et al. (2016) |
| Chlortetracycline | Cattle manure | 48.3–6.8 µg/kg | Xie et al. (2016) |
| | Poultry manure | 185.9–57.6 µg/kg | Xie et al. (2016) |
| | Swine manure | 80.8–31.6 µg/kg | Xie et al. (2016) |
| | Pig manure | 2–0.8 mg/kg | Kim et al. (2012) |
| Doxycycline | Broiler manure | 13.64–0.027 mg/kg | Ho et al. (2013) |
| | Cattle manure | 71.8–22.8 µg/kg | Xie et al. (2016) |
| | Poultry manure | 33.1–30.1 µg/kg | Xie et al. (2016) |
| | Swine manure | 94.8–24.3 µg/kg | Xie et al. (2016) |
| Enrofloxacin | Broiler manure | 36.70–0.014 mg/kg | Ho et al. (2013) |
| | Poultry manure | 18.2–3633.0 µg/kg | Xie et al. (2016) |
| Erythromycin | Broiler manure | 15.58–<BDL mg/kg | Ho et al. (2013) |

*(Continued)*

**TABLE 6.4 (*Continued*)**
**Change in the Concentrations of Antibiotics and Heavy Metals in Composting Process**

| Contaminant | Feedstock | Concentration (Initial–Final) | References |
|---|---|---|---|
| Flumequine | Broiler manure | 42.03–<BDL mg/kg | Ho et al. (2013) |
| Norfloxacin | Broiler manure | 5.09–<BDL mg/kg | Ho et al. (2013) |
| | Cattle manure | 22.5–39.9 μg/kg | Xie et al. (2016) |
| | Poultry manure | 38.5–110.2 μg/kg | Xie et al. (2016) |
| | Swine manure | 8.0–491.9 μg/kg | Xie et al. (2016) |
| Tilmicosin | Broiler manure | 5.24–< BDL mg/kg | Ho et al. (2013) |
| Trimethoprim | Broiler manure | 21.98–0.022 mg/kg | Ho et al. (2013) |
| Tylosin | Broiler manure | 62.57–<BDL mg/kg | Ho et al. (2013) |
| | Pig manure | 20–1.0 mg/kg | Kim et al. (2012) |
| Progesterone | Broiler manure | 6.31–< BDL mg/kg | Ho et al. (2013) |
| Heavy Metals | | | |
| Copper | Cow manure | 94.67–112.23 mg/kg | Pinto et al. (2020) |
| | Dairy : beef : swine manure | 185.2–465.7 mg/kg | Ko et al. (2008) |
| | Poultry manure | 92.59–138.96 mg/kg | Tiquia and Tam (2000) |
| | Corn stalk residue : activated sewage sludge | 9,430–11,490 mg/kg | Kharrazi et al. (2014) |
| | Poultry manure | 14.4–9.2 mg/kg | Irshad et al. (2014) |
| Zinc | Cow manure | 313–466.0 mg/kg | Pinto et al. (2020) |
| | Dairy : beef : swine manure | 194.4–656.7 mg/kg | Ko et al. (2007) |
| | Poultry manure | 268.7–642.08 mg/kg | Tiquia and Tam (2000) |
| | Corn stalk residue : activated sewage sludge | 382–614 mg/kg | Kharrazi et al. (2014) |
| | Poultry manure | 22.3–14.8 mg/kg | Irshad et al. (2014) |
| Cadmium | Cow manure | 0.34–1.38 mg/kg | Smith (2021) |
| | Dairy : beef : swine manure | 0.52–1.09 mg/kg | Ko et al. (2007) |
| | Poultry manure | 1.1–0.9 mg/kg | Irshad et al. (2014) |
| Nickel | Cow manure | 3.13–14.37 mg/kg | Pinto et al. (2020) |
| | Dairy : beef : swine manure | 9.2–10.9 mg/kg | Ko et al. (2007) |
| | Corn stalk residue : activated sewage sludge | 43.29–52.29 mg/kg | Kharrazi et al. (2014) |
| Chromium | Cow manure | 6.33–35.33 mg/kg | Pinto et al. (2020) |
| | Dairy : beef : swine manure | 8–12.9 mg/kg | Ko et al. (2007) |
| | Corn stalk residue : activated sewage sludge | 48.29–45.29 mg/kg | Kharrazi et al. (2014) |
| Mercury | Cow manure | 0.06–0.38 mg/kg | Pinto et al. (2020) |
| Selenium | Cow manure | 1.41–1.55 mg/kg | Pinto et al. (2020) |
| Molybdenum | Cow manure | 4.27–3.23 mg/kg | Pinto et al. (2020) |
| Lead | Cow manure | 9.43–73.23 mg/kg | Pinto et al. (2020) |
| | Dairy : beef : swine manure | 11.1–38.2 mg/kg | Ko et al. (2007) |
| | Corn stalk residue : activated sewage sludge | 43.75–72.10 mg/kg | Kharrazi et al. (2014) |
| | Poultry manure | 4.5–3.8 mg/kg | Irshad et al. (2014) |
| Arsenic | Cow manure | 8.83–29.67 mg/kg | Pinto et al. (2020) |
| | Dairy : beef : swine manure | 28–21.8 mg/kg | Ko et al. (2007) |
| Manganese | Corn stalk residue : activated sewage sludge | 348–444 mg/kg | Kharrazi et al. (2014) |
| | Poultry manure | 23.8–17.2 mg/kg | Irshad et al. (2014) |
| Iron | Poultry manure | 30.7–24.8 mg/kg | Irshad et al. (2014) |

and ecosystems. Nitrous oxide ($N_2O$) has a global warming potential of 273 times that of $CO_2$ for a century. It also plays a significant role in the stratospheric ozone hole (Griffis et al. 2017). As of 2021, the atmospheric concentration of $N_2O$ reached 334 ppb, a matter of great concern. Moreover, agricultural activities have been identified as the primary source of $N_2O$ (Smith 2021). About 60% of all human-made $N_2O$ emissions come from farming fields (Reay et al. 2012). Therefore, reducing these emissions of $N_2O$ is a crucial goal to slow down the pace of CC. In this part of the chapter, we focused on identifying the effects of CC and emerging pollutants on nitrogen dynamics during composting.

Apart from synthetic chemical fertilizers, compost is the primary source of plant nutrients used across the globe. Nitrogen is an indispensable element in the composting microenvironment, and microbial communities primarily mediate its interconversion between inorganic to organic and vice versa throughout composting (Shi et al. 2018; Jiang et al. 2015). In a composting process, the three major stages involved in nitrogen transformation are ammonification, nitrification, and denitrification (Cáceres et al. 2018; Meng et al. 2016). The process of nitrogen transformation in compost is summarized in Figure 6.5.

Ammonification is the key and the first step in N mineralization of composting, during which microorganisms degrade organic molecules containing amine or amide groups ($R\text{-}NH_2$) to ammonium ions ($NH_4^+$) (Reddy et al. 1984). In general, ammonification occurs at the mesophilic phase of composting, overlapping with the early phase of the thermophilic phase. Ammonium ions produced are further assimilated by microorganisms and used for metabolic purposes. When these ammonium ions are produced more than their requirement, they will be available for nitrification and denitrification processes. However, under a situation where the nitrification rate is lower than ammonification, $NH_4^+$ ions accumulate in compost to various degrees. Under such conditions, depending on the conditions of microenvironmental, the compost equilibrium between $NH_4^+$ and $NH_3$ gets altered. As the pH of the compost rises, establishing alkaline conditions, the conversion of $NH_4^+$to $NH_3$ is expedited, thereby causing the loss of nitrogen in the form of $NH_3$ (Wong et al. 2017; Cáceres et al. 2018). An enormous and diverse variety of microorganisms participate in the ammonification of the nitrogen cycle. Important bacterial species involved in this process include *Bacillus*, *Streptomyces*, and *Proteus*. Ectotrophic mycorrhizal fungi also participate in nitrogen transformation by taking part in the ammonification of organic nitrogen (Lakhanpal 2000; Giri et al. 2005; Kaviya et al. 2019).

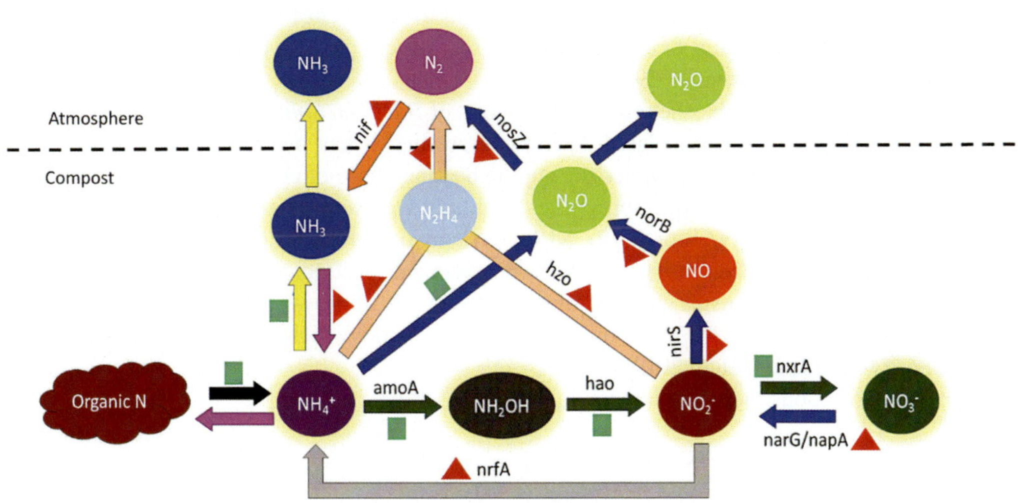


**FIGURE 6.5** Biotransformation of nitrogen under composting environment. (nif, nitrogenase; *nosZ*, nitrous oxide reductase; *norB*, nitric oxide reductase; *nirS*, nitrite reductase genes; *nxrA*, nitrite oxidoreductase; *nrfA*, nitrite reductase; *hzo*, hydrazine oxidoreductase; *narG/napA,* nitrate reductase; *amoA*, ammonia monooxygenase; *hao*, hydroxylamine oxidoreductase.)

Biochemical reaction in ammonification:

$$\text{Organic N} \rightarrow NH_4^+$$

Eventually, microorganisms responsible for nitrification convert the ammonia ($NH_4^+$) into nitrate ($NO_3^-$) via the formation of nitrite ($NO_2^-$) (Tanikawa et al. 2018). Different microorganisms mediate the conversion of ammonia to nitrite and nitrite to nitrate. In the first stage, ammonia is oxidized to nitrite by ammonia-oxidizing bacteria (AOB) and/or archaea (AOA). In the second stage, nitrite is oxidized to nitrate by nitrite-oxidizing bacteria (NOB). Further, microbes can use this nitrate as the nitrogen source in various metabolic processes. Nitrogen assimilation has been reported to occur throughout the curing stage of the composting process (Zeng et al. 2012). The two most abundant AOB and NOB in the environment are the beta-proteobacteria, *Nitrosomonas* and *Nitrospira* (Fujitani et al. 2015; Sun et al. 2019). The biochemical process of nitrification takes place through the following steps (Kuypers et al. 2018):

Biochemical reactions in nitrification:

$$NH_4^+ + O_2 \rightarrow NO + e^- + 2H^+ + H_2O$$

$$NO + H_2O \rightarrow NO_2^- + e^- + 2H^+$$

$$NO_2^- + H_2O \rightarrow NO_3^- + 2e^- + 2H^+$$

According to some recent studies, some species of *Nitrospira* were found to oxidize the ammonia to nitrate within a single cell. The process is called comammox (Daims et al. 2015; Van Kessel et al. 2015) and is represented using the following equation:

$$NH_4^+ + O_2 \rightarrow NO_3^- + H_2O + 2H^+$$

Denitrification is a microbial-mediated biochemical process carried out under anoxic/anaerobic conditions during which nitrite and nitrate are reduced to $N_2$. Further, under partial anaerobic conditions, $N_2O$, NO, and $NO_2$ are also produced, which are lost as gas to the atmosphere (Maeda et al. 2010; Wang et al. 2015). During the mesophilic and thermophilic phases of composting, denitrification takes place. Pseudomonas and Bacillus were found to have genes for nitrite reductase (NIR) (*nirK*, *nirS*, and *nirZ*) and nitrous oxide reductase (*nosZ*), which speed up the process of denitrification (Liu et al. 2019b; Yang et al. 2019).

Several denitrifying microbes belonging to the class of *Bacillus*, *Sphingomonas*, *Pseudomonas*, *Flavobacterium*, etc. have been isolated during composting. It has been found that denitrifying microbes are generally low in abundance during the heating phase of composting but rise steadily during the thermophilic and maturation phases. A Study by Lycus et al. (2017) reports that the enzymes nitrate reductases (NAR or NAP), NIRs, nitric oxide reductases (NOR), and nitrous oxide reductases ($N_2OR$) are required to speed up the biochemical reactions of denitrification at different stages. These microorganisms can turn nitrate into $N_2$ in a step-by-step process by temporarily storing intermediates like nitrite, nitric oxide, and nitrous oxide, which can be used to make fully functional enzymes (Zhang et al. 2019).

Biochemical reactions in denitrification:

$$NO_3^- + 2e^- + 2H^+ \rightarrow NO_2^- + H_2O$$

$$NO_2^- + e^- + 2H^+ \rightarrow NO + H_2O$$

$$2NO + 2e^- + 2H^+ \rightarrow N_2O + H_2O$$

$$N_2O + 2e^- + 2H^+ \rightarrow N_2 + H_2O$$

$N_2$ formed can further be converted to $NH_3$ by several nitrogen-fixing bacteria with the aid of the nitrogenase enzyme (Zhang et al. 2020). Nitrogen fixation promotes composting and minimizes nitrogen loss. Only bacteria (*Azotobacter vinelandii*, *Azospirillum brasilense*, *Rhizobium* spp., and archaea) with the nitrogenase metalloenzyme can convert dinitrogen into ammonia and will participate in this process. Understanding and enhancing the process of nitrogen fixation during composting can effectually prevent nitrogen loss and are crucial for enhancing composting quality (Shi et al. 2020). The $N_2$ fixation is represented using the following equation (Kuypers et al. 2018):

$$N_2 + 8e^- + 8H^+ + 2NH_3 + H_2$$

The dissimilarity nitrate reduction to ammonia (DNRA) is a nitrate reduction method that preserves nitrogen and was first discovered in the soil bacterium *Clostridium welchii* (Wood 1938). The end product from the DNRA reaction is ammonium (Tiedje 1988). A variety of heterotrophic and anaerobic chemolithoautotrophic bacteria engage in the two-step DNRA reaction (Dalsgaard and Bak 1994). According to Tiedje (1988), under anoxic conditions, $NO_3^-$ is first converted to $NO_2^-$ and then to $NH_4^+$ without its uptake by microbial cells, while it is still easily accessible to primary producers as a source of N (Pandey et al. 2009; Silver et al. 2001; Minick et al. 2016). The initial step is known as $NO_3^-$ respiration, similar to denitrification, while the final phase can either be respiratory or fermentative (Cole and Brown 1980). The equations of DNRA are as follows (Kuypers et al. 2018):

$$NO_3^- + 2e^- + 2H^+ \rightarrow NO_2^- + H_2O$$
$$NO_2^- + 6e^- + 8H^+ \rightarrow NH_4^+ + 2H_2O$$

In addition to denitrification, the process of anaerobic ammonium oxidation, also known as anammox, is responsible for the emission of fixed nitrogen into the atmosphere in the form of dinitrogen gas. In the early 1990s, wastewater sludge AOB were found to convert ammonium and nitrite into dinitrogen gas in the absence of oxygen for autotrophic growth. The AOB grew oxygen-free (Richards 1965).

$$NO_3^- + 2H^+ + e^- \rightarrow NO + H_2O$$
$$NO + NH_4^+ + 2H^+ + 3e^- \rightarrow N_2H_4 + H_2O$$
$$N_2H_4 \rightarrow N_2 + 4H^+ + 4e^-$$
$$NH_4^+ + NO_2^- \rightarrow N_2 + 2H_2O$$

Anaerobic microenvironments provide ideal circumstances for anammox bacteria to grow. The presence of anammox bacteria in the composted waste mass indicates typical activity in the thermophilic phase of the process, during which the concentration of ammonium peaks under anaerobic conditions formed by increasing breakdown of organic matter (González et al. 2019; Stegenta et al. 2022).

## 6.7 EFFECT OF CLIMATE CHANGE AND EMERGING CONTAMINANTS ON NITROGEN DYNAMICS OF COMPOSTING

There are several factors that simultaneously affect the nitrogen transformations in compost. Hence, it is cumbersome to monitor the individual's effect. However, several earlier studies attempted to analyze these factors affecting the nitrogen processing in soil, water, and compost. The feedstock influences microbial community architectures, which play a vital part in composting (Villar et al. 2016). Therefore, the microbiome of a particular organic waste is influenced by the physical and chemical qualities of the feedstock. According to studies, pH, total organic carbon (TOC), and $NH_4^+$ are among the most significant factors influencing bacterial populations' diversity. De Guardia et al. (2010) found that certain feedstock, such as spent mushroom compost, can provide a suitable micro-environment for nitrification.

Also, nitrogen loss in the composting process largely depends on the C/N ratio of feedstock. In a feedstock with a low C/N ratio, the high nitrogen content contributes to the faster conversion of organic nitrogen to $NH_4^+$, which accumulates in compost that gets lost in the form of $NH_3$. During composting of organic feedstocks, nitrogen loss was estimated to be as high as 50%, among which $NH_3$ alone contributes to 90% of total nitrogen loss (Tiquia and Tam 2000; Wong et al. 2009). Such nitrogen loss can be reduced by amending lignocellulosic biomass to increase the C/N ratio. However, how the lignocellulosic biomass grown at altered climate conditions with varied cell wall content behaves during composting is yet to be studied. Further, if the C/N ratio of feedstock is too high, both composting efficiency and compost quality are affected due to the inadequate nitrogen supply for microbial activity.

As the concentration of $CO_2$ increases in the atmosphere, changes in the ammonification and nitrification processes take place, which in turn alter the compost quality. Disrupting intracellular electron transport and impairing denitrification ability could result from the dissolved $CO_2$ combining with iron ions to create a precipitate. Shi et al. (2020) suggested that recirculating $CO_2$ during composting could be controlled by denitrification and lower GHG ($CO_2$ and $N_2O$) emissions, improving composting process efficiency.

As observed in a meta-analysis of published data performed by Barnard et al. (2005), the activities of nitrifying and denitrifying enzymes were found to decrease in response to elevated $CO_2$ concentration. However, no significant changes in gross nitrification were observed. The suppression of nitrifying enzyme activity by the elevated level of $CO_2$ is primarily through decreasing soil $O_2$ and by increasing soil water content (Hungate et al. 1997). In several similar studies, a significant decrease in soil $NO_3^-$ was recorded under higher $CO_2$ levels, correlated with reduced denitrifying enzyme activity (Barnard et al. 2004, 2005). On the contrary, denitrification activity, being an anaerobic process, was found to increase under higher soil water content or reduced $CO_2$ level (Ruser et al. 2006; Escuer-Gatius et al. 2020). Horz et al. (2004) recorded a decline in abundance in AOB under elevated $CO_2$ levels in grassland soils. Further, a synergistic effect was recorded with elevated $CO_2$ and precipitation treatments, where the decrease in abundance of AOB was much more significant than $CO_2$ alone. However, these variations in microbial abundance did not affect nitrification potential. One of the most important elements governing denitrification is oxygen availability, which is controlled by soil moisture, directly reliant on precipitation (Firestone et al. 1989). Increased soil moisture fosters anaerobic conditions and plays a more significant role in denitrification (Duan et al. 2017). Several earlier investigations have established a favorable correlation between increased soil water content and rising $N_2O$ emission from both laboratory and field experiments (Flessa et al. 1995; Dobbie et al. 1999; Cantarel et al. 2011).

Excessive $NH_3$ and interstitial $CO_2$ can inhibit nitrification during composting stages (Brito et al. 2012). Elevated $CO_2$ alters gross mineralization, which affects $NH_4^+$ availability and nitrification. $NH_3$ at high pH levels limits the second step of nitrification, causing nitrite to temporarily accumulate during compost maturation (Eklind and Kirchmann 2000; Chen et al. 2015). Low $NH_4^+/NH_3$ ratios inhibit nitrification (Zeng et al. 2012). $NH_3$ liberation raises compost water pH, which can inhibit nitrification (Wang et al. 2015). In addition, it is reported that loss of moisture, ammonification, and a rise in temperature are the primary reasons for excess $NH_3$ emissions. The optimum pH range for nitrification is typically between 7.5 and 8.0; however, Kim et al. (2007) proposed that the process also relies on the material's temperature. Specifically, some authors contend that low pH inhibits bacteria that oxidize ammonia (Zou et al. 2016). $NH_4^+$ oxidation and $CO_2$ stripping alter the pH during wastewater nitrification (Ge et al. 2015). Throughout the bio-oxidative phase of composting, the pH might rise to between 8.5 and 9.0 as a result of the ammonification process.

Nevertheless, an increase in pH during the bio-oxidative phase can enhance the volatilization of $NH_3$, thereby removing any excess $NH_4^+$-N that could limit nitrification.

At mesophilic temperatures (20°C–35°C), organic matter decomposes slowly and nitrifying bacteria grow aerobically (Sánchez-Monedero et al. 2001; Zeng et al. 2012). Temperature affects denitrifying microorganisms and enzymes, which release $N_2O$. Denitrifying microbes prefer 25°C–35°C (Braker et al. 2010; Hao et al. 2022). According to Braker et al. (2010), the community composition of the denitrifier (nirK and nirS types) altered during incubation at various temperatures. Their research indicated that higher or lower temperatures inhibit the growth and metabolism of denitrifying microorganisms and that high temperatures inhibit denitrifying enzyme activities and reduce denitrification efficiency. It is suggested that the composting temperature be adjusted to reduce $N_2O$ emissions.

At higher or lower temperatures, the nitrification process would cease (De Guardia et al. 2010). The $NO_3$-N levels in all treatments increased steadily during green waste composting with waste mushroom compost and biochar at thermophilic to mesophilic temperatures. Nitrifying bacteria cannot survive at thermophilic temperatures (Brito et al. 2008; Sánchez-Monedero 2001). Nevertheless, nitrification has been observed in the thermophilic composting stage, particularly in the later thermophilic stage (Cáceres et al. 2015; Li et al. 2005; Guo et al. 2016). According to Zeng et al. (2011), archaea are essential to the first stage of the nitrification process because they are able to acquire a competitive advantage in environments that are high-stress due to temperature and low-available oxygen. As global warming is a gradual process, it is presumed that this will give enough time for bacteria to develop mechanisms of nitrification under high temperatures. However, further studies are required to understand these mechanisms of high-temperature nitrifying bacteria.

Cáceres et al. (2018) reported increased levels of $NO_2$ and $NO_3$ during the thermophilic phase of composting chicken manure and rice straw residue, indicating the persistence of some ammonia oxidation microbes at high temperatures. Studies suggest the resistance of nitrifiers to high temperatures favors their ammonia oxidation potential (Jiang et al. 2015; Zhang et al. 2016). The nitrification process largely depends on temperature and pH (Kim et al. 2007). However, the initial nitrification rates are low at lower pH and do not vary with temperature fluctuations. At higher temperatures, nitrification rates are faster and strongly depend on pH. Additionally, AOB and NOB react differently to temperature, and their respective minimum residence times in wastewater treatment change as a function of temperature. For instance, AOB has multiplication rates higher than NOB at temperatures over 15°C (Ge et al. 2015).

Zheng et al. (2020) stated that warming, higher $CO_2$, and increased precipitation are some factors that promote overall terrestrial N fixation. Nitrogen fixation is inhibited by soil physicochemical changes, plant and microbial characteristics, and soil resources. Since many N-fixing microorganisms are anaerobes and high oxygen levels inhibit nitrogenase production, increased precipitation (or wetness) favors them by reducing oxygen levels in saturated soils. Nevertheless, water deficiency (or dryness) frequently prevents N fixation in forests and deserts (Reed et al. 2011). Nitrogen fixation rates are impacted by global warming since nitrogenase works best at a temperature of about 25°C (Houlton et al. 2008). In composting, enhanced nitrogen fixation may reduce nitrogen loss. Enhanced water content or decreased $CO_2$ level promotes denitrification. As per our understanding, the conservation of existing nitrogen is preferred over fixing gaseous nitrogen during composting.

The nitrogen cycle has been significantly disrupted by the extensive use of fertilizers in agricultural activities, as seen by the increasing concentration of nitrogen oxides in the air, nitrates in soils, and water (Olivares et al. 2013). HMs affect the nitrification rates in paddy soils, while excess usage of chemical fertilizers and pesticides affects nitrifying archaea/bacteria, denitrifying bacteria, and anammox bacteria (Rahman et al. 2021). A review by DeVries and Zhang (2016) reported a variable response of the nitrification process exposed to different AB at different concentrations in soil and wastewater sludge. AB include ciprofloxacin, NF, ofloxacin, erythromycin, virginiamycin, roxarsone, colistin, sulfadiazine, sulfadimethoxine, sulfamethoxazole, OTC, and TC were reported to inhibit the nitrification process to a varying extent. Chloramphenicol, ivermectin, difloxacin, monensin, bacitracin, and chlortetracycline were ineffective in the nitrification process. Similarly, the denitrification process was inhibited by gentamicin, amoxicillin, difloxacin, flumequine, norsasin, clarithromycin, erythromycin, bacitracin, sulfadiazine, sulfamethazine, sulfamethoxazole, chlortetracycline, and roxarsone at varying levels. Other AB tested, such as ciprofloxacin, TC, and virginiamycin, had no effect on denitrification at tested concentrations. In paddy soil, dose-related effects of TC and sulfamethazine, separately or in combination, were observed with 19.8%–75.5%, 11.3%–30.5%, and 22.1%–86.0% inhibition of DNRA rates, respectively, at 0.6–6,000 μg/kg exposure (Shan et al. 2018). Hou et al. (2015) observed a drastic increase in $N_2O$ flux (300%) when sediment was treated with AB (sulfamethoxazole, >50 mg/L). Further, a coincidence was observed between increased $N_2O$ flux and a reduction in the denitrification process. The author proposed that the increase in $N_2O$ is because of AB, which inhibited the reaction of $N_2O$ reduction to $N_2$. Similar observations were reported by DeVries et al. (2015) with narasin antibiotic at 11,000 ng/kg concentrations.

Chloramphenicol was most widely detected in sludge where the anammox activity was inhibited up to 95% at 200 mg/L concentration (Fernandez et al. 2009; Van de Graaf et al. 1995). The anammox process is susceptible to temperature and pH variations (Tomaszewski 2017). Depending on pH and temperature, the process may also be repressed by its substrates, ammonium and nitrite (Fernández et al. 2012). However, the exact mechanism underlying this phenomenon remains unknown.

The oxygen and nitrite in the air affect the anammox bacteria very much. Oxygen levels as low as 2 mM and nitrite levels between 5 and 10 mM completely stop anammox from working, but this can be undone (Jetten et al. 1998). Anammox bacteria are predominantly employed in wastewater treatment where HMs show a hormetic effect, promoting anammox activity during short-term exposure at low concentrations (up to 5 mg/L) (Li et al. 2018), while Cd(II) and Cr(VI) often inhibit anammox even at low concentrations (Jiang et al. 2018). Afzal et al. (2019) reported that the cadmium contamination (10 mg/kg) in the paddy field significantly reduced the microbial population involved in nitrogen transformation. Further, cadmium was found to decrease the activity of nitrifying and denitrifying bacteria, which was evident by decreased $NO_2$ emission. According to their observations, the effect of cadmium was higher in non-flooded soil than in flooded soil. As a similar group of bacteria, enzymes and reactions are also found to play a crucial role in composting; it is presumed that the above-discussed EC may similarly affect the composting process.

## 6.8 EFFECT OF HUMIC SUBSTANCES ON FEEDSTOCK AND COMPOSTING PARAMETERS

In the biodegradation process of composting, organic wastes are degraded, and stable organic final products are produced. The finished products are outstanding soil amendments and stable, safe, high-quality fertilizers. The most prominent composting by-products are pungent humic compounds (HS) (Stevenson 1994), which can be used as a marker of a completed composting process (Jednak et al. 2017). The primary functional groups found in HS are carboxyl, quinone, and hydroxyl (Guo et al. 2019). It plays a multifactorial role in the environment through the process of ion exchange, redox, adsorption, complexation, and others. They improve soil structure and fertility, stimulate plant growth, and even help the plant to resist disease. HS can be complex with various metals, inorganic, and organic pollutants, thereby immobilizing the toxic pollutants (Guo et al. 2019; Canellas et al. 2015; Jurado et al. 2015).

The parental materials for HS production are primarily the lignin component of lignocellulosic biomass and other polyphenolic compounds (Wang et al. 2021). Peptides, amino acids, and amino sugars are some of the additional organic matter constituents that become a part of HS, making it more heterogeneous than lignin (Maier et al. 2015; Sánchez–Monedero et al. 1999; Zhang et al. 2019). This makes the HS more recalcitrant for microbial degradation, which is widely evident by its slow turnover rate of 2%–5% per year (Wagner and Wolf 1999). During composting, HS production happens in two stages: (1) formation of reactive monomers during the degradation of organic matter and (2) spontaneous polymerization of reactive monomers to form complex HS (Maier 2015; Guo et al. 2019). During heating and thermophilic stages of aerobic composting, HS precursors are primarily produced. The compost becomes more stable as additional precursors are polymerized throughout the cooling and maturing processes (Roig et al. 2006). According to investigations, the more prolonged thermophilic phase during composting was favorable for the synthesis of low micromolecular HS precursors (Guo et al. 2019; Yu et al. 2018). Numerous studies found a connection between microbial abundance and diversity and their ability to break down plant matter and produce HS (Jurado et al. 2015; Zhao et al. 2016). Microorganisms first break down large organic molecules, and then low-molecular-weight compounds like polyphenols, carboxyls, and amino acids are used in biochemical synthesis. During the microbial attack, phenolic aldehydes and acids are liberated from lignin and are converted into quinones, which then polymerize in the presence or absence of amino compounds to produce HS (Paradelo et al. 2013; Zhang et al. 2019; Kulikowska 2016; Stevenson 1994; Wu et al. 2017). The major microorganisms involved in humification are identified as Proteobacteria, Actinobacteria, Bacteroidetes, and Firmicutes (Kulikova and Perminova 2021).

The factors affecting the development of HS in compost include raw materials, microorganisms and enzymes involved, oxygen concentration, temperature during the process of composting, C/N ratio, substrate pH, moisture, and EC (Swift et al. 1979; Binner et al. 2011). In addition, humification, being a microbes-mediated process affected by all those factors affecting microbial abundance, diversity, and function, will have an impact on the composting process and quality.

According to our earlier discussions, plants exposed to various stresses respond by modifying their cell wall structure primarily by accumulating more lignin. When such biomass is used for composting along with easily degradable organic substances, the possibility of enhanced HS production needs to be explored. On the other hand, the lignin-rich feedstock may also alter the C/N ratio, affecting the composting process and quality. In a study conducted by Wang et al. (2021), the introduction of alkali lignin and sodium lignosulfonate (a by-product of the paper-making industry) in composting enhanced the HS content by 44.42 mg/g and 45.87 mg/g, respectively. The observations were further supported by an increased degree of polymerization in the final compost. Here, the feedstock used was already depolymerized to various extents, which may speed up the process of HS formation.

Shan et al. (2013) attempted to optimize the quantity of easily degradable organic matter (Kitchen waste) during straw composting. They observed that the right proportion of kitchen waste and straw (1:2) was required to obtain a high yield of HS with reduced GHG emissions. According to the authors, higher microbial activity is related to the loss of compost mass, leading to reduced HS. Also, treatment with a higher straw ratio recorded increased β-glucosidase activity and lower microbial activity, leading to the production of a good quantity of HS. Supporting this, Smidt et al. (2008) analyzed the 174 composts prepared from various combinations and proportions of organic wastes and found that a well-balanced mixture of biowaste comprising kitchen, market, and yard waste produced a higher quantity of HS, which was correlated with the most prolonged period of microbial activity. In a similar study, Zhao et al. (2022a) emphasized that the role of lignin and microbial activity is crucial to obtain a good quantity of HS while composting different combinations of straw, cabbage, chicken manure, and pig manure.

Though several attempts to understand the CC consequences on soil humus were made, a similar effort in the research area of composting is required. A large group of scientists suspected that the humus stock of arable land is declining due to the influence of CC. Increased temperature due to global warming leads to the degradation of HS, thereby releasing GHG due to enhanced microbial activity or an increase in dissolved organic matter leaching out of the soil and polluting water bodies (Ewa 2018; Lorenzoni et al. 2007; Ritson et al. 2017; Knorr et al. 2005). Niklinska et al. (1999) studied the impact of global warming on nitrogen degradation and the rate of humus respiration at high altitudes in the northern hemisphere (Scots pine stands located in Europe). According to their estimation, under 2°C and 4°C temperature-increase situations, the soil respiration was found to increase in the range of $0.07 \times 10^{15}$ to $0.13 \times 10^{15}$ g $CO_2$. In general, the N mineralization was not significantly affected up to an increase in temperature up to 15°C; above, N mineralization was found to have a significant increment.

The characteristics of binding between HS and pollutants are also likely to change due to climate-induced changes in HS structural makeup due to their microbial transformations. This could decrease HS ability to immobilize metals and organic pollutants, which could further affect the quality and process of composting (Lipczynska-Kochany 2018; Balser et al. 2010; Zogg et al. 1997). Such compost, upon field application, may underperform, reducing the reliability of compost use. Similar behavior of HS formation or degradation or altered properties in composting under varied climatic factors and emerging pollutants may be expected. However, more research studies are needed on composting to unravel the impact of CC and emerging pollutants.

## 6.9 CONCLUSIONS AND PERSPECTIVES

Composts produced from different organic residues vary considerably in their quality and stability, which further tend to depend upon the composition of feedstock used for compost production, along with the diverse chemical and biological variations that occur during the composting process. As discussed in the previous sections, raw material and compost processes are anthropogenically influenced by diverse CC factors and EC. The measurable properties that define compost quality include parameters such as the temperature, $CO_2$ concentration, C/N ratio, the amount of organic matter, salinity, total nitrogen, total phosphorus, stability, pH, maturity, particle size, HM concentration, and pathogen concentrations and phytotoxicity. According to our observations, the available research studies to prove the impact of CC on compost processing and quality are insufficient. However, the impact of CC on plant growth and health and, subsequently, agro-residue-based composting feedstocks is primarily evidenced by the plant cell wall composition change. Microbes, being highly versatile organisms in nature, may already start evolving to enhance their fitness under changing compost environments. Among the various CC factors, temperature, moisture, and $CO_2$ were found to play a vital role in determining the composting process by altering the course of aerobic respiration. At the same time, compost quality is primarily determined by the feedstock composition (C/N ratio). On the other hand, compost microbes learned to overcome the adverse impact of EC, such as AB and HMs. Ample evidence is available to support the compost as the best process that removes AB and reduces the adverse effects of HMs in agricultural fields. However, one of the major concerns still is the possibility of origin and dissemination of antibiotic resistance genes in/through compost to the environment, which poses a potential risk to human and animal health. This chapter provides an insight into how CC and EC impact the feedstock, composting process, and compost quality. Providing an opportunity to take suitable measures harnesses the benefits of compost as a part of sustainable agriculture in the era of CC.

## REFERENCES

Abou Elezz, Ahmed, Ahmed Easa, Fathy Atia, and Talaat Ahmed. "The potential impact data of tylosin and enrofloxacin veterinary antibiotics on germination and accumulation in barley seed as a forage crop and good dietary sources using LC/MS-MS." *Data in Brief* 25 (2019): 104326.

Abu Qdais, Hani, and Mohamad Al-Widyan. "Evaluating composting and co-composting kinetics of various agro-industrial wastes." *International Journal of Recycling of Organic Waste in Agriculture* 5, no. 3 (2016): 273–280.

Achuba, Felix I. "The effect of sublethal concentrations of crude oil on the growth and metabolism of cowpea (Vigna unguiculata) seedlings." *Environmentalist* 26 (2006): 17–20.

Aemig, Quentin, Nicolas Doussiet, Alice Danel, Nadine Delgenès, Julie Jimenez, Sabine Houot, and Dominique Patureau. "Organic micropollutants' distribution within sludge organic matter fractions explains their dynamic during sewage sludge anaerobic digestion followed by composting." *Environmental Science and Pollution Research* 26 (2019): 5820–5830.

Afzal, Muhammad, Mengjie Yu, Caixian Tang, Lujun Zhang, Niaz Muhammad, Haochun Zhao, Jiayin Feng, Lu Yu, and Jianming Xu. "The negative impact of cadmium on nitrogen transformation processes in a paddy soil is greater under non-flooding than flooding conditions." *Environment International* 129 (2019): 451–460.

Alengebawy, Ahmed, Sara Taha Abdelkhalek, Sundas Rana Qureshi, and Man-Qun Wang. "Heavy metals and pesticides toxicity in agricultural soil and plants: Ecological risks and human health implications." *Toxics* 9, no. 3 (2021): 42.

Alkoaik, Fahad N. "Integrating aeration and rotation processes to accelerate composting of agricultural residues." *PLoS One* 14, no. 7 (2019): e0220343.

Alvarez, Sophie, Ellen L. Marsh, Steve G. Schroeder, and Daniel P. Schachtman. "Metabolomic and proteomic changes in the xylem sap of maize under drought." *Plant, Cell & Environment* 31, no. 3 (2008): 325–340.

Arias, Paola A., Nicolas Bellouin, Erika Coppola, Richard G. Jones, Gerhard Krinner, Jochem Marotzke, Vaishali Naik et al. "Technical summary." In *Climate Change 2021: The Physical Science Basis*, edited by Valérie Masson-Delmotte, Panmao Zhai, Anna Pirani, Sarah L. Connors, Clotilde Péan, Sophie Berger, Nada Caud, et al., 33–144. Cambridge University Press, 2021.

Arikan, Osman A., Lawrence J. Sikora, Walter Mulbry, Shahamat U. Khan, and Gregory D. Foster. "Composting rapidly reduces levels of extractable oxytetracycline in manure from therapeutically treated beef calves." *Bioresource Technology* 98, no. 1 (2007): 169–176.

Azanu, David, Christiana Mortey, Godfred Darko, Johan Juhl Weisser, Bjarne Styrishave, and Robert Clement Abaidoo. "Uptake of antibiotics from irrigation water by plants." *Chemosphere* 157 (2016): 107–114.

Azim, Khalil, Benaissa Soudi, Samira Boukhari, Claude Perissol, Sevastianos Roussos, and Imad Thami Alami. "Composting parameters and compost quality: A literature review." *Organic Agriculture* 8, no. 2 (2018): 141–158.

Baligar, Virupax C., Nand Kumar Fageria, and Zhenli He. "Nutrient use efficiency in plants." *Communications in Soil Science and Plant Analysis* 32, no. 7–8 (2001): 921–950.

Balser, Teri C., Jessica L. M. Gutknecht, and Chao Liang. "How will climate change impact soil microbial communities?" In *Soil Microbiology and Sustainable Crop Production*, edited by Geoffrey R. Dixon and David E. Tilston, 373–397. Springer, 2010.

Bao, Jianfeng, Xiaojuan Wang, Jie Gu, Xiaoxia Dai, Kaiyu Zhang, Qianzhi Wang, Jiyue Ma, and Huiling Peng. "Effects of macroporous adsorption resin on antibiotic resistance genes and the bacterial community during composting." *Bioresource Technology* 295 (2020): 121997.

Barnard, Romain, Laure Barthes, Xavier Le Roux, and Paul W. Leadley. "Dynamics of nitrifying activities, denitrifying activities and nitrogen in grassland mesocosms as altered by elevated $CO_2$." *New Phytologist* 162, no. 2 (2004): 365–376.

Barnard, Romain, Paul W. Leadley, and Bruce A. Hungate. "Global change, nitrification, and denitrification: A review." *Global Biogeochemical Cycles* 19, no. 1 (2004): GB1007.

Barnard, Romain, Paul W. Leadley, Robert Lensi, and Laure Barthes. "Plant, soil microbial and soil inorganic nitrogen responses to elevated $CO_2$: A study in microcosms of Holcus lanatus." *Acta Oecologica* 27, no. 3 (2005): 171–178.

Başlar, Süleyman, Yunus Doğan, Hüseyin Bağ, and Aydan Elçi. "Trace element biomonitoring by needles of Pinus brutia Ten. from western Anatolia, Turkey." *Fresenius Environmental Bulletin* 12 (2003): 451.

Bernal, María P., Juan A. Alburquerque, and Raúl Moral. "Composting of animal manures and chemical criteria for compost maturity assessment. A review." *Bioresource Technology* 100, no. 22 (2009): 5444–5453.

Bhatia, Shashi Kant, Sujit Sadashiv Jagtap, Ashwini Ashok Bedekar, Ravi Kant Bhatia, Anil Kumar Patel, Deepak Pant, Jeyakumar Rajesh Banu, Christopher V. Rao, Yun-Gon Kim, and Yung-Hun Yang. "Recent developments in pretreatment technologies on lignocellulosic biomass: Effect of key parameters, technological improvements, and challenges." *Bioresource Technology* 300 (2020): 122724.

Bhattacharjya, Sudeshna, Asha Sahu, Manas Chandra Manna, and Ashok Kumar Patra. "Potential of surplus crop residues, horticultural waste and animal excreta as a nutrient source in the central and western regions of India." *Current Science* 116, no. 314 (2019): 314–318.

Bhuiyan, Nazmul H., Weiping Liu, Guosheng Liu, Gopalan Selvaraj, Yangdou Wei, and John King. "Transcriptional regulation of genes involved in the biosynthesis pathways and supply of methyl units in response to powdery mildew attack and abiotic stresses in wheat." *Plant Molecular Biology* 64, no. 3 (2007): 305–318.

Binner, Erwin, Ena Smidt, Johannes Tintner, Katharina Böhm, and Peter Lechner. "How to enhance humification during composting of separately collected biowaste: Impact of feedstock and processing." *Waste Management & Research* 29, no. 11 (2011): 1153–1163.

Bishop, Paul L. "Nitrogen variations during sludge composting." *Biocycle* 24 (1983): 34–39.

Bita, Craita E., and Tom Gerats. "Plant tolerance to high temperature in a changing environment: Scientific fundamentals and production of heat stress-tolerant crops." *Frontiers in Plant Science* 4 (2013): 273.

Bona, Cleusa, Igor Mendonça de Rezende, Gedir de Oliveira Santos, and Luiz Antônio de Souza. "Effect of soil contaminated by diesel oil on the germination of seeds and the growth of Schinus terebinthifolius Raddi (Anacardiaceae) seedlings." *Brazilian Archives of Biology and Technology* 54 (2011): 1379–1387.

Braker, Gesche, Julia Schwarz, and Ralf Conrad. "Influence of temperature on the composition and activity of denitrifying soil communities." *FEMS Microbiology Ecology* 73, no. 1 (2010): 134–148.

Brito, Luís Miguel, Jorge Coutinho, and Stephen R. Smith. "Methods to improve the composting process of the solid fraction of dairy cattle slurry." *Bioresource Technology* 99, no. 18 (2008): 8955–8960.

Brito, Luís Miguel, Isabel de Mourão, João Coutinho, and Stephen R. Smith. "Simple technologies for on-farm composting of cattle slurry solid fraction." *Waste Management* 32, no. 7 (2012): 1332–1340.

Budíková, Svatava. "Structural changes and aluminium distribution in maize root tissues." *Biologia Plantarum* 42, no. 2 (1999): 259–266.

Burritt, David John. "The polycyclic aromatic hydrocarbon phenanthrene causes oxidative stress and alters polyamine metabolism in the aquatic liverwort Riccia fluitans L." *Plant, Cell & Environment* 31, no. 10 (2008): 1416–1431.

Cáceres, Rafaela, Narcís Coromina, Krystyna Malińska, and Oriol Marfà. "Evolution of process control parameters during extended co-composting of green waste and solid fraction of cattle slurry to obtain growing media." *Bioresource Technology* 179 (2015): 398–406.

Cáceres, Rafaela, Krystyna Malińska, and Oriol Marfà. "Nitrification within composting: A review." *Waste Management* 72 (2018): 119–137.

Canellas, Luciano P., Fábio L. Olivares, Natália O. Aguiar, Davey L. Jones, Antonio Nebbioso, Pierluigi Mazzei, and Alessandro Piccolo. "Humic and fulvic acids as biostimulants in horticulture." *Scientia Horticulturae* 196 (2015): 15–27.

Cantarel, Amélie AM, Juliette MG Bloor, Nicolas Deltroy, and Jean-François Soussana. "Effects of climate change drivers on nitrous oxide fluxes in an upland temperate grassland." *Ecosystems* 14, no. 2 (2011): 223–233.

Cayuela, María Luz, Miguel Ángel Sánchez-Monedero, and Asunción Roig. "Evaluation of two different aeration systems for composting two-phase olive mill wastes." *Process Biochemistry* 41, no. 3 (2006): 616–623.

Central Research Institute for Dryland Agriculture (CRIDA). *Vermicompost from Wastes*, 2 pp. CRIDA, 2003.

Chaplygin, Victor A., Tatiana M. Minkina, Saglara S. Mandzhieva, Olga G. Nazarenko, Irina V. Zimulina, Tatiana V. Bauer, Yuri A. Litvinov, and Vishnu D. Rajput. "Heavy metals in agricultural crops of Rostov region through the example of soft wheat (Triticum aestivum)." *IOP Conference Series: Earth and Environmental Science*, 624, no. 1 (2021): 012204.

Chauhan, Prakram Singh. "Role of various bacterial enzymes in complete depolymerization of lignin: A review." *Biocatalysis and Agricultural Biotechnology* 23 (2020): 101498.

Chen, Fu-Sheng, Guohua Geoff Wang, Xiang-Min Fang, Song-Ze Wan, Yang Zhang, and Chao Liang. "Nitrogen deposition effect on forest litter decomposition is interactively regulated by endogenous litter quality and exogenous resource supply." *Plant and Soil* 437, no. 1 (2019): 413–426.

Chen, Jun, Wen-Jing Deng, You-Sheng Liu, Li-Xin Hu, Liang-Ying He, Jian-Liang Zhao, Tuan-Tuan Wang, and Guang-Guo Ying. "Fate and removal of antibiotics and antibiotic resistance genes in hybrid constructed wetlands." *Environmental Pollution* 249 (2019): 894–903.

Chen, Ruirui, Yiming Wang, Wei Wang, Shiping Wei, Zhongwang Jing, and Xiangui Lin. "$N_2O$ emissions and nitrogen transformation during windrow composting of dairy manure." *Journal of Environmental Management* 160 (2015): 121–127.

Chen, Zhou, Yanzeng Li, Chengsong Ye, Xin He, and Shenghua Zhang. "Fate of antibiotics and antibiotic resistance genes during aerobic co-composting of food waste with sewage sludge." *Science of the Total Environment* 784 (2021): 146950.

Choi, Jong-Myung, Chun-Ho Pak, and Chiwon W. Lee. "Micro nutrient toxicity in French marigold." *Journal of Plant Nutrition* 19, no. 6 (1996): 901–916.

Cole, Jeffrey A., and Christopher M. Brown. "Nitrite reduction to ammonia by fermentative bacteria: A short circuit in the biological nitrogen cycle." *FEMS Microbiology Letters* 7, no. 2 (1980): 65–72.

Daims, Holger, Elena V. Lebedeva, Petra Pjevac, Ping Han, Craig Herbold, Mads Albertsen, Nico Jehmlich et al. "Complete nitrification by Nitrospira bacteria." *Nature* 528, no. 7583 (2015): 504–509.

Dalsgaard, Tage, and Friedhelm Bak. "Nitrate reduction in a sulfate-reducing bacterium, Desulfovibrio desulfuricans, isolated from rice paddy soil: Sulfide inhibition, kinetics, and regulation." *Applied and Environmental Microbiology* 60, no. 1 (1994): 291–297.

Davey, Matthew P., David N. Bryant, Ian Cummins, Trevor W. Ashenden, Phillip Gates, Robert Baxter, and Robert Edwards. "Effects of elevated $CO_2$ on the vasculature and phenolic secondary metabolism of Plantago maritima." *Phytochemistry* 65, no. 15 (2004): 2197–2204.

De Guardia, Amandine, Pierre Mallard, Cécile Teglia, Alain Marin, Christophe Le Pape, Mickael Launay, Jean-Claude Benoist, and Charles Petiot. "Comparison of five organic wastes regarding their behaviour during composting: Part 2, nitrogen dynamic." *Waste Management* 30, no. 3 (2010): 415–425.

Delmer, Deborah P., and Candace H. Haigler. "The regulation of metabolic flux to cellulose, a major sink for carbon in plants." *Metabolic Engineering* 4, no. 1 (2002): 22–28.

Deng, Wenwen, Anyun Zhang, Shujuan Chen, Xueping He, Lei Jin, Xiumei Yu, Shengzhi Yang et al. "Heavy metals, antibiotics and nutrients affect the bacterial community and resistance genes in chicken manure composting and fertilized soil." *Journal of Environmental Management* 257 (2020): 109980.

DeVries, Stephanie L., and Pengfei Zhang. "Antibiotics and the terrestrial nitrogen cycle: A review." *Current Pollution Reports* 2, no. 1 (2016): 51–67.

DeVries, Stephanie L., Madeline Loving, Xiqing Li, and Pengfei Zhang. "The effect of ultralow-dose antibiotics exposure on soil nitrate and $N_2O$ flux." *Scientific Reports* 5, no. 1 (2015): 1–8.

Directorate of Agriculture, Assam. *Method of Composting and Preparation of Vermicompost.* Government of Assam, 2003. https://diragri.assam.gov.in/sites/default/files/swf_utility_folder/departments/diragri_medhassu_in_oid_4/portlet/level_2/11%28B%29.10.pdf.

Dobbie, K. E., I. P. McTaggart, and K. A. Smith. "Nitrous oxide emissions from intensive agricultural systems: variations between crops and seasons, key driving variables, and mean emission factors." Journal of Geophysical Research: Atmospheres 104, no. D21 (1999): 26891–26899.

Dolliver, Holly, Kuldip Kumar, and Satish Gupta. "Sulfamethazine uptake by plants from manure-amended soil." *Journal of Environmental Quality* 36, no. 4 (2007): 1224–1230.

Donlan, Rodney M., and John William Costerton. "Biofilms: Survival mechanisms of clinically relevant microorganisms." *Clinical Microbiology Reviews* 15, no. 2 (2002): 167–193.

Duan, Yun-Feng, Xian-Wang Kong, Andreas Schramm, Rodrigo Labouriau, Jørgen Eriksen, and Søren O. Petersen. "Microbial N transformations and $N_2O$ emission after simulated grassland cultivation: effects of the nitrification inhibitor 3, 4-dimethylpyrazole phosphate (DMPP)." Applied and environmental microbiology 83, no. 1 (2017): e02019–16.

Duxbury, Cheryl L., Douglass George Dixon, and Bruce M. Greenberg. "Effects of simulated solar radiation on the bioaccumulation of polycyclic aromatic hydrocarbons by the duckweed Lemna gibba." *Environmental Toxicology and Chemistry: An International Journal* 16, no. 8 (1997): 1739–1748.

Ebbs, Stephen D., and Leon V. Kochian. *Toxicity of Zinc and Copper to Brassica Species: Implications for Phytoremediation*. Vol. 26, no. 3. American Society of Agronomy, Crop Science Society of America, and Soil Science Society of America, 1997.

Ederli, Luisa, Lara Reale, Francesco Ferranti, and Stefania Pasqualini. "Responses induced by high concentration of cadmium in Phragmites australis roots." *Physiologia Plantarum* 121, no. 1 (2004): 66–74.

Eisa, Musaed, Ramesh Chand, and Arun K. Joshi. "Biochemical and histochemical traits: A promising way to screen resistance against spot blotch (Bipolaris sorokiniana) of wheat." *European Journal of Plant Pathology* 137 (2013): 805–820.

Eklind, Ylva, and Holger Kirchmann. "Composting and storage of organic household waste with different litter amendments. II: Nitrogen turnover and losses." *Bioresource Technology* 74, no. 2 (2000): 125–133.

Ellis, Reginald John. "Molecular chaperones: The plant connection." *Science* 250, no. 4983 (1990): 954–959.

Escuer-Gatius, Jordi, Merrit Shanskiy, Kaido Soosaar, Alar Astover, and Henn Raave. "High-temperature hay biochar application into soil increases $N_2O$ fluxes." *Agronomy* 10, no. 1 (2020): 109.

FAO, WHO. "Principles and methods for the risk assessment of chemicals in food." *Environmental Health Criteria* 240 (2009).

Feng, Wei, Daniel Kita, Alexis Peaucelle, Heather N. Cartwright, Vinh Doan, Qiaohong Duan, Ming-Che Liu et al. "The FERONIA receptor kinase maintains cell-wall integrity during salt stress through $Ca^{2+}$ signaling." *Current Biology* 28, no. 5 (2018): 666–675.

Fernandez, Ismael, Anuska Mosquera-Corral, José Luis Campos, and Ramón Méndez. "Operation of an Anammox SBR in the presence of two broad-spectrum antibiotics." *Process Biochemistry* 44, no. 4 (2009): 494–498.

Fernandez, Ismael, Anuska Mosquera-Corral, José Luis Campos, and Ramón Méndez. "Short-and long-term effects of ammonium and nitrite on the Anammox process." *Journal of Environmental Management* 95 (2012): S170–S174.

Firestone, M. K., and E. A. Davidson. "Microbiological basis of NO and N 2 O production and consumption in soil." *Exchange of Trace Gases Between Terrestrial Ecosystems and the Atmosphere* 47 (1989): 7–21.

Flessa, H., P. Dörsch, and F. Beese. "Seasonal variation of $N_2O$ and CH4 fluxes in differently managed arable soils in southern Germany." *Journal of Geophysical Research: Atmospheres* 100, no. D11 (1995): 23115–23124.

Flexas, Jaume, Miquel Ribas-Carbó, Antonio Díaz-Espejo, Jeroni Galmés, and Hipólito Medrano. "Mesophyll conductance to $CO_2$: Current knowledge and future prospects." *Plant, Cell & Environment* 31, no. 5 (2008): 602–621.

Fogarty, Andrew M., and Olli H. Tuovinen. "Microbiological degradation of pesticides in yard waste composting." *Microbiological Reviews* 55, no. 2 (1991): 225–233.

Fujitani, Hirotsugu, Asami Kumagai, Norisuke Ushiki, Kengo Momiuchi, and Satoshi Tsuneda. "Selective isolation of ammonia-oxidizing bacteria from autotrophic nitrifying granules by applying cell-sorting and sub-culturing of microcolonies." *Frontiers in Microbiology* 6 (2015): 1159.

Füleky, György, and Szilveszter Benedek. "Composting to recycle biowaste." In *Sociology, Organic Farming, Climate Change and Soil Science*, edited by Eric Lichtfouse, 319–346. Sustainable Agriculture Reviews 3. Springer, 2010.

Gall, Hyacinthe Le, Florian Philippe, Jean-Marc Domon, Françoise Gillet, Jérôme Pelloux, and Catherine Rayon. "Cell wall metabolism in response to abiotic stress." *Plants* 4, no. 1 (2015): 112–166.

Garg, Payal, Asha Gupta, and Santosh Satya. "Vermicomposting of different types of waste using Eisenia foetida: A comparative study." *Bioresource Technology* 97, no. 3 (2006): 391–395.

Ge, Shijian, Shanyun Wang, Xiong Yang, Shuang Qiu, Baikun Li, and Yongzhen Peng. "Detection of nitrifiers and evaluation of partial nitrification for wastewater treatment: A review." *Chemosphere* 140 (2015): 85–98.

Gehlot, Priyanka, and Pundi Hariprasad. "Computational and data mining studies to understand the distribution and dynamics of temoneria (TEM) β-lactamase and their interaction with β-lactam and β-lactamase inhibitors." *Environmental Pollution* 314 (2022): 120289.

Geng, Jiagen, Xiaoying Liu, Jie Wang, and Si Li. "Accumulation and risk assessment of antibiotics in edible plants grown in contaminated farmlands: A review." *Science of the Total Environment* 852 (2022): 158616.

Ghanati, Faezeh, Akio Morita, and Hiromi Yokota. "Deposition of suberin in roots of soybean induced by excess boron." *Plant Science* 168, no. 2 (2005): 397–405.

Ghosh, Sudeshna, and Timothy M. LaPara. "The effects of subtherapeutic antibiotic use in farm animals on the proliferation and persistence of antibiotic resistance among soil bacteria." *The ISME Journal* 1, no. 3 (2007): 191–203.

Giri, Bhoopander, Pham Huong Giang, Rina Kumari, Ram Prasad, and Ajit Varma. "Microbial diversity in soils." In *Microorganisms in Soils: Roles in Genesis and Functions*, edited by Friedhelm Buscot and Ajit Varma, 19–55. Springer, 2005.

Gliniak, Maciej, Łukasz Grabowski, and Daria Polek. "Ozone aeration impact on the maturation phase in the process of green waste composting." *BIO Web of Conferences* 10 (2018): 01005.

Gondek, Matthew, David C. Weindorf, Carmen Thiel, and Greg Kleinheinz. "Soluble salts in compost and their effects on soil and plants: A review." *Compost Science & Utilization* 28, no. 2 (2020): 59–75.

González, Daniel, Joan Colón, David Gabriel, and Antoni Sánchez. "The effect of the composting time on the gaseous emissions and the compost stability in a full-scale sewage sludge composting plant." *Science of the Total Environment* 654 (2019): 311–323.

González-Lamothe, Rocío, Gabriel Mitchell, Mariza Gattuso, Moussa S. Diarra, François Malouin, and Kamal Bouarab. "Plant antimicrobial agents and their effects on plant and human pathogens." *International Journal of Molecular Sciences* 10, no. 8 (2009): 3400–3419.

Gray, Sharon B., and Siobhan M. Brady. "Plant developmental responses to climate change." *Developmental Biology* 419, no. 1 (2016): 64–77.

Griffis, Timothy J., Zichong Chen, John M. Baker, Jeffrey D. Wood, Dylan B. Millet, Xuhui Lee, Rodney T. Venterea, and Peter A. Turner. "Nitrous oxide emissions are enhanced in a warmer and wetter world." *Proceedings of the National Academy of Sciences* 114, no. 45 (2017): 12081–12085.

Guo, Xiaobo, Yanyu Lu, and Qunliang Li. "Effect of adding flue gas desulphurization gypsum on the transformation and fate of nitrogen during composting." *Compost Science & Utilization* 24, no. 4 (2016): 230–237.

Guo, Xiao-xia, Hong-tao Liu, and Shu-biao Wu. "Humic substances developed during organic waste composting: Formation mechanisms, structural properties, and agronomic functions." *Science of the Total Environment* 662 (2019): 501–510.

Habermann, Eduardo, Eduardo A. Dias de Oliveira, Gustavo Delvecchio, Rafael Belisario, Rafael Ferreira Barreto, Dilier Olivera Viciedo, Nadia Oliveira Rossingnoli et al. "How does leaf physiological acclimation impact forage production and quality of a warmed managed pasture of *Stylosanthes capitata* under different conditions of soil water availability?." *Science of the Total Environment* 759 (2021): 143505.

Hajar, Erna Wati Ibnu, Ahmad Ziad Bin Sulaiman, and Abdul Munaim Mimi Sakinah. "Assessment of heavy metals tolerance in leaves, stems and flowers of *Stevia rebaudiana* plant." *Procedia Environmental Sciences* 20 (2014): 386–393.

Hao, Zhen-Le, Amjad Ali, Yi Ren, Jun-Feng Su, and Zhao Wang. "A mechanistic review on aerobic denitrification for nitrogen removal in water treatment." *Science of The Total Environment* 848 (2022): 157452.

Harindintwali, Jean Damascene, Jianli Zhou, and Xiaobin Yu. "Lignocellulosic crop residue composting by cellulolytic nitrogen-fixing bacteria: A novel tool for environmental sustainability." *Science of the Total Environment* 715 (2020): 136912.

Harrison, Matthew T., Everard J. Edwards, Graham D. Farquhar, Adrienne B. Nicotra, and John R. Evans. "Nitrogen in cell walls of sclerophyllous leaves accounts for little of the variation in photosynthetic nitrogen-use efficiency." *Plant, Cell & Environment* 32, no. 3 (2009): 259–270.

Hasan, Golam Mohammad Monowar Anwarul, Anuj Kumer Das, and Mohammed Abdus Satter. "Accumulation of heavy metals in rice (Oryza sativa. L) grains cultivated in three major industrial areas of Bangladesh." *Journal of Environmental and Public Health* 2022 (2022): 4113032.

Haug, Roger T. *The Practical Handbook of Compost Engineering*. Routledge, 2018.

Hernandez-Blanco, Camilo, Dong Xin Feng, Jian Hu, Andrea Sanchez-Vallet, Laurent Deslandes, Francisco Llorente, Marta Berrocal-Lobo et al. "Impairment of cellulose synthases required for Arabidopsis secondary cell wall formation enhances disease resistance." *Plant Cell* 19, no. 3 (2007): 890–903.

Hobman, Jon L., and Lisa C. Crossman. "Bacterial antimicrobial metal ion resistance." *Journal of Medical Microbiology* 64, no. 5 (2015): 471–497.

Horwath, William R., and Lloyd F. Elliott. "Ryegrass straw component decomposition during mesophilic and thermophilic incubations." *Biology and Fertility of Soils* 21, no. 4 (1996): 227–232.

Horz, Hans-Peter, Adrian Barbrook, Christopher B. Field, and Brendan J. M. Bohannan. "Ammonia-oxidizing bacteria respond to multifactorial global change." *Proceedings of the National Academy of Sciences* 101, no. 42 (2004): 15136–15141.

Hou, Lijun, Guoyu Yin, Min Liu, Junliang Zhou, Yanling Zheng, Juan Gao, Haibo Zong, Yi Yang, Lei Gao, and Chunfu Tong. "Effects of sulfamethazine on denitrification and the associated $N_2O$ release in estuarine and coastal sediments." *Environmental Science & Technology* 49, no. 1 (2015): 326–333.

Houlton, Benjamin Z., Ying-Ping Wang, Peter M. Vitousek, and Christopher B. Field. "A unifying framework for dinitrogen fixation in the terrestrial biosphere." *Nature* 454, no. 7202 (2008): 327–330.

Hu, Xiangang, Qixing Zhou, and Yi Luo. "Occurrence and source analysis of typical veterinary antibiotics in manure, soil, vegetables and groundwater from organic vegetable bases, northern China." *Environmental Pollution* 158, no. 9 (2010): 2992–2998.

Huhe, Jiang C., Yanpei Wu, and Yunxiang Cheng. "Bacterial and fungal communities and contribution of physicochemical factors during cattle farm waste composting." *MicrobiologyOpen* 6: e518." (2017).

Hungate, Bruce A., Christopher P. Lund, Holly L. Pearson, and F. STUART CHAPIN III. "Elevated CO2 and nutrient addition after soil N cycling and N trace gas fluxes with early season wet-up in a California annual grassland." *Biogeochemistry* 37, no. 2 (1997): 89–109.

Huo, Meixia, Wenjin Ma, Kaixiang Zhou, Xiangyue Xu, Zhenli Liu, and Lingli Huang. "Migration and toxicity of toltrazuril and its main metabolites in the environment." *Chemosphere* 302 (2022): 134888.

Insam, Heribert, and Marco de Bertoldi. "Microbiology of the composting process." In *Waste Management Series*, vol. 8, pp. 25–48. Elsevier, 2007.

Iranzo, María, Miguel Gamón, Rafael Boluda, and Salvador Mormeneo. "Analysis of pharmaceutical biodegradation of WWTP sludge using composting and identification of certain microorganisms involved in the process." *Science of the Total Environment* 640 (2018): 840–848.

Irshad, M., Shazia Gul, A. Egrinya Eneji, Zobia Anwar, and M. Ashraf. "Extraction of heavy metals from manure and their bioavailability to spinach (Spinacia oleracea L.) after composting." *Journal of Plant Nutrition* 37, no. 10 (2014): 1661–1675.

Jajoo, Anjana, Nageswara Rao Mekala, Rupal Singh Tomar, Michele Grieco, Mikko Tikkanen, and Eva-Mari Aro. "Inhibitory effects of polycyclic aromatic hydrocarbons (PAHs) on photosynthetic performance are not related to their aromaticity." *Journal of Photochemistry and Photobiology B: Biology* 137 (2014): 151–155.

Jednak, Tanja, Jelena Avdalović, Srđan Miletić, Latinka Slavković-Beškoski, Dalibor Stanković, Jelena Milić, Mila Ilić, Vladimir Beškoski, Gordana Gojgić-Cvijović, and Miroslav M. Vrvić. "Transformation and synthesis of humic substances during bioremediation of petroleum hydrocarbons." *International Biodeterioration & Biodegradation* 122 (2017): 47–52.

Jetten, Mike SM, Marc Strous, Katinka T. Van de Pas-Schoonen, Jos Schalk, Udo GJM van Dongen, Astrid A. van de Graaf, Susanne Logemann, Gerard Muyzer, Mark CM van Loosdrecht, and J. Gijs Kuenen. "The anaerobic oxidation of ammonium." *FEMS Microbiology Reviews* 22, no. 5 (1998): 421–437.

Jiang, Jishao, Xueling Liu, Yimei Huang, and Hua Huang. "Inoculation with nitrogen turnover bacterial agent appropriately increasing nitrogen and promoting maturity in pig manure composting." *Waste Management* 39 (2015): 78–85.

Jiang, Xiao-Yan, Ya-Fei Cheng, Wei-Qin Zhu, Yu-Hui Bai, Xue-Qi Wu, and Ren-Cun Jin. "Effect of chromium on granule-based anammox processes." *Bioresource Technology* 260 (2018): 1–8.

Jurado, Marina. M., Francisca Suárez-Estrella, María José López, María del Carmen Vargas-García, José Antonio López-González, and Joaquín Moreno. "Enhanced turnover of organic matter fractions by microbial stimulation during lignocellulosic waste composting." *Bioresource Technology* 186 (2015): 15–24.

Kalra, Naveen, D. Chakraborty, Anil Sharma, Hari Krishna Rai, Monica Jolly, Subhash Chander, Ponduru Ramesh Kumar et al. "Effect of increasing temperature on yield of some winter crops in northwest India." *Current Science* 94 (2008): 82–88.

Karmakar, Ruma, Shefali Bindiya, and Pundi Hariprasad. "Convergent evolution in bacteria from multiple origins under antibiotic and heavy metal stress, and endophytic conditions of host plant." *Science of the Total Environment* 650 (2019): 858–867.

Katsou, Evina, Teresa Alvariño, Simos Malamis, Sonia Suárez, Nicola Frison, Francisco Omil, and Francesco Fatone. "Effects of selected pharmaceuticals on nitrogen and phosphorus removal bioprocesses." *Chemical Engineering Journal* 295 (2016): 509–517.

Kaur, Simardeep, Mahesh Kumar Samota, Manoj Choudhary, Mukesh Choudhary, Abhay K. Pandey, Anshu Sharma, and Julie Thakur. "How do plants defend themselves against pathogens-Biochemical mechanisms and genetic interventions." *Physiology and Molecular Biology of Plants* 28, no. 2 (2022): 485–504.

Kaviya, N., Viabhav K. Upadhayay, Jyoti Singh, Amir Khan, Manisha Panwar, and Ajay Veer Singh. "Role of microorganisms in soil genesis and functions." In *Mycorrhizosphere and Pedogenesis*, edited by Ajay Kumar and Narendra Tuteja, 25–52. Springer, 2019.

Khadra, Ahmed, Amine Ezzariai, Georges Merlina, Marion-Justine Capdeville, Hélène Budzinski, Hassan Hamdi, Eric Pinelli, and Mohamed Hafidi. "Fate of antibiotics present in a primary sludge of WWTP during their co-composting with palm wastes." *Waste Management* 84 (2019): 13–19.

Khan, Kiran Yasmin, Barkat Ali, Shuang Zhang, Peter Joseph Stoffella, Shan Yuan, Qian Xia, Hongjun Qu, Yu Shi, Xiaoqiang Cui, and Ya Guo. "Effects of antibiotics stress on growth variables, ultrastructure, and metabolite pattern of *Brassica rapa ssp. chinensis*." *Science of the Total Environment* 778 (2021): 146333.

Khan, Sardar, Qing Cao, Yuanming Zheng, Yizong Huang, and Yongguan Zhu. "Health risks of heavy metals in contaminated soils and food crops irrigated with wastewater in Beijing, China." *Environmental Pollution* 152, no. 3 (2008a): 686–692.

Khan, Shad Ali, Lajbar Khan, Iqbal Hussain, Khan Bahadar Marwat, and Naveed Akhtar. "Profile of heavy metals in selected medicinal plants." *Pakistan Journal of Weed Science Research* 14, no. 1–2 (2008b): 101–110.

Kharrazi, Seyede Maryam, Habibollah Younesi, and Javad Abedini-Torghabeh. "Microbial biodegradation of waste materials for nutrients enrichment and heavy metals removal: An integrated composting-vermicomposting process." *International Biodeterioration & Biodegradation* 92 (2014): 41–48.

Kim, Jin-Wook, Young-Kyu Hong, Jae-E. Yang, Oh-Kyung Kwon, and Sung-Chul Kim. "Bioaccumulation and mass balance analysis of veterinary antibiotics in an agricultural environment." *Toxics* 10, no. 5 (2022): 213.

Kim, Young Mo, Donghee Park, Dae Sung Lee, and Jong Moon Park. "Instability of biological nitrogen removal in a cokes wastewater treatment facility during summer." *Journal of Hazardous Materials* 141, no. 1 (2007): 27–32.

Kirchmann, Holger, and Ernst Witter. "Ammonia volatilization during aerobic and anaerobic manure decomposition." *Plant and Soil* 115 (1989): 35–41.

Knorr, Wolfgang, Iain Colin Prentice, Joanna I. House, and Elisabeth A. Holland. "Long-term sensitivity of soil carbon turnover to warming." *Nature* 433, no. 7023 (2005): 298–301.

Ko, Han J., Ki Y. Kim, Hyeon T. Kim, Chi N. Kim, and M. Umeda. "Evaluation of maturity parameters and heavy metal contents in composts made from animal manure." *Waste Management* 28, no. 5 (2008): 813–820.

Kong, Xiu-Xiu, Jian-Lan Jiang, Bin Qiao, Hong Liu, Jing-Sheng Cheng, and Ying-Jin Yuan. "The biodegradation of cefuroxime, cefotaxime and cefpirome by the synthetic consortium with probiotic *Bacillus clausii* and investigation of their potential biodegradation pathways." *Science of the Total Environment* 651 (2019): 271–280.

Košnář, Zdeněk, Lucie Wiesnerová, Tereza Částková, Stanislava Kroulíková, Jiří Bouček, Filip Mercl, and Pavel Tlustoš. "Bioremediation of polycyclic aromatic hydrocarbons (PAHs) present in biomass fly ash by co-composting and co-vermicomposting." *Journal of Hazardous Materials* 369 (2019): 79–86.

Kuhlman, Lawrence R. "Windrow composting of agricultural and municipal wastes." *Resources, Conservation and Recycling* 4, no. 1–2 (1990): 151–160.

Kulhari, Alpana, Arun Sheorayan, Somvir Bajar, Susheel Sarkar, Ashok Chaudhury, and Rajwant K. Kalia. "Investigation of heavy metals in frequently utilized medicinal plants collected from environmentally diverse locations of north western India." *SpringerPlus* 2, no. 1 (2013): 1–9.

Kulikova, Natalia A., and Irina V. Perminova. "Interactions between humic substances and microorganisms and their implications for nature-like bioremediation technologies." *Molecules* 26, no. 9 (2021): 2706.

Kulikowska, Dorota. "Kinetics of organic matter removal and humification progress during sewage sludge composting." *Waste Management* 49 (2016): 196–203.

Kumar, Arvind, Ashok Kumar, Vipin Kumar, Mukesh Kumar, B. Singh, and P. Cjouhan. "Recycling of harmful weeds through NADEP composting." *Vegetos* 25 (2012): 315–318.

Kumar, K., Satish C. Gupta, S. K. Baidoo, Y. Chander, and Carl J. Rosen. "Antibiotic uptake by plants from soil fertilized with animal manure." *Journal of Environmental Quality* 34, no. 6 (2005): 2082–2085.

Kummerová, Marie, and Lucie Vanova. "Chlorophyll fluorescence as an indicator of fluoranthene phototoxicity." *Plant Soil and Environment* 53, no. 10 (2007): 430.

Kuypers, Marcel M. M., Hannah K. Marchant, and Boran Kartal. "The microbial nitrogen-cycling network." *Nature Reviews Microbiology* 16, no. 5 (2018): 263–276.

Lakhanpal, Tej Nath. "Ectomycorrhiza: An overview." In *Mycorrhizal Biology*, edited by K. G. Mukerji, B. P. Chamola, and J. Singh, 101–118. Springer, 2000.

Li, Changsheng, Steve Frolking, and Klaus Butterbach-Bahl. "Carbon sequestration in arable soils is likely to increase nitrous oxide emissions, offsetting reductions in climate radiative forcing." *Climatic Change* 72, no. 3 (2005): 321–338.

Li, Huayu, Hong Yao, Dayi Zhang, Lushen Zuo, Jia Ren, Jinyuan Ma, Jin Pei, Yaru Xu, and Chengyong Yang. "Short-and long-term effects of manganese, zinc and copper ions on nitrogen removal in nitritation-anammox process." *Chemosphere* 193 (2018): 479–488.

Li, Yingbin, T. Martijn Bezemer, Junjie Yang, Xiaotao Lü, Xinyu Li, Wenju Liang, Xingguo Han, and Qi Li. "Changes in litter quality induced by N deposition alter soil microbial communities." *Soil Biology and Biochemistry* 130 (2019): 33–42.

Lima, Rogério Barbosa, Tiago Benedito dos Santos, Luiz Gonzaga Esteves Vieira, Maria de Lourdes Lúcio Ferrarese, Osvaldo Ferrarese-Filho, Lucélia Donatti, Maria Regina Torres Boeger, and Carmen Lúcia de Oliveira Petkowicz. "Heat stress causes alterations in the cell-wall polymers and anatomy of coffee leaves (Coffea arabica L.)." *Carbohydrate Polymers* 93, no. 1 (2013): 135–143.

Lipczynska-Kochany, Ewa. "Effect of climate change on humic substances and associated impacts on the quality of surface water and groundwater: A review." *Science of the Total Environment* 640 (2018): 1548–1565.

Liu, Binyang, Weiqiu Liu, Xiangping Nie, Chao Guan, Yufeng Yang, Zhaohui Wang, and Wei Liao. "Growth response and toxic effects of three antibiotics on Selenastrum capricornutum evaluated by photosynthetic rate and chlorophyll biosynthesis." *Journal of Environmental Sciences* 23, no. 9 (2011): 1558–1563.

Liu, Feng, Guang-Guo Ying, Ran Tao, Jian-Liang Zhao, Ji-Feng Yang, and Lan-Feng Zhao. "Effects of six selected antibiotics on plant growth and soil microbial and enzymatic activities." *Environmental Pollution* 157, no. 5 (2009): 1636–1642.

Liu, Lin, Jie Li, Hongyong Fan, Xu Huang, Lili Wei, and Chaoxiang Liu. "Fate of antibiotics from swine wastewater in constructed wetlands with different flow configurations." *International Biodeterioration & Biodegradation* 140 (2019a): 119–125.

Liu, Lin, Yu-hong Liu, Chao-xiang Liu, Zhen Wang, Jian Dong, Ge-fu Zhu, and Xu Huang. "Potential effect and accumulation of veterinary antibiotics in Phragmites australis under hydroponic conditions." *Ecological Engineering* 53 (2013): 138–143.

Liu, Ning, Tao Hou, Hongjie Yin, Lujia Han, and Guangqun Huang. "Effects of amoxicillin on nitrogen transformation and bacterial community succession during aerobic composting." *Journal of Hazardous Materials* 362 (2019b): 258–265.

Liu, Qingquan, Le Luo, and Luqing Zheng. "Lignins: Biosynthesis and biological functions in plants." *International Journal of Molecular Sciences* 19, no. 2 (2018): 335.

Lobell, David B., Wolfram Schlenker, and Justin Costa-Roberts. "Climate trends and global crop production since 1980." *Science* 333, no. 6042 (2011): 616–620.

Lorenzoni, Irene, Sophie Nicholson-Cole, and Lorraine Whitmarsh. "Barriers perceived to engaging with climate change among the UK public and their policy implications." *Global Environmental Change* 17, no. 3–4 (2007): 445–459.

Löser, Christian, Holm Ulbricht, Petra Hoffmann, and Heinz Seidel. "Composting of wood containing polycyclic aromatic hydrocarbons (PAHs)." *Compost Science & Utilization* 7, no. 3 (1999): 16–32.

Lucintel. (2020). *Compost Market Report: Trends, Forecast and Competitive Analysis*. Value Market Report (ID: 5817475). Lucintel, 2020.

Lu, Yintao, Hong Yao, Dan Shan, Yichen Jiang, Shichao Zhang, and Jun Yang. "Heavy Metal Residues in Soil and Accumulation in Maize at Long-Term Wastewater Irrigation Area in Tongliao, China." *Journal of Chemistry* 2015, no. 1 (2015): 628280.

Lycus, Pawel, Kari Lovise Bøthun, Linda Bergaust, James Peele Shapleigh, Lars Reier Bakken, and Åsa Frostegård. "Phenotypic and genotypic richness of denitrifiers revealed by a novel isolation strategy." *ISME Journal* 11, no. 10 (2017): 2219–2232.

Maeda, Koki, Dai Hanajima, Riki Morioka, and Takashi Osada. "Characterization and spatial distribution of bacterial communities within passively aerated cattle manure composting piles." *Bioresource Technology* 101, no. 24 (2010): 9631–9637.

Maier, Raina M. "Biogeochemical cycling." In *Environmental Microbiology*, edited by Raina M. Maier, Ian L. Pepper, and Charles P. Gerba, 339–373. Academic Press, 2015.

Maila, Mphekgo P., and Thomas E. Cloete. "Germination of Lepidium sativum as a method to evaluate polycyclic aromatic hydrocarbons (PAHs) removal from contaminated soil." *International Biodeterioration & Biodegradation* 50, no. 2 (2002): 107–113.

Manwani, Seema, Narain Bhoot, Harsh Pandey, and Garima Awasthi. "Mitigation of lead and Cadmium from vegetable crops through different biochemical adsorbent treatments at Sanganer industrial area, Jaipur." *Materials Today: Proceedings* 62 (2022): 6172–6177.

Mao, Chuanzao, Keke Yi, Ling Yang, Bingsong Zheng, Yunrong Wu, Feiyan Liu, and Ping Wu. "Identification of aluminium-regulated genes by cDNA-AFLP in rice (*Oryza sativa L.*): Aluminium-regulated genes for the metabolism of cell wall components." *Journal of Experimental Botany* 55, no. 394 (2004): 137–143.

Mattei, Paola, Alessandra Cincinelli, Tania Martellini, Riccardo Natalini, Ersilia Pascale, and Giancarlo Renella. "Reclamation of river dredged sediments polluted by PAHs by co-composting with green waste." *Science of the Total Environment* 566 (2016): 567–574.

Mawar, Ritu, and Tanu Mathur. "Enhancing survival and multiplication of *Trichoderma harzianum* by *Prosopis juliflora* based compost in Indian arid region." *Indian Phytopathology* 75 (2022): 807–815.

Meng, Liqiang, Weiguang Li, Shumei Zhang, Chuandong Wu, and Ke Wang. "Effects of sucrose amendment on ammonia assimilation during sewage sludge composting." *Bioresource Technology* 210 (2016): 160–166.

Michel, Frederick, Tim O'Neill, Robert Rynk, Jane Gilbert, Steven Wisbaum, and Thomas Halbach. "Passively aerated composting methods, including turned windrows." In *The Composting Handbook*, edited by Robert Rynk, 159–196. Academic Press, 2022.

Michelini, Luciana R. R., Ruediger Reichel, Wilfried Werner, Rossella Ghisi, and Sören Thiele-Bruhn. "Sulfadiazine uptake and effects on *Salix fragilis L.* and *Zea mays L.* plants." *Water, Air, & Soil Pollution* 223, no. 8 (2012): 5243–5257.

Migliore, Luciana, Alice Rotini, Nadia L. Cerioli, Salvatore Cozzolino, and Maurizio Fiori. "Phytotoxic antibiotic sulfadimethoxine elicits a complex hormetic response in the weed lythrum salicaria L." *Dose-Response* 8, no. 4 (2010): 414–427.

Minden, Vanessa, Andrea Deloy, Anna Martina Volkert, Sara Diana Leonhardt, and Gesine Pufal. "Antibiotics impact plant traits, even at small concentrations." *AoB Plants* 9, no. 2 (2017).

Minick, Kevyn J., Chandra B. Pandey, Thomas R. Fox, and Santosh Subedi. "Dissimilatory nitrate reduction to ammonium and $N_2O$ flux: Effect of soil redox potential and N fertilization in loblolly pine forests." *Biology and Fertility of Soils* 52, no. 5 (2016): 601–614.

Montealegre, Constanza M., Chris Van Kessel, Michael P. Russelle, and Michael J. Sadowsky. "Changes in microbial activity and composition in a pasture ecosystem exposed to elevated atmospheric carbon dioxide." *Plant and Soil* 243, no. 2 (2002): 197–207.

Moura, Jullyana Cristina Magalhães Silva, Cesar Augusto Valencise Bonine, Juliana de Oliveira Fernandes Viana, Marcelo Carnier Dornelas, and Paulo Mazzafera. "Abiotic and biotic stresses and changes in the lignin content and composition in plants." *Journal of Integrative Plant Biology* 52, no. 4 (2010): 360–376.

Narendra, Savitha, Sujatha Venkataramani, Guoxin Shen, Jing Wang, Vijaya Pasapula, Yun Lin, Dmytro Kornyeyev, A. Scott Holaday, and Hong Zhang. "The arabidopsis ascorbate peroxidase 3 is a peroxisomal membrane-bound antioxidant enzyme and is dispensable for arabidopsis growth and development." *Journal of Experimental Botany* 57, no. 12 (2006): 3033–3042.

National Aeronautics and Space Administration. Climate change. NASA Science. (2026, June 3)https://science.nasa.gov/climate-change/

Neves, Geraldo Y. S., Rogério Marchiosi, Maria de Lourdes L. Ferrarese, Rita de Cássia Siqueira-Soares, and Oswaldo Ferrarese-Filho. "Root growth inhibition and lignification induced by salt stress in soybean." *Journal of Agronomy and Crop Science* 196, no. 6 (2010): 467–473.

Niklińska, Maria, Maciej Maryański, and Ryszard Laskowski. "Effect of temperature on humus respiration rate and nitrogen mineralization: Implications for global climate change." *Biogeochemistry* 44, no. 3 (1999): 239–257.

Oguntimehin, Ilemobayo, Fawzy Eissa, and Hiroshi Sakugawa. "Negative effects of fluoranthene on the ecophysiology of tomato plants (Lycopersicon esculentum Mill): Fluoranthene mists negatively affected tomato plants." *Chemosphere* 78, no. 7 (2010): 877–884.

Oksanen, Elina, Johanna Riikonen, Seija Kaakinen, Toini Holopainen, and Elina Vapaavuori. "Structural characteristics and chemical composition of birch (Betula pendula) leaves are modified by increasing $CO_2$ and ozone." *Global Change Biology* 11, no. 5 (2005): 732–748.

Olivares, José, Eulogio J. Bedmar, and Juan Sanjuán. "Biological nitrogen fixation in the context of global change." *Molecular Plant-Microbe Interactions* 26, no. 5 (2013): 486–494.

Pal, Chandan, Johan Bengtsson-Palme, Erik Kristiansson, and Daniel G. Larsson. "Co-occurrence of resistance genes to antibiotics, biocides and metals reveals novel insights into their co-selection potential." *BMC Genomics* 16, no. 1 (2015): 1–14.

Pan, Xiangliang, Chunnuan Deng, Daoyong Zhang, Jianlong Wang, Guijin Mu, and Ying Chen. "Toxic effects of amoxicillin on the photosystem II of *Synechocystis sp.* characterized by a variety of in vivo chlorophyll fluorescence tests." *Aquatic Toxicology* 89, no. 4 (2008): 207–213.

Panda, Sanjib Kumar, František Baluška, and Hideaki Matsumoto. "Aluminum stress signaling in plants." *Plant Signaling & Behavior* 4, no. 7 (2009): 592–597.

Pandey, Bhawna, Surindra Suthar, and Vineet Singh. "Accumulation and health risk of heavy metals in sugarcane irrigated with industrial effluent in some rural areas of Uttarakhand, India." *Process Safety and Environmental Protection* 102 (2016): 655–666.

Pandey, Chandra B., Ramesh Chandra Srivastava, and Ravi Kumar Singh. "Soil nitrogen mineralization and microbial biomass relation, and nitrogen conservation in humid-tropics." *Soil Science Society of America Journal* 73, no. 4 (2009): 1142–1149.

Paradelo, Remigio, Ana Belén Moldes, and María Teresa Barral. "Evolution of organic matter during the mesophilic composting of lignocellulosic winery wastes." *Journal of Environmental Management* 116 (2013): 18–26.

Peralta-Videa, Jose R., Jorge L. Gardea-Torresdey, Elizabeth Gomez, Kirk J. Tiemann, Jason G. Parsons, and Gustavo Carrillo. "Effect of mixed cadmium, copper, nickel and zinc at different pHs upon alfalfa growth and heavy metal uptake." *Environmental Pollution* 119, no. 3 (2002): 291–301.

Pugnaire, Francisco I., José A. Morillo, Josep Peñuelas, Peter B. Reich, Richard D. Bardgett, Aurora Gaxiola, David A. Wardle, and Wim H. Van Der Putten. "Climate change effects on plant-soil feedbacks and consequences for biodiversity and functioning of terrestrial ecosystems." *Science Advances* 5, no. 11 (2019): eaaz1834.

Purwantisari, Susiana, Achmadi Priyatmojo, Retno Peni Sancayaningsih, Rina Sri Kasiamdari, and Kadarwati Budihardjo. "Lignification on potatoes by application of trichoderma viride." *IOP Conference Series: Earth and Environmental Science* 518, no. 1 (2020): 012075.

Rahman, M. Mizanur, Azmerry Khanom, and Shudhangshu Kumar Biswas. "Effect of pesticides and chemical fertilizers on the nitrogen cycle and functional microbial communities in paddy soils: Bangladesh perspective." *Bulletin of Environmental Contamination and Toxicology* 106, no. 2 (2021): 243–249.

Ramaswamy, Jayashree, Shiv O. Prasher, Ramanbhai M. Patel, Syed A. Hussain, and Suzelle F. Barrington. "The effect of composting on the degradation of a veterinary pharmaceutical." *Bioresource Technology* 101, no. 7 (2010): 2294–2299.

Rao, Mruthyumjaya Meda, and Ajay KumarMeena. "Detection of toxic heavy metals and pesticide residue in herbal plants which are commonly used in the herbal formulations." *Environmental Monitoring and Assessment* 181, no. 1 (2011): 267–271.

Ray, Partha, Chaoqi Chen, Katharine F. Knowlton, Amy Pruden, and Kang Xia. "Fate and effect of antibiotics in beef and dairy manure during static and turned composting." *Journal of Environmental Quality* 46, no. 1 (2017): 45–54.

Reay, Dave S., Eric A. Davidson, Keith A. Smith, Pete Smith, Jerry M. Melillo, Frank Dentener, and Paul J. Crutzen. "Global agriculture and nitrous oxide emissions." *Nature Climate Change* 2, no. 6 (2012): 410–416.

Reddy, K. Ramesh, William H. Patrick, and Francis E. Broadbent. "Nitrogen transformations and loss in flooded soils and sediments." *Critical Reviews in Environmental Science and Technology* 13, no. 4 (1984): 273–309.

Reed, Sasha C., Cory C. Cleveland, and Alan R. Townsend. "Functional ecology of free-living nitrogen fixation: A contemporary perspective." *Annual Review of Ecology, Evolution and Systematics* 42, no. 1 (2011): 489–512.

Richards, Francis A. "Anoxic basins and fjords." *Chemical Oceanography* 1 (1965): 611–643.

Ritson, Jonathan P., Richard E. Brazier, Nigel J. D. Graham, Chris Freeman, Michael R. Templeton, and Joanna M. Clark. "The effect of drought on dissolved organic carbon (DOC) release from peatland soil and vegetation sources." *Biogeosciences* 14, no. 11 (2017): 2891–2902.

Robertson, Elizabeth J., and Rachel M. Leech. "Significant changes in cell and chloroplast development in young wheat leaves (Triticum aestivum cv Hereward) grown in elevated $CO_2$." *Plant Physiology* 107, no. 1 (1995): 63–71.

Rogers, Hugo H., Stephen A. Prior, and Elizabeth G. O'Neill. "Cotton root and rhizosphere responses to free-air $CO_2$ enrichment." *Critical Reviews in Plant Sciences* 11, no. 2–3 (1992): 251–263.

Roig, Asunción, María Luz Cayuela, and M. A. Sánchez-Monedero. "An overview on olive mill wastes and their valorisation methods." *Waste Management* 26, no. 9 (2006): 960–969.

Rolz, Carlos, Ricardo De Leon, R. Cifuentes, and C. Porres. "Windrow composting of sugarcane and coffee byproducts." *Sugar Tech* 12, no. 1 (2010): 15–20.

Ruiz-Vera, Ursula M., Matthew Siebers, Sharon B. Gray, David W. Drag, David M. Rosenthal, Bruce A. Kimball, Donald R. Ort, and Carl J. Bernacchi. "Global warming can negate the expected $CO_2$ stimulation in photosynthesis and productivity for soybean grown in the Midwestern United States." *Plant Physiology* 162, no. 1 (2013): 410–423.

Runion, G. Brett, Elroy A. Curl, Hugo H. Rogers, Paul A. Backman, Rodrigo Rodriguez-Kabana, and Bobby E. Helms. "Effects of free-air $CO_2$ enrichment on microbial populations in the rhizosphere and phyllosphere of cotton." *Agricultural and Forest Meteorology* 70, no. 1–4 (1994): 117–130.

Ruser, Reiner, Holger Flessa, Roland Russow, G. Schmidt, Friedhelm Buegger, and Jean Charles Munch. "Emission of $N_2O$, $N_2$ and $CO_2$ from soil fertilized with nitrate: Effect of compaction, soil moisture and rewetting." *Soil Biology and Biochemistry* 38, no. 2 (2006): 263–274.

Saha, Koel, Poulami Dwibedi, Ankita Ghosh, Jaya Sikder, Sudip Chakraborty, and Stefano Curcio. "Extraction of lignin, structural characterization and bioconversion of sugarcane bagasse after ionic liquid assisted pretreatment." *3 Biotech* 8, no. 8 (2018): 1–12.

Sánchez-Monedero, Miguel A., Asunción Roig, Juan Cegarra, and Maria Pilar Bernal. "Relationships between water-soluble carbohydrate and phenol fractions and the humification indices of different organic wastes during composting." *Bioresource Technology* 70, no. 2 (1999): 193–201.

Sánchez-Monedero, Miguel A., Asunción Roig, Concepción Paredes, and Maria Pilar Bernal. "Nitrogen transformation during organic waste composting by the Rutgers system and its effects on pH, EC and maturity of the composting mixtures." *Bioresource Technology* 78, no. 3 (2001): 301–308.

Sasaki, Masao, Yoko Yamamoto, and Hideaki Matsumoto. "Lignin deposition induced by aluminum in wheat (*Triticum aestivum*) roots." *Physiologia Plantarum* 96, no. 2 (1996): 193–198.

Saxe, Henrik. "Physiological and biochemical tools in diagnosis of forest decline and air pollution injury to plants." In *Plant Response to Air Pollution*, edited by Mohamed Yunus and Mohammad Iqbal, 449–487. John Wiley & Sons, 1996.

Schortemeyer, Marcus, Paul Dijkstra, Dale W. Johnson, and Bert G. Drake. "Effects of elevated atmospheric $CO_2$ concentration on C and N pools and rhizosphere processes in a Florida scrub oak community." *Global Change Biology* 6, no. 4 (2000): 383–391.

Schortemeyer, Marcus, Ueli A. Hartwig, George R. Hendrey, and Michael J. Sadowsky. "Microbial community changes in the rhizospheres of white clover and perennial ryegrass exposed to free air carbon dioxide enrichment (FACE)." *Soil Biology and Biochemistry* 28, no. 12 (1996): 1717–1724.

Schutzendubel, Andres, and Andrea Polle. "Plant responses to abiotic stresses: Heavy metal-induced oxidative stress and protection by mycorrhization." *Journal of Experimental Botany* 53, no. 372 (2002): 1351–1365.

Scott-Craig, John S., Katherine B. Kerby, Barbara D. Stein, and Shauna C. Somerville. "Expression of an extracellular peroxidase that is induced in barley (Hordeum vulgare) by the powdery mildew pathogen (Erysiphe graminis f. sp. hordei)." *Physiological and Molecular Plant Pathology* 47, no. 6 (1995): 407–418.

Segura, Marisa S. I., Olavo A. da Silva, Mitsue Nikaido, Tânia Maria B. Trevilato, Antoni Bocio, Angela Maria M. Takayanagui, and José L. Domingo. "Metals levels in sugar cane (*Saccharum* spp.) samples from an area under the influence of a municipal landfill and a medical waste treatment system in Brasil." *Enviroment International* 32, no. 1 (2006): 52–57.

Seiler, Cornelia, and Thomas U. Berendonk. "Heavy metal driven co-selection of antibiotic resistance in soil and water bodies impacted by agriculture and aquaculture." *Frontiers in Microbiology* 3 (2012): 399.

Sellami, Faten, Rim Jarboui, Sami Hachicha, Khaled Medhioub, and Emna Ammar. "Co-composting of oil exhausted olive-cake, poultry manure and industrial residues of agro-food activity for soil amendment." *Bioresource Technology* 99, no. 5 (2008): 1177–1188.

Selvam, Ammaiyappan, Delin Xu, Zhenyong Zhao, and Jonathan WC Wong. "Fate of tetracycline, sulfonamide and fluoroquinolone resistance genes and the changes in bacterial diversity during composting of swine manure." *Bioresource Technology* 126 (2012): 383–390.

Sierra, M. J., R. Millán, A. I. Cardona, and T. Schmid. "Potential cultivation of Hordeum vulgare L. in soils with high mercury background concentrations." *International Journal of Phytoremediation* 13, no. 8 (2011): 765–778.

Shah, Zahir, Yaakob Mohd Jani, and Farmanullah Khan. "Evaluation of organic wastes for composting." *Communications in Soil Science and Plant Analysis* 45, no. 3 (2014): 309–320.

Shak, Katrina Pui Yee, Ta Yeong Wu, Su Lin Lim, and Chieh Ai Lee. "Sustainable reuse of rice residues as feedstocks in vermicomposting for organic fertilizer production." *Environmental Science and Pollution Research* 21, no. 2 (2014): 1349–1359.

Shan, Jun, Pinpin Yang, M. Mizanur Rahman, Xiaoxia Shang, and Xiaoyuan Yan. "Tetracycline and sulfamethazine alter dissimilatory nitrate reduction processes and increase $N_2O$ release in rice fields." *Environmental Pollution* 242 (2018): 788–796.

Shan, Yi N., Jin H. Chen, Lei Wang, Fan Li, Xiao H. Fu, and Yi Q. Le. "Influences of adding easily degradable organic waste on the minimization and humification of organic matter during straw composting." *Journal of Environmental Science and Health, Part B* 48, no. 5 (2013): 384–392.

Sharma, Kanika, and Vinod Kumar Garg. Recycling of lignocellulosic waste as vermicompost using earthworm Eisenia fetida. *Environmental Science and Pollution Research*, 26, no. 14 (2019): 14024–14035.

Shi, Mingzi, Yue Zhao, Longji Zhu, Xinyu Song, Yu Tang, Haishi Qi, Hongjie Cao, and Zimin Wei. "Denitrification during composting: Biochemistry, implication and perspective." *International Biodeterioration & Biodegradation* 153 (2020): 105043.

Shi, Mingzi, Zimin Wei, Liqin Wang, Junqiu Wu, Duoying Zhang, Dan Wei, Yu Tang, and Yue Zhao. "Response of humic acid formation to elevated nitrate during chicken manure composting." *Bioresource Technology* 258 (2018): 390–394.

Shrivas, Vijay Laxmi, Anil Kumar Choudhary, Puttaswamy Hariprasad, and Shilpi Sharma. "Nutrient concentrations affect the antimicrobial resistance profiles of cattle manures." *Environmental Science and Pollution Research* 29 (2021): 1–7.

Silver, Whendee L., Donald J. Herman, and Mary K. Firestone. "Dissimilatory nitrate reduction to ammonium in upland tropical forest soils." *Ecology* 82, no. 9 (2001): 2410–2416.

Smidt, Ena, Johannes Tintner, Katharina Meissl, and Erwin Binner. "Influence of input materials and composting operation on humification of organic matter." In *Compost I: Dynamic Soil, Dynamic Plant*, edited by Michael Insam, Ika Franke-Whittle, and Manfred Goberna, 209–224. Springer, 2008.

Smith, Joel B. "Climate Change Risks to the United States and Its Security Interests." *Our Warming Planet: Climate Change Impacts and Adaptation* 2 (2021): 459.

Snowden, Kimberley C., and Richard C. Gardner. "Five genes induced by aluminum in wheat (*Triticum aestivum L.*) roots." *Plant Physiology* 103, no. 3 (1993): 855–861.Sriprachote, Anongnat, Pornthiwa Kanyawongha, Kumiko Ochiai, and Toru Matoh. "Current situation of cadmium-polluted paddy soil, rice and soybean in the Mae Sot District, Tak Province, Thailand." *Soil Science and Plant Nutrition* 58, no. 3 (2012): 349–359.

Stegenta-Dąbrowska, Sylwia, Karolina Sobieraj, Joanna Rosik, Robert Sidełko, Marvin Valentine, and Andrzej Białowiec. "The development of anammox and chloroflexi bacteria during composting of sewage sludge." *Science of the Total Environment* 827 (2022): 154308.

Stevenson, Frank J. *Humus Chemistry: Genesis, Composition, Reactions*. John Wiley & Sons, 1994.

Stocker, Thomas F., Dahe Qin, Gian-Kasper Plattner, Melinda M. B. Tignor, Simon K. Allen, Judith Boschung, Alexander Nauels, Yu Xia, Vincent Bex, and Pauline M. Midgley, eds. *Climate Change 2013: The Physical Science Basis. Contribution of Working Group I to the Fifth Assessment Report of the Intergovernmental Panel on Climate Change*. Cambridge University Press, 2014.

Sulaiman, Fazrul Razman, and Huda Asilah Hamzah. "Heavy metals accumulation in suburban roadside plants of a tropical area (Jengka, Malaysia)." *Ecological Processes* 7, no. 1 (2018): 1–11.

Sun, Yu, Liping Zhu, Xiuhong Xu, Qingxin Meng, Mengqi Men, Benshu Xu, and Liting Deng. "Correlation between ammonia-oxidizing microorganisms and environmental factors during cattle manure composting." *Revista Argentina de Microbiología* 51, no. 4 (2019): 371–380.

Sweeten, John M., and Brent W. Auvermann, *Brent Composting Manure and Sludge*, 1–7. Texas AgriLife Extension Service, 2006.

Swift, Michael John, Oliver W. Heal, Jonathan Michael Anderson, and J. M. Anderson. *Decomposition in Terrestrial Ecosystems*. Vol. 5. Univ of California Press, 1979.

Tahara, Ko, Mariko Norisada, Taizo Hogetsu, and Katsumi Kojima. "Aluminum tolerance and aluminum-induced deposition of callose and lignin in the root tips of Melaleuca and Eucalyptus species." *Journal of Forest Research* 10, no. 4 (2005): 325–333.

Tanikawa, Daisuke, Yuki Nakamura, Hiroaki Tokuzawa, Yuka Hirakata, Masashi Hatamoto, and Takashi Yamaguchi. "Effluent treatment in an aquaponics-based closed aquaculture system with single-stage nitrification–denitrification using a down-flow hanging sponge reactor." *International Biodeterioration & Biodegradation* 132 (2018): 268–273.

Tasho, Reep Pandi, and Jae Yong Cho. "Veterinary antibiotics in animal waste, its distribution in soil and uptake by plants: A review." *Science of the Total Environment* 563 (2016): 366–376.

Tchounwou, Paul B., Clement G. Yedjou, Anita K. Patlolla, and Dwayne J. Sutton. "Heavy metal toxicity and the environment." In *Molecular, Clinical and Environmental Toxicology*, edited by Andreas Luch, 133–164. Springer, 2012.

Tiedje, James M. "Ecology of denitrification and dissimilatory nitrate reduction to ammonium." In *Biology of Anaerobic Microorganisms*, edited by Alexander J. B. Zehnder, 179–244. John Wiley & Sons, 1988.

Tiquia, Sonia M., and Nora F. Y. Tam. "Fate of nitrogen during composting of chicken litter." *Environmental Pollution* 110, no. 3 (2000): 535–541.

Tiquia, Sonia M., Tom L. Richard, and Mark S. Honeyman. "Carbon, nutrient, and mass loss during composting." *Nutrient Cycling in Agroecosystems* 62 (2002): 15–24.

Tiwari, Garima, P. Duraivadivel, and Satyawati Sharm. "1-Aminocyclopropane-1-carboxylic acid deaminase producing beneficial rhizobacteria ameliorate the biomass characters of Panicum maximum Jacq. by mitigating drought and salt stress." *Scientific Reports* 8, no. 1 (2018): 1–12.

Tomaszewski, Mariusz, Grzegorz Cema, and Aleksandra Ziembińska-Buczyńska. "Influence of temperature and pH on the anammox process: A review and meta-analysis." *Chemosphere* 182 (2017): 203–214.

Tomati, Umberto, Elisabetta Galli, L. Pasetti, and Enrica Volterra. "Bioremediation of olive-mill wastewaters by composting." *Waste Management & Research* 13, no. 5 (1995): 509–518.

Tsui, Lo, and William R. Roy. "Effect of compost age and composition on the atrazine removal from solution." *Journal of Hazardous Materials* 139, no. 1 (2007): 79–85.

Van de Graaf, Astrid A., Arnold Mulder, Peter de Bruijn, Mike S. M. Jetten, Lesley A. Robertson, and J. Gijs Kuenen. "Anaerobic oxidation of ammonium is a biologically mediated process." *Applied and Environmental Microbiology* 61, no. 4 (1995): 1246–1251.

Van Kessel, Maartje A. H. J., Daan R. Speth, Mads Albertsen, Per H. Nielsen, Huub J. M. Op den Camp, Boran Kartal, Mike S. M. Jetten, and Sebastian Lücker. 2015. "Complete nitrification by a single microorganism." *Nature* 528(7583), pp. 555–559.

Verma, Rajhans, W. P. Badole, Parvati Deewan, and Vijay Singh Meena. "Carbon and weight loss during composting of wheat straw by different methods." *Annals of Biology* 30, no. 2 (2014): 354–357.

Villar, Iria, David Alves, Josefina Garrido, and Salustiano Mato. "Evolution of microbial dynamics during the maturation phase of the composting of different types of waste." *Waste Management* 54 (2016): 83–92.

Wagner, George H., and Duane C. Wolf. "Carbon transformations and soil organic matter formation." *Principles and Applications of Soil Microbiology*, edited by David M. Sylvia, Jeffry J. Fuhrmann, Peter G. Hartel, and David A. Zuberer, 218–258. Prentice Hall, 1999.

Waksman, Selman A., Theone C. Cordon, and Neculai J. Hulpoi. "Influence of temperature upon the microbiological population and decomposition processes in composts of stable manure." *Soil Science* 47, no. 2 (1939): 83–114.

Wang, Hang, Naseer Sangwan, Hong-Yi Li, Jian-Qiang Su, Wei-Yin Oyang, Zhi-Jian Zhang, Jack A. Gilbert, Yong-Guan Zhu, Fan Ping, and Han-Luo Zhang. "The antibiotic resistome of swine manure is significantly altered by association with the *Musca domestica* larvae gut microbiome." *ISME Journal* 11, no. 1 (2017): 100–111.

Wang, Ke, Weiguang Li, Xiangkun Li, and Nanqi Ren. "Spatial nitrifications of microbial processes during composting of swine, cow and chicken manure." *Scientific Reports* 5, no. 1 (2015): 1–8.

Wang, Weiwu, Yi Hou, Wenfeng Huang, Xiaoming Liu, Ping Wen, Yueqiang Wang, Zhen Yu, and Shungui Zhou. "Alkali lignin and sodium lignosulfonate additives promote the formation of humic substances during paper mill sludge composting." *Bioresource Technology* 320 (2021): 124361.

Wang, Xueqin, Hongyang Cui, Jianhong Shi, Xinyu Zhao, Yue Zhao, and Zimin Wei. "Relationship between bacterial diversity and environmental parameters during composting of different raw materials." *Bioresource Technology* 198 (2015): 395–402.

Wang, Yongcui, Lei Chu, Jian Ma, Guangyu Chi, Caiyan Lu, and Xin Chen. "Effects of multiple antibiotics residues in broiler manure on composting process." *Science of the Total Environment* 817 (2022): 152808.

Wei, Haiying, Shanjuan Song, Hongling Tian, and Ting Liu. "Effects of phenanthrene on seed germination and some physiological activities of wheat seedling." *Comptes Rendus Biologies* 337, no. 2 (2014): 95–100.

Westerman, Philip W., and José R. Bicudo. "Management considerations for organic waste use in agriculture." *Bioresource Technology* 96, no. 2 (2005): 215–221.

Williams, Mark A., Charles W. Rice, and Clenton E. Owensby. "Carbon dynamics and microbial activity in tallgrass prairie exposed to elevated $CO_2$ for 8 years." *Plant and Soil* 227, no. 1 (2000): 127–137.

Wong, Jonathan W.-C., Shun On Fung, and Ammaiyappan Selvam. "Coal fly ash and lime addition enhances the rate and efficiency of decomposition of food waste during composting." *Bioresource Technology* 100, no. 13 (2009): 3324–3331.

Wong, Jonathan W. C., Xia Wang, and Ammaiyappan Selvam. "Improving compost quality by controlling nitrogen loss during composting." In *Current Developments in Biotechnology and Bioengineering*, edited by Ashok Pandey, 59–82. Elsevier, 2017.

Woods, Donald Devereux. "The reduction of nitrate to ammonia by Clostridium welchii." *Biochemical Journal* 32, no. 11 (1938): 2000.

Wu, Junqiu, Yue Zhao, Wei Zhao, Tianxue Yang, Xu Zhang, Xinyu Xie, Hongyang Cui, and Zimin Wei. "Effect of precursors combined with bacteria communities on the formation of humic substances during different materials composting." *Bioresource Technology* 226 (2017): 191–199.

Wu, Leilei, and Lena Qiying Ma. "Relationship between compost stability and extractable organic carbon." *Journal of Environmental Quality* 31, no. 4 (2002): 1323–1328.

Xie, Wan-Ying, Xin-Ping Yang, Qian Li, Long-Hua Wu, Qi-Rong Shen, and Fang-Jie Zhao. "Changes in antibiotic concentrations and antibiotic resistome during commercial composting of animal manures." *Environmental Pollution* 219 (2016): 182–190.

Yan, Qing, Yuan Xu, Long Chen, Zhaoyun Cao, Yafang Shao, Yufeng Xu, Yonghong Yu et al. "Irrigation with secondary municipal-treated wastewater: Potential effects, accumulation of typical antibiotics and grain quality responses in rice (Oryza sativa L.)." *Journal of Hazardous Materials* 410 (2021): 124655.

Yang, Fan, Yun Li, Yuhua Han, Wentao Qian, Guoxue Li, and Wenhai Luo. "Performance of mature compost to control gaseous emissions in kitchen waste composting." *Science of the Total Environment* 657 (2019): 262–269.

Yang, Huizhen, He Zhang, Huizhen Qiu, Dominic Kwadwo Anning, Mengchan Li, Youling Wang, and Chunhong Zhang. "Effects of C/N ratio on lignocellulose degradation and enzyme activities in aerobic composting." *Horticulturae* 7, no. 11 (2021): 482.

Yang, Xinping, Qian Li, Zhu Tang, Wenwen Zhang, Guanghui Yu, Qirong Shen, and Fang-Jie Zhao. "Heavy metal concentrations and arsenic speciation in animal manure composts in China." *Waste Management* 64 (2017): 333–339.

Yin, Yanan, Wen Song, Jie Gu, Kaiyu Zhang, Xun Qian, Xin Zhang, Yajun Zhang, Yang Li, and Xiaojuan Wang. "Effects of copper on the abundance and diversity of ammonia oxidizers during dairy cattle manure composting." *Bioresource Technology* 221 (2016): 181–187.

Yu, Yao, Yanan Wan, Aboubacar Younoussa Camara, and Huafen Li. "Effects of the addition and aging of humic acid-based amendments on the solubility of Cd in soil solution and its accumulation in rice." *Chemosphere* 196 (2018): 303–310.

Zagoskina, Natalia V., Elena A. Goncharuk, and Anna K. Alyavina. "Effect of cadmium on the phenolic compounds formation in the callus cultures derived from various organs of the tea plant." *Russian Journal of Plant Physiology* 54, no. 2 (2007): 237–243.

Zekker, Ivar, Gourav Dhar Bhowmick, Hans Priks, Dibyojyoty Nath, Ergo Rikmann, Madis Jaagura, Taavo Tenno, Kaido Tämm, and Makarand Madhao Ghangrekar. "ANAMMOX-denitrification biomass in microbial fuel cell to enhance the electricity generation and nitrogen removal efficiency." *Biodegradation* 31, no. 4 (2020): 249–264.

Zeng, Guangming, Jiachao Zhang, Yaoning Chen, Zhen Yu, Man Yu, Hui Li, Zhifeng Liu, Ming Chen, Lunhui Lu, and Chunxiao Hu. "Relative contributions of archaea and bacteria to microbial ammonia oxidation differ under different conditions during agricultural waste composting." *Bioresource Technology* 102, no. 19 (2011): 9026–9032.

Zeng, Yang, Amaury De Guardia, Christine Ziebal, Flávia Junqueira De Macedo, and Patrick Dabert. "Nitrification and microbiological evolution during aerobic treatment of municipal solid wastes." *Bioresource Technology* 110 (2012): 144–152.

Zhang, Bo, Meng Wang, Bing Wang, Yanjun Xin, Jiaqi Gao, and Huiling Liu. "The effects of bio-available copper on macrolide antibiotic resistance genes and mobile elements during tylosin fermentation dregs co-composting." *Bioresource Technology* 251 (2018): 230–237.

Zhang, Jin, Hui Lin, Junwei Ma, Wanchun Sun, Yuyi Yang, and Xin Zhang. "Compost-bulking agents reduce the reservoir of antibiotics and antibiotic resistance genes in manures by modifying bacterial microbiota." *Science of the Total Environment* 649 (2019): 396–404.

Zhang, Junya, Meixue Chen, Qianwen Sui, Rui Wang, Juan Tong, and Yuansong Wei. "Fate of antibiotic resistance genes and its drivers during anaerobic co-digestion of food waste and sewage sludge based on microwave pretreatment." *Bioresource Technology* 217 (2016): 28–36.

Zhang, Ming-Kui, Zhao-Yun Liu, and Huo Wang. "Use of single extraction methods to predict bioavailability of heavy metals in polluted soils to rice." *Communications in Soil Science and Plant Analysis* 41, no. 7 (2010): 820–831.

Zhang, Ranran, Xiaojuan Wang, Jie Gu, and Yajun Zhang. "Influence of zinc on biogas production and antibiotic resistance gene profiles during anaerobic digestion of swine manure." *Bioresource Technology* 244 (2017): 63–70.

Zhang, Xuezhen, Xiaxiang Li, Deliang Chen, Huijuan Cui, and Quansheng Ge. "Overestimated climate warming and climate variability due to spatially homogeneous $CO_2$ in climate modeling over the Northern Hemisphere since the mid-19th century." *Scientific Reports* 9, no. 1 (2019): 1–9.

Zhang, Yun, Yue Zhao, Yanni Chen, Qian Lu, Mingxiao Li, Xueqin Wang, Yuquan Wei, Xinyu Xie, and Zimin Wei. "A regulating method for reducing nitrogen loss based on enriched ammonia-oxidizing bacteria during composting." *Bioresource Technology* 221 (2016): 276–283.

Zhang, Zhechao, Yue Zhao, Tianxue Yang, Zimin Wei, Yingjun Li, Yuquan Wei, Xiaomeng Chen, and Liqin Wang. "Effects of exogenous protein-like precursors on humification process during lignocellulose-like biomass composting: Amino acids as the key linker to promote humification process." *Bioresource Technology* 291 (2019): 121882.

Zhang, Zhengzhe, Yu Zhang, and Yinguang Chen. "Recent advances in partial denitrification in biological nitrogen removal: From enrichment to application." *Bioresource Technology* 298 (2020): 122444.

Zhao, Bing, Yuyun Wang, Huarong Sun, and Zhi Xu. "Analysis of humus formation and factors for driving the humification process during composting of different agricultural wastes." *Frontiers in Environmental Science* 23 (2022a): 1364.

Zhao, Fangkai, Lei Yang, Liding Chen, Shoujuan Li, and Long Sun. "Bioaccumulation of antibiotics in crops under long-term manure application: Occurrence, biomass response and human exposure." *Chemosphere* 219 (2019): 882–895.

Zhao, Shuaixiang, Susanne Schmidt, Hongjian Gao, Tingyu Li, Xinping Chen, Yong Hou, Dave Chadwick et al. "A precision compost strategy aligning composts and application methods with target crops and growth environments can increase global food production." *Nature Food* 3, no. 9 (2022b): 741–752.

Zhao, Yue, Qian Lu, Yuquan Wei, Hongyang Cui, Xu Zhang, Xueqin Wang, Si Shan, and Zimin Wei. "Effect of actinobacteria agent inoculation methods on cellulose degradation during composting based on redundancy analysis." *Bioresource Technology* 219 (2016): 196–203.

Zheng, Mianhai, Zhenghu Zhou, Ping Zhao, Yiqi Luo, Qing Ye, Kerong Zhang, Liang Song, and Jiangming Mo. "Effects of human disturbance activities and environmental change factors on terrestrial nitrogen fixation." *Global Change Biology* 26, no. 11 (2020): 6203–6217.

Zhu, Chunwu, Lewis Ziska, Jianguo Zhu, Qing Zeng, Zubing Xie, Haoyan Tang, Xiaodong Jia, and Toshihiro Hasegawa. "The temporal and species dynamics of photosynthetic acclimation in flag leaves of rice (*Oryza sativa*) and wheat (*Triticum aestivum*) under elevated carbon dioxide." *Physiologia Plantarum* 145, no. 3 (2012): 395–405.

Ziółkowska, Aleksandra, Agnieszka I. Piotrowicz-Cieślak, Małgorzata Margas, Barbara Adomas, and Grzegorz Nałęcz-Jawecki. "Accumulation of tetracycline, oxytetracycline and chlortetracycline in pea (*Pisum sativum L.*)." *Fresenius Environmental Bulletin* 24, no. 4a (2015): 1386–1391.

Zogg, Gregory P., Donald R. Zak, David B. Ringelberg, David C. White, Neil W. MacDonald, and Kurt S. Pregitzer. "Compositional and functional shifts in microbial communities due to soil warming." *Soil Science Society of America Journal* 61, no. 2 (1997): 475–481.

Zou, Yina, Zhen Hu, Jian Zhang, Huijun Xie, Christophe Guimbaud, and Yingke Fang. "Effects of pH on nitrogen transformations in media-based aquaponics." *Bioresource Technology* 210 (2016): 81–87.

# 7 Biodetoxification of Lignocellulose

*Wankui Jiang*[1], *Wenming Zhang*[1,2], *Carol S.K. Lin*[3], *Min Jiang*[1,2], *and Fengxue Xin*[1,2]
[1]State Key Laboratory of Materials-Oriented Chemical Engineering, College of Biotechnology and Pharmaceutical Engineering, Nanjing Tech University, Nanjing, P.R. China
[2]Jiangsu National Synergetic Innovation Center for Advanced Materials, Nanjing Tech University, Nanjing, P.R. China
[3]School of Energy and Environment, City University of Hong Kong, Hong Kong

## 7.1 INTRODUCTION

Lignocellulose is the most abundant and renewable resource on earth, consisting of cellulose, hemicellulose and lignin, which has been recognized as a potential source to produce biofuels and biochemicals such as ethanol, succinic acid and lactic acid (Dong et al., 2017; Hu et al., 2016; He et al., 2016; Wang et al., 2016; Kurosawa et al., 2015). However, the direct utilization of lignocellulose is difficult due to its solid crystalline structure. Generally, pretreatment is necessary to destroy the biorecalcitrance of lignocellulose and release reducing sugars (Jørgensen et al., 2007). Various methods such as chemical (dilute acid, alkaline or alkaline/oxidation treatment), physical (high-temperature pyrolysis, microwave or crushing) and biological methods can be used for pretreatment (Talebnia et al., 2010; Uppugundla et al., 2014; Guo et al., 2011; Toquero and Bolado, 2014). During the pretreatment process, various toxins derived from lignocellulose will be generated, including lignin-derived aromatic compounds (vanillin, p-hydroxybenzoate, ferulate, etc.), furan derivatives (furfural and 5-hydroxymethylfurfural (HMF)) and low-molecular-weight organic acids (acetic acid and formic acid) (Palmqvist et al., 2000; Klinke et al., 2004; Jönsson et al., 2013). These toxins severely inhibit the microbial growth and activity, which in turn hinders the industrial production of biofuels and biochemicals (Ko et al., 2015; Kim et al., 2011; Qin et al., 2016). Therefore, detoxification is another vital step to remove these toxic substances for the following fermentation process.

Several detoxification strategies have been explored to mitigate the effects of inhibitors on the microbial fermentation, including water washing, evaporation, organic liquid–liquid extraction, ion exchange adsorption and activated carbon adsorption (Ko et al., 2015; Kim et al., 2011; Qin et al., 2016; Zhu et al., 2011; Florencio et al., 2016; Cao et al., 2013; Kunasundari et al., 2017; Silva et al., 2013). However, these methods also lead to some negative outcomes such as the extensive use of fresh water and generation of wastewater, high energy consumption and incomplete removal of inhibitors (Dong and Bao, 2010). Recently, biodetoxification has attracted great attention, which uses microorganisms or their enzymes to remove or transform inhibitory compounds into less toxic ones, thereby improving the fermentability of the pretreated lignocellulosic biomass (Wierckx et al., 2010). Biodetoxification shows several advantages over other methods, including mild reaction conditions, low consumption of fresh water and minimal wastewater production, and low energy consumption (Ko et al., 2015; Zhang et al., 2010). There have been some reports on the

DOI: 10.1201/9781032676890-9

use of microorganisms for the biological detoxification of lignocellulosic hydrolysates. For instance, *Kurthia huakuii* LAM0618 was capable of detoxifying phenolic inhibitors in lignocellulosic hydrolysates. The detoxified hydrolysate can be directly used for 33.47 g/L of lactic acid production, which was 1.87 times higher than that of the non-detoxified hydrolysate (Xie et al., 2018). *Issatchenkia occidentalis* CCTCC M206097 can remove 67% of syringaldehyde, 73% of ferulate, 62% of furfural and 85% of HMF after 24 hours (Fonseca et al., 2011). *Coniochaeta ligniaria* NRRL30616 was able to remove more than 95% of acetate and 65% of furfural, hydroxymethylfurfural and phenolic compounds generated during the pretreatment of corn stover (Cao et al., 2013). *Aspergillus oryzae* ZN1 can also degrade furfural/HMF produced during the dilute acid pretreatment of corn straw. The detoxified hydrolysate can be used for 104.5 g/L of lactic acid production (Liu et al., 2015).

In the bioconversion of lignocellulose, pretreatment is necessary to maximize the accessibility of cellulose, which often results in the generation of excessive inhibitory substances, leading to a decrease in the efficiency of subsequent fermentation processes.

This chapter first provides an overview of the pretreatment processes for lignocellulose. The inhibitory substances, including furans, furfural, phenolics, formic acid and acetic acid, produced from lignocellulose and their mechanisms of action are enumerated. Subsequently, strategies to alleviate the adverse effects of these inhibitors on enzymatic hydrolysis and microbial activity are summarized. It is highlighted that the process of biodegradation shows promise and warrants further research. Finally, an overview of the challenges and prospects of biodegradation techniques in lignocellulosic bioconversion is provided.

## 7.2 LIGNOCELLULOSE PRETREATMENT AND INHIBITORS

### 7.2.1 Pretreatment of Lignocellulosic Biomass

The pretreatment of lignocellulose can be classified into physical, chemical, physicochemical and biological methods (Table 7.1). Physical pretreatment, also referred to as mechanical pretreatment, involves the use of mechanical methods such as grinding, cutting, milling and shredding to decrease particle size and enhance the surface area of biomass. The increased surface area of biomass enhances the availability of enzymes, thereby promoting the disintegration of biomass (Agbor et al., 2011). Additionally, this pretreatment can partially modify the structure of biomass, decrease the crystallinity of cellulose and break chemical bonds (Zakaria et al., 2014). For example, after ball milling for 120 minutes, the crystallinity index of oil palm empty fruit bunch decreased from 56% to 9% (Zakaria et al., 2014). Overall, physical pretreatment methods are relatively simple and enable the reduction of biomass particle size and enhancement of surface area.

Chemical pretreatment involves the hydrolysis and decomposition of lignocellulosic biomass using acids, alkalis, organic solvents or ionic liquids. Acid pretreatment is commonly performed under specific conditions using concentrated or dilute acids to break the glycosidic bonds between cellulose and hemicellulose, thereby disrupting the structure of lignocellulose. Acid pretreatment offers advantages of rapid reaction rates and effective treatment, making it widely applied. However, it presents severe corrosion issues and necessitates complex acid recovery equipment, resulting in high operational and maintenance costs. Moreover, the formation of inhibitors during the hydrolysis process makes the utilization of hydrolysate in subsequent fermentation processes less ideal (Luo et al., 2022). Studies have shown that acid pretreatment of corn stover produces phenolic inhibitors such as ferulic acid (Hou et al., 2017a). Alkaline pretreatment typically involves the use of sodium hydroxide and potassium hydroxide reagents at relatively low temperatures, but the high cost is a limiting factor (Hou et al., 2017a). This pretreatment leads to the expansion of lignocellulosic biomass, dissolution of lignin and hemicellulose, and de-esterification of intermolecular ester bonds, resulting in reduced polymerization degree and crystallinity. The disruption of lignin structure increases the exposure of internal surfaces, rendering it more amenable to enzymatic digestion and thus improving hydrolysis efficiency (Park and Kim, 2012). Nevertheless, alkaline pretreatment

**TABLE 7.1**
**Inhibitors Generated During Different Pretreatment Processes and Their Mechanisms of Action**

| Pretreatment | Inhibitor | Source | Action Mechanism of Inhibitor | References |
|---|---|---|---|---|
| Hydrothermal pretreatment | Phenols, monosaccharides, furans, carboxylic acids | Lignin<br>Hemicellulose<br>Cellulose | Highly toxic to fermenting microorganisms | Baruah et al. (2018) and Maitra and Singh (2021) |
| Mechanical grinding | Phenols<br>Furfural | Hemicellulose<br>Lignin | Inhibition of enzyme activity | Agrawal et al. (2021), Cheng et al. (2021), Kumar et al. (2020) |
| Microwave radiation | Furfural<br>HMF | Hemicellulose | Inhibition of microbial and enzyme activity | Anh et al. (2021), Bundhoo (2018), Jaffur et al. (2021), Rios-Gonzalez et al. (2021) |
| Acoustic degradation method (ultrasonic) | Monosaccharide<br>Furfural<br>HMF | Hemicellulose | Inhibition of enzyme activity | Bhatia et al. (2020), Fadimu et al. (2022), Kong et al. (2021), Lu et al. (2022), Zhang et al. (2022) |
| Acid hydrolysis | Acetic acid<br>Furfural<br>Monosaccharide<br>Formic acid<br>HMF<br>Pseudo-lignin | Cellulose<br>Hemicellulose | Inhibition of cellulase activity | Baruah et al. (2018), Du et al. (2022), Sun et al. (2021) |
| Alkali pretreatment | Furan<br>Acetic acid<br>Formic acid | Hemicellulose | Disrupt DNA, proteins and membrane structure of yeast cells | Lu et al. (2022), Solowski et al. (2020), Wu et al. (2022) |
| Oxidation and ozonation | Acetaldehyde<br>Phenols | Hemicellulose | Inhibition of yeast cells | Alvira et al. (2010), Lu et al. (2022), Minmunin et al. (2018) |
| Ionic liquid handling | Furfural<br>HMF<br>Weak acid | Hemicellulose<br>Lignin | Toxic to enzymes and microorganisms | Bhatia et al. (2020) |
| Treatment with solvents (organic and other) | Weak acid<br>Furan | Hemicellulose | Inhibits enzyme and microbial activity | Chin et al. (2021), Rajendran et al. (2018), Tramontina et al. (2020) |
| Supercritical fluids | Furfural<br>HMF<br>Weak acid | Hemicellulose<br>Lignin | Inhibition of enzyme activity | Bhatia et al. (2020), Rosales-Calderon and Arantes (2019) |
| Steam blasting | Organic acid<br>Phenolic furan | Hemicellulose | Inhibition of cellulose and enzyme activity | Brandt et al. (2021), Jaffur et al. (2021), Satari and Karimi (2018) |
| Microorganisms | Furfural<br>Phenols<br>HMF<br>Organic acids | Hemicellulose<br>Cellulose<br>Lignin | Inhibits the growth of microorganisms | Moreno et al. (2015), Xu et al. (2022) |
| Ammonia fiber blasting | Phenols<br>Furans | Lignin<br>Hemicellulose | Inhibits enzyme and microbial activity | Baruah et al. (2018), Jaffur et al. (2021) |
| Electrochemical | Ferulic acid | Lignin | Inhibition of enzyme activity | Taherzadeh and Karimi (2008) |

often necessitates high temperatures to enhance the extent of lignin dissolution. Additionally, alkaline pretreatment is more suitable for the pretreatment of biomass with lower lignin content (Rizal et al., 2018). Ionic liquids are an inexpensive pretreatment process that can be conducted at low/moderate temperatures (Bhatia et al., 2020). They are non-toxic to enzymes or cells during saccharification and fermentation processes, and they can selectively depolymerize lignin. Hence, this pretreatment process is attractive for industrial-scale applications (Elgharbawy et al., 2016; Financie et al., 2016). Furthermore, ionic liquids can be recycled and reused (Elgharbawy et al., 2016). However, the recovery process involves a step to remove inhibitors, which incurs additional costs (Sen et al., 2012; Domínguez de María, 2014). Research reports that approximately 11% of the lignin component is removed from oil palm fronds after treatment with ionic liquids, which is higher than enzymatic delignification by laccase (Financie et al., 2016).

Physicochemical pretreatment methods encompass steam explosion, ammonia fiber explosion (AFEX), and liquid hot water (LHW) processes. Steam explosion is among the extensively researched and rapidly evolving pretreatment methods for the physicochemical transformation of lignocellulosic feedstocks. Steam explosion offers significant advantages, including cost-effectiveness, minimal environmental impact, low energy demand, and limited or no reliance on chemicals (Alvira et al., 2010; Avellar and Glasser, 1998). Nevertheless, this method can generate adverse inhibitory by-products that have the potential to interfere with downstream processes (Mosier et al., 2005). Research has indicated that furfural produced from the steam explosion of wheat straw ranges from 0.16 to 2.14 g/L, with an inhibitory concentration of furfural at 0.02 g/L (Kim 2018). AFEX, akin to steam explosion, induces sudden expansion of ammonia due to high pressure and temperature, leading to the expansion and physical disintegration of biomass structure, and partial reduction of crystalline cellulose and lignin (Laureano-Perez et al., 2005). AFEX offers numerous advantages, including the absence of inhibitor generation, high sugar recovery (up to 99%), elimination of the need for particle size reduction prior to pretreatment, absence of post-treatment washing, ammonia recyclability and no requirement of nitrogen source for subsequent microbial fermentation (Zheng et al., 2009; Teymouri et al., 2005; Sendich et al., 2008a, 2008b). However, compared with biomass with high lignin content, AFEX is more advantageous for herbaceous, agricultural crop residues and low lignin feedstocks, as only a small amount of hemicellulose and lignin is dissolved during the pretreatment process (Wyman et al., 2005). The AFEX method is prone to generating phenolic inhibitors when applied to biomass with high lignin content. Poplar, after undergoing pretreatment with AFEX, can produce 2.1 mg/g of phenolic inhibitors (Hou et al., 2017a). LHW pretreatment is a process that utilizes liquid water at various temperatures (160°C–240°C) to disrupt the structure of hemicellulose or lignin and adequately expose cellulose (Kim et al., 2018). During pretreatment, sugars derived from hemicellulose predominantly exist in the liquid fraction as oligomers, leading to minimal formation of undesired degradation compounds (Mosier et al., 2005a, Mosier et al., 2005b). In comparison with other pretreatment techniques, LHW offers advantages such as not requiring additional catalysts or chemicals, moderate process temperatures, and reduced sugar loss and inhibitory compound formation (Gírio et al., 2010; Hu et al., 2008; Kim et al., 2009). Although LHW has higher C5 sugar recovery and lower production of sugar-derived inhibitory compounds, compared with steam explosion, it is not yet economically feasible on an industrial scale due to higher energy costs and water requirements.

Biological pretreatment utilizes fungi, bacteria or other microorganisms to degrade lignin, hemicellulose and/or cellulose and produce sugars (Solowski et al., 2020). Biological pretreatment usually takes place under mild reaction conditions, making it widely regarded as environmentally friendly and safe (Kumar et al., 2020). Biological pretreatment is dependent on the enzymatic activity of functional microorganisms, primarily lignin peroxidases, laccases and hemicellulases, to remove lignin and hemicellulose from the lignocellulosic substrate (Reginatto and Antonio, 2015). The reaction rate of biological pretreatment is typically slower compared with chemical pretreatment. Consequently, they are often used in combination with other processes to enhance the production of fermentable sugars while reducing processing time and inhibitor formation (Xu et al., 2022).

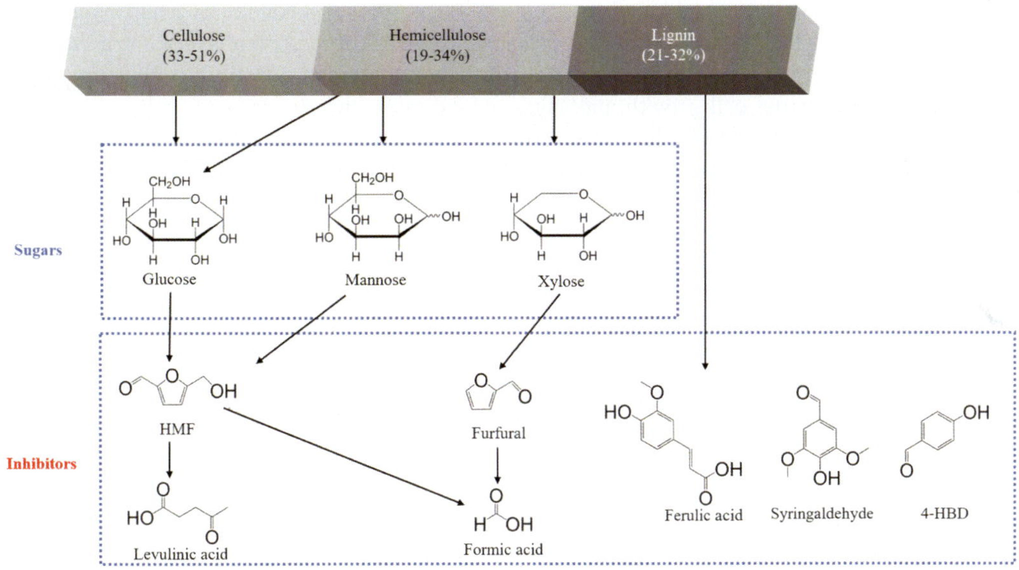


**FIGURE 7.1** Model diagram of the average chemical composition and main inhibitory compound formation of lignocellulosic biomass.

## 7.2.2 Pretreatment-Derived Inhibitors

The hydrolysis products of the primary components of lignocellulosic materials are glucose, mannose, xylose, phenols and acetic acid (Figure 7.1). The first two can undergo isomerization to form fructose, which can then undergo dehydration to produce HMF. Xylose can be enzymatically converted to xylulose, which can subsequently undergo dehydration to yield furfural. The resulting HMF and furfural can undergo hydrolysis to produce formic acid. The specific inhibitors recognized during pretreatment include HMF, furfural, phenols, formic acid and acetic acid.

Multiple phenolic compounds with different molecular weights (MWs), polarities and side chains are generated during lignin degradation in biomass pretreatment. Certain aromatic hydrocarbon molecules present in lignocellulosic biomass may also be released as extractives during the sugar generation process. Phenolic compounds have been demonstrated to be strong inhibitors of cellulase enzymes. For example, the presence of 10 mg/mL vanillin reduced the cellulose hydrolysis conversion rate to 26%, which is half of the 53% conversion rate observed in the control group (Qin et al., 2016). Research has also revealed that p-coumaric acid and ferulic acid decreased the rate of cellulose conversion to glucose by approximately 30% and 16%, respectively (Qin et al., 2016). Research has also shown that phenolic compounds exhibit greater toxicity than other potent inhibitory molecules (such as furfural, weak acids and other degradation by-products) even at lower concentrations. This is due to their low MW, which enables them to penetrate cell membranes, disrupt internal structures and cause alterations in cell morphology (Jönsson et al., 2016; Palmqvist et al., 2000; Gurram et al., 2011). Ferulic acid and *p*-coumaric acid have been reported as the most toxic phenolic acids against the *Clostridium beijerinckii* BA101, with these two compounds at a concentration of 1 g/L inhibiting cell growth by as much as 74% (Ezeji et al., 2007). Investigating the toxicity levels of ferulic acid (1.8 mM) and p-coumaric acid (9.7 mM) on a *Saccharomyces cerevisiae* revealed that compared with the control, ferulic acid and *p*-coumaric acid inhibited cell growth by up to 80% (Adeboye et al., 2015). Phenolic compounds can also enhance the fluidity of cell membranes, potentially resulting in a significant decrease in intracellular potassium levels (Fitzgerald et al., 2004). Additionally, phenolic compounds can facilitate the loss of membrane integrity, resulting in decreased cell growth and impaired sugar assimilation. They may also cause DNA degradation, thereby inhibiting RNA and protein synthesis (Gurram et al., 2011; Fitzgerald et al., 2004).

Furfural and HMF are furan derivatives produced during the degradation of pentose and hexose sugars and are typically present in hydrolysates. Although it is currently unknown whether these molecules significantly inhibit enzyme performance, they can negatively affect microbial fermentation by inhibiting cell growth and sugar uptake rates, thereby reducing ethanol production (Palmqvist et al., 2000a). Studies have reported that cellulase activity is affected when the concentrations of acetic acid and furfural reach 13 g/L and 4 g/L, respectively (Kim et al., 2011). The inhibitory effect of furans can slow down the overall fermentation process by prolonging the lag phase of cells, but it usually does not have a significant impact on the final ethanol production in the fermentation of *S. cerevisiae* and *Zymomonas mobilis* (Behera et al., 2014). Furthermore, the inhibitory effect of furfural on fermentation can be mitigated by increasing the initial cell inoculation of the yeast (Taherzadeh et al., 1999). However, high concentrations of furans or their mixture with other components (such as acetic acid, furfural and lignin derivatives) can result in adverse microbial growth and fermentation. For instance, 0.5 g/L of furfural has no impact on the cell growth of *Scheffersomyces stipitis*, whereas 2 g/L of furfural is harmful to cell growth (Roberto et al., 1991). Similarly, in the process of ethanol production by *S. stipitis* from wheat straw hydrolysate, 0.25 g/L of furfural has no impact on cell growth and ethanol production, whereas 1.5 g/L of furfural inhibits ethanol yield and production by 90.4% and 85.1%, respectively (Nigam, 2001). Importantly, they also observed a synergistic inhibitory effect between acetic acid, furfural and lignin derivatives, leading to lower yield and productivity compared with the combined inhibitory effects of individual compounds (Nigam et al., 2001). Furan derivatives also exert negative effects on microbial growth, metabolism and cell wall formation (Liu, 2011; Zaldivar et al., 1999; Hadi and Rehman, 1989). Furfural inhibits primary carbon-degrading enzymes such as aldehyde dehydrogenase, alcohol dehydrogenase (ADH), aldehyde reductase (TAL), glycolaldehyde-3-phosphate dehydrogenase and pyruvate dehydrogenase, making it more toxic to ethanol fermentation than other inhibitory compounds like HMF (Guo et al., 2008). Furan derivatives also have an impact on the assimilation of sulfur-containing amino acids such as cysteine and methionine. Furfural is also associated with increased reactive oxygen species (ROS) levels, which can damage mitochondria and vacuolar membranes (cellular scaffolding and nuclear chromatin) (Allen et al., 2010). In certain cases, furfural can be transformed by specific yeast into other inhibitory compounds, such as furfuryl alcohol and furoic acid (Allen et al., 2010). In comparison with furfural, HMF has a less pronounced inhibitory effect on microbial activity but can prolong the lag phase and consume cell growth. It also has a longer duration compared with furfural, as the conversion rate of furfural is four times faster than that of HMF (Liu and Blaschek, 2010).

## 7.3 BIODETOXIFICATING INHIBITORS

The abundance of by-products generated during lignocellulosic pretreatment results in inhibition of cellulase activity and microbial fermentation. Extensive research and efforts have been dedicated to combating inhibitors and minimizing the inhibitory issues before and after pretreatment (Alves-Ferreira et al., 2022). Although various physical (evaporation and membrane separation) and chemical detoxification (neutralization, filtration, activated carbon and ion exchange) processes have been studied, the use of microbial or enzymatic bioprocessing methods offers several advantages, such as mild reaction conditions, absence of chemical additives, minimal side reactions and low energy requirements (Parawira and Tekere, 2011). This section primarily discusses the utilization of biological approaches to counteract the formation of inhibitors and their adverse effects. The utilization of specific microorganisms for detoxification of pretreated lignocellulose represents an environmentally friendly approach, avoiding energy-intensive processing conditions, harsh chemicals and costly processing materials. Within this approach, microbial pretreatment can mitigate and/or eliminate inhibitors originating from lignocellulosic materials prior to enzymatic hydrolysis and subsequent fermentation (Jin et al., 2010; Cannella et al., 2014; Nichols et al., 2010; Nichols et al., 2014).

### 7.3.1 Microbial Detoxification Processes

Microbial detoxification involves the utilization of microorganisms to mitigate the inhibitory effects of degradation compounds formed during the physicochemical pretreatment of lignocellulose. Important factors to consider in an effective microbial detoxification process include the addition of nutrients for the growth of detoxifying microorganisms, pH, temperature, treatment time, inoculum size and microbial strains. A viable microbial detoxification strategy is to employ fungi, bacteria or yeast for additional detoxification steps before product fermentation. Among various microorganisms, fungi such as *Trichoderma reesei* and *Coniochaeta ligniaria* have been extensively studied. Larsson et al. (1999) utilized *T. reesei* to enhance the fermentability of dilute acid hydrolysate from spruce and observed the removal of furans and a small portion of phenolics without impacting the weak acid concentration (Larsson et al., 1999). A newly isolated fungus, *C. ligniaria* NRRL30616, was found to possess the ability to metabolize furfural, HMF, as well as aromatic acids and aldehydes, making it an ideal candidate for hydrolysate detoxification (López et al., 2004). Strain *C. ligniaria* exhibits effectiveness in reducing inhibitors formed during the dilute acid pretreatment of different biomass sources, including switchgrass, reed, foxtail, alfalfa stems, corn stover and rice husks, thereby reducing fermentation lag phase (Nichols et al., 2005a, 2005b, 2008, 2010, 2014). It is worth mentioning that the recombinant bacterium *Escherichia coli* FBR5 can utilize C5 and C6 sugars for ethanol production. However, due to its sensitivity to inhibitors such as furfural, HMF and acetic acid, achieving efficient synthesis in the process of utilizing pretreated hydrolysates for chemical production is challenging. However, strain FBR5 can effectively utilize C5 and C6 sugars when diluted acid-pretreated corn stover hydrolysate detoxified by *C. ligniaria* is employed as a substrate for microbial fermentation, whereas it is unable to do so in non-biologically detoxified hydrolysate (Nichols et al., 2008). The fungal *Amorphotheca resinae* ZN1 has the ability to degrade various inhibitory compounds present in steam-exploded corn stover, as well as dilute acid-pretreated corn stover, rice straw, wheat straw, and rapeseed straw (Zhang et al., 2010). Furthermore, *Aspergillus nidulans* FLZ10 can fully degrade furfural and HMF and partially eliminate formic acid and acetic acid when utilized on steam-exploded corn stover (Yu et al., 2011).

Bacteria and yeast are also utilized for the detoxification of lignocellulosic hydrolysis, although there is relatively less research compared with fungi. *Ureibacillus thermophaercus* is employed for the removal of furfural and HMF, as well as for the elimination of phenolic compounds in dilute acid hydrolysates of waste wood (Okuda et al., 2008). López et al. (2004) isolated five bacteria, *Methylobacterium extorquens*, *Pseudomonas* sp., *Flavobacterium indologenes*, *Acinetobacter* sp. and *Arthrobacter aurescens*, using specific mineral media, and they are capable of degrading ferulic acid, hydroxymethylfurfural and furfural (López et al., 2004). Fonseca et al. (2011) reported that *Issatchenkia occidentalis* CCTCC M206097 can remove vanillin, ferulic acid, furfural and HMF from sugarcane bagasse hemicellulosic hydrolysate, and the detoxified hydrolysate was utilized for yeast fermentation (Fonseca et al., 2011).

*In situ* microbial detoxification is a feasible method that utilizes the natural abilities of various fermentation microorganisms to remove inhibitory compounds. For example, most strains of *Saccharomyces cerevisiae* can convert furan derivatives into compounds with lower antibacterial activity (Ferreira et al., 2011; Palmqvist et al., 2000a; Schneider, 1996; Thomsen et al., 2009). For example, furfural can be reduced to furfuryl alcohol, and HMF can be reduced to 2,5-bis-hydroxymethylfuran (Liu et al., 2005). *Saccharomyces cerevisiae* also has the ability to metabolize certain phenolic compounds due to the presence of cinnamic acid decarboxylase, an enzyme that catalyzes the decarboxylation reaction, converting aromatic carboxylic acids into the corresponding vinyl derivatives (Goodey and Tubbs, 1982). Schneider (1996) reported the selective removal of acetic acid from sulfite liquor of hardwood waste using a mutated yeast strain, *S. cerevisiae* YGSCD 308.3, thereby converting all sugar biomass into ethanol (Schneider, 1996). Thomsen et al. (2009) described the ability of *S. cerevisiae* to detoxify water-pretreated wheat straw hydrolysate by degrading furfural and phenolic aldehydes (Thomsen et al., 2009). In addition to *S. cerevisiae*, *Pichia stipitis* has also been reported to have the ability to remove various inhibitory compounds. For example, *P. stipitis* CBS5773 has been reported to remove furfural and HMF from lignocellulosic acid hydrolysate (Ferreira et al., 2011). Co-culture systems have also been used for detoxification and chemical synthesis of lignocellulosic biomass. To increase the yield of bioethanol, Yu et al. (2011) co-cultivated *Aspergillus nidulans* FLZ10 with *S. cerevisiae* for simultaneous detoxification and fermentation of steam-exploded corn stover. By employing these two microorganisms, the final ethanol production increased more than threefold, reaching a concentration of 34 g/L (Yu et al., 2011).

### 7.3.2 Adaptation of Microbial

Alternative approaches to tackle inhibition issues involve subjecting fermenting microorganisms to adaptive evolution in culture media containing inhibitory substances. This approach can cultivate microorganisms with high tolerance to phenols, furfural and other small organic molecules, which positively impacts productivity and yield in the fermentation process (Koppram et al., 2012; Almario et al., 2013). Liu et al. (2005) developed novel evolved strains of *S. cerevisiae* and *P. stipitis* through multiple cultivations in synthetic media with gradually increasing inhibitor concentrations. These strains demonstrated increased tolerance to furfural and HMF and exhibited faster growth and glucose metabolism compared with control strains (Liu et al., 2005). Tian et al. (2010) utilized evolved strain Y5 of brewing Saccharomyces cerevisiae, which exhibited good inhibitor tolerance and furfural metabolism capability while maintaining high ethanol production capacity (Tian et al., 2010). The inhibitor-tolerant and thermotolerant *Saccharomyces cerevisiae* (Fm13) selected through adaptive evolution showed favorable fermentation performance under high inhibitor concentrations, including acetic acid (7.2 g/L), formic acid (2.44 g/L), lactic acid (6.89 g/L), furfural (2.77 g/L) and HMF (3.75 g/L), achieving an ethanol fermentation yield of around 90% of the theoretical maximum (Favaro et al., 2013).

### 7.3.3 Genetic/Metabolic Engineering

The utilization of genetic/metabolic engineering technologies to construct recombinant strains for overcoming fermentation inhibition issues has gained significant acceptance among research scholars. In the engineered *Saccharomyces cerevisiae* strains, the overexpression of specific enzymes, such as ADH and TAL, enables the enhancement of ethanol production during fermentation with high concentrations of furfural hydrolysate (Hasunuma et al., 2014). Similar investigations have also been performed on the *Trametes versicolor* and *Escherichia coli*, demonstrating their increased tolerance to hydrolysates of spruce wood and sugarcane bagasse, respectively (Larsson et al., 2001a; Wang et al., 2013). Research has reported that overexpression of the glucose-6-phosphate dehydrogenase (ZWF1), an enzyme in the pentose phosphate pathway (PPP) of *Saccharomyces cerevisiae*, also enhances its tolerance to furfural derivatives (Gorsich et al., 2006). The design of a genetically modified *Saccharomyces cerevisiae* strain resistant to phenolic compounds has also been studied. Larsson et al. (2001a) reported that overexpression of the Pad1p gene, encoding phenylacrylic acid decarboxylase, in brewing yeast enhances its tolerance to phenylacrylic acid (Larsson et al. 2001a). The engineered brewing yeast strain was cultured in a synthetic basal medium supplemented with ferulic acid, cinnamic acid, and spruce hydrolysate. In comparison with the wild-type strain, the recombinant strain with Pad1p protein overexpression can convert these two acids at a higher rate, demonstrating faster cell growth and increased ethanol production. Investigations have also been conducted to explore the expression of exogenous laccases in brewing yeast for increased degradation of phenolic compounds. Larsson et al. (2001b) designed a recombinant brewing yeast strain that harbors the laccase gene from the white fungus *T. versicolor* and overexpresses the homologous t-SNARE Sso2p, a membrane protein involved in protein secretion mechanisms. The strain exhibits higher laccase activity compared with the laccase gene-only carrying *Saccharomyces cerevisiae* and can convert coniferyl aldehyde at a faster rate. Moreover, this transformed strain can ferment dilute acid spruce hydrolysate, leading to higher ethanol production compared with the control group (Larsson et al., 2001b).

## 7.4 CHALLENGES AND PROSPECTS

The effective detoxification/elimination of inhibitors generated during lignocellulosic pretreatment to mitigate their adverse effects often encounters economic challenges (Carpita and McCann, 2020; Espro et al., 2018). The challenges worthy of further research include the following.

### 7.4.1 Reducing Costs of Biological Detoxification

Compared with physical and chemical detoxification, biological detoxification has environmental and economic advantages (Sodre et al., 2021). However, the high cost of detoxification enzyme preparations is the main obstacle to their widespread industrial-scale application (Bhatia et al., 2020). For example, the price of laccase is 36.1 US dollars per kilogram, and the price of peroxidase is 12.5 US dollars per kilogram (https://www.sunsonenzymes.com/). Reducing enzyme costs can be achieved by improving enzyme production by separating new isolates or constructing new secreted strains. White rot fungus is currently the only strain that can completely degrade lignin into soluble products, which has attracted widespread research attention. For example, *Pleurotus ostreatus* and *Cerrena* sp. produce highly active laccase, while *Phanerochaete chrysosporium* and *Trametes versicolor* produce highly active peroxidase (Suryadi et al., 2022). A novel *Paecilomyces variotii* FN89 isolated from corn straw pretreated with sulfuric acid can completely degrade furan and phenolic inhibitors derived from lignocellulose, while retaining most of the fermentable sugars. In addition to exploring more excellent detoxification strains, constructing new strains or microbial communities is also an effective way to reduce the cost of biological detoxification (Zhang et al., 2021). Ali et al. (2021) constructed a novel bacterial complex SST-4, which can degrade over 90% of phenolic substances and exhibit high detoxification efficiency (Ali et al., 2021). It is considered a promising biological complex for detoxifying lignocellulosic hydrolysates. *Pseudomonas putida* KT2440 and *Bacillus coagulans* NL01 have also been constructed to form a microbial consortium, which can not only remove organic acids and furaldehyde, but also remove 90% of aromatic compounds (Zou et al., 2021). Overall, although biological detoxification has advantages over physical chemical detoxification, experimental research is still needed in the future to achieve more significant economic benefits of biological detoxification and enable large-scale commercialization of lignocellulose (Sodre et al., 2021).

### 7.4.2 Minimizing Accumulation of Inhibitory Compounds

The minimum accumulation of inhibitors can lead to high performance in subsequent fermentation stages, maximizing the profitability of biological refining. Emerging technologies are applied to pretreatment processes to reduce or eliminate the production of inhibitory compounds. For example, by combining mechanocatalytic $P_2O_5$ pretreatment with strain domestication, the conversion rate of lactic acid synthesis by *Pediococcus penttosaceus* using the pretreated hydrolysate increased by 22.4% after domestication, and no phenolic inhibitors were produced (Liu et al., 2021b). Reducing the formation of inhibitors or enhancing their removal can reduce their accumulation, while inhibitors are also valuable chemicals. The biological detoxification method of ketoglutaric acid can provide intermediates for the TCA cycle (tricarboxylic acid cycle). Phenols can be converted into phenolic resins through catalytic reactions, and acetic acid and formic acid purified by physical methods can be used as renewable chemicals. However, all of these strategies require reducing accumulation and increasing conversion at an affordable cost. Considering the diversity of lignocellulosic biomass and the diversity of pretreatment processes, the relative detoxification methods should not be universally applicable (Ghazali and Razak, 2021). All new technologies being developed should be considered as candidate technologies for specific scenarios, and economic evaluation and large-scale testing based on all steps of biological refining require commercial scale detoxification in biological refining.

### 7.4.3 Development of Inhibitor-Tolerant Microbial Species or Enzymes

Although physical and chemical detoxification methods have been widely used in the detoxification process of lignocellulosic biorefineries, there are still issues with high costs. Therefore, it is important to develop research on inhibitor-tolerant microorganisms in biological detoxification (Chang et al., 2021). Different methods have been developed in microorganisms to address the inhibitory compounds produced in the biological refining process of lignocellulose. For example, a microbial co-culture system was constructed by synthesizing a novel microbial community with strong tolerance, in which *Pseudomonas putrescens* KT2440 and *Bacillus coagulans* NL01 can quickly remove inhibitors produced during the biological refining process of lignocellulose. The experimental results showed that the degradation rate of organic acids and furaldehyde by this microbial community was 100%, demonstrating the enormous potential of this process in the biotransformation of lignocellulose (Sjulander and Kikas, 2020). Adaptive evolution is also an effective way to obtain highly tolerant bacteria. Some studies have shown that adaptive evolution of formic acid can lead to the evolution of strains that are tolerant to formic acid and acetic acid (Du et al., 2022). Understanding the mechanism of action between strains and inhibitory compounds is also essential for biological detoxification. For example, understanding the membrane potential and membrane permeability of cell membranes can better respond to the inhibitory mechanisms of inhibitors and provide useful information in developing new tolerant strains (López et al., 2021).

## 7.5 CONCLUSIONS AND PERSPECTIVES

Considering the sustainable development of the environment and society, future biorefineries must not only be able to produce large quantities of low-cost bio-based chemicals but also reduce the generation of downstream waste. In order to reduce the stubborn structure of lignocellulose, pretreatment is a key step to ensure the effective utilization of these raw materials and plays an important role in the total cost. As mentioned above, multiple microorganisms and lignin-degrading enzymes have the potential to be used as a single pretreatment method or in combination with other cost-effective physicochemical techniques. After the biological detoxification step, the pretreated lignocellulose hydrolysate retains less inhibitory compounds, saving the fresh water required for washing materials and reducing wastewater. In addition, lower inhibitors allow fermentation microorganisms to work at higher rates at higher substrate concentrations, reducing fermentation time and increasing final product concentration, thereby reducing costs and ensuring green production processes. To promote the proper utilization of lignocellulosic raw materials, the biological detoxification of lignocellulose can be combined with different physical and chemical technologies to increase global production. In this case, various microorganisms and enzymes have been successfully applied to the detoxification process of lignocellulose hydrolysate, promoting saccharification and fermentation steps, although there are still no reported data showing microorganisms or enzymes capable of detoxifying all types of phenolic and furfural inhibitors.

This chapter summarizes the research progress in the formation, mechanism of action and biological detoxification of hydrolysates from lignocellulosic biomass pretreatment. Reducing the formation of inhibitors or enhancing their removal can reduce the accumulation of inhibitors and improve the performance of lignocellulose biorefinery. Biological detoxification is an important step in achieving the commercialization of lignocellulose biological refining. At the same time, biological detoxification is still in the stage of technological exploration, requiring a large amount of technological development to adapt to different sources and application scenarios of lignocellulose. In the future, developing new inhibitor-resistant microorganisms remains a major challenge, and developing more advanced biotechnology and comprehensively understanding the inhibitory mechanisms of microbial inhibitors still require continuous exploration.

## REFERENCES


Adeboye, P.T., Bettiga, M., Aldaeus, F., Larsson, P.T., Olsson, L., 2015. Catabolism of coniferyl aldehyde, ferulic acid, and p-coumaric acid by Saccharomyces cerevisiae yields less toxic products. *Microb. Cell Fact.* 14, 149.

Agbor, V.B., Cicek, N., Sparling, R., Berlin, A., Levin, D.B., 2011. Biomass pretreatment: Fundamentals toward application. *Biotechnol. Adv.* 29, 675–685.

Agrawal, R., Verma, A., Singhania, R.R., Varjan, S., Dong, C.D., Patel, A.K., 2021. Current understanding of the inhibition factors and their mechanism of action for the lignocellulosic biomass hydrolysis. *Bioresour. Technol.* 332, 125042.

Ali, S.S., Jiao, H., Mustafa, A.M., Koutra, E., El-Sapagh, S., Kornaros, M., Elsamahy, T., Khalil, M., Bulgariu, L., Sun, J., 2021. Construction of a novel *microbial consortium* valued for the effective degradation and detoxification of creosote-treated sawdust along with enhanced methane production. *J. Hazard. Mater.* 418, 126091.

Allen, S.A., Clark, W., McCaffery, J.M., Cai, Z., Lanctot, A., Slininger, P.J., Liu, Z.L., Gorsich, S.W., 2010. Furfural induces reactive oxygen species accumulation and cellular damage in Saccharomyces cerevisiae. *Biotechnol. Biofuels.* 3, 2.

Almario, M.P., Reyes, L.H., Kao, K.C., 2013. Evolutionary engineering of *Saccharomyces cerevisiae* for enhanced tolerance to hydrolysates of lignocellulosic biomass. *Biotechnol. Bioeng.* 110, 2616–2623.

Alves-Ferreira, J., Cavalheiro, F., Duarte, L.C., Ferreira, A.R.P., Martinez, A., Pereira, H., Fernandez, M.C., 2022. D-Lactic acid production from Cistus ladanifer residues: Cofermentation of pentoses and hexoses by *Escherichia coli* JU15. *Ind. Crop. Prod.* 177, 114519.

Alvira, P., Tomás-Pejó, E., Ballesteros, M., Negro, M.J., 2010. Pretreatment technologies for an efficient bioethanol production process based on enzymatic hydrolysis: A review. *Bioresour. Technol.* 101, 4851–4861.

Anh, T.H., Nizetic, S., Ong, H.C., Mofijur, M., Ahmed, S.F., Ashok, B., Van The Vinh, B., Minh, Q.C., 2021. Insight into the recent advances of microwave pretreatment technologies for the conversion of lignocellulosic biomass into sustainable biofuel. *Chemosphere* 281, 130878.

Avellar, B.K., Glasser, W.G., 1998. Steam-assisted biomass fractionation. I. Process considerations and economic evaluation. *Biomass Bioenerg.* 14, 205–218.

Baruah, J., Nath, B.K., Sharma, R., Kumar, S., Deka, R.C., Baruah, D.C., Kalita, E., 2018. Recent trends in the pretreatment of lignocellulosic biomass for value-added products. *Front. Energy Res.* 6, 141.

Behera, S., Arora, R., Nandhagopal, N., Kumar, S., 2014. Importance of chemical pretreatment for bioconversion of lignocellulosic biomass. *Renew. Sust. Energ. Rev.* 36, 91–106.

Bhatia, S.K., Jagtap, S.S., Bedekar, A.A., Bhatia, R.K., Patel, A.K., Pant, D., Rajesh Banu, J., Rao, C.V., Kim, Y.-G., Yang, Y.-H., 2020. Recent developments in pretreatment technologies on lignocellulosic biomass: Effect of key parameters, technological improvements, and challenges. *Bioresour. Technol.* 300, 122724.

Brandt, B.A., Garcia-Aparicio, M.D.P., Gorgens, J.F., van Zyl, W.H., 2021. Rational engineering of *Saccharomyces cerevisiae* towards improved tolerance to multiple inhibitors in lignocellulose fermentations. *Biotechnol. Biofuels.* 14, 173.

Bundhoo, Z.M.A., 2018. Microwave-assisted conversion of biomass and waste materials to biofuels. *Renew. Sustain. Energy Rev.* 82, 1149–1177.

Cannella, D.; Sveding, P.V.; Jørgensen, H., 2014. PEI detoxification of pretreated spruce for high solids ethanol fermentation. *Appl. Energ.* 132, 394–403.

Cao, G., Ximenes, E., Nichols, N.N., Zhang, L., Ladisch, M., 2013. Biological abatement of cellulase inhibitors. *Bioresour. Technol.* 146, 604–610.

Carpita, N.C., McCann, M.C., 2020. Redesigning plant cell walls for the biomass-based bioeconomy. *J. Biol. Chem.* 295, 15144–15157.

Chang, D., Wang, C., Ndayisenga, F., Yu, Z., 2021. Mutations in adaptively evolved *Escherichia coli LGE2* facilitated the cost-effective upgrading of undetoxified bio-oil to bioethanol fuel. *Bioresour. Bioprocess.* 8, 105.

Cheng, Y., Quan, W., Qu, T., He, Y., Wang, Z., Zeng, M., Qin, F., Chen, J., He, Z., 2021. Effects of 60Co-irradiation and superfine grinding wall disruption pretreatment on phenolic compounds in pine (*Pinus yunnanensis*) pollen and its antioxidant and α-glucosidase-inhibiting activities. *Food Chem.* 345, 128808.

Chin, D.W.K., Lim, S., Pang, Y.L., Lim, C.H., Shuit, S.H., Lee, K.M., Chong, C.T., 2021. Effects of organic solvents on the organosolv pretreatment of degraded empty fruit bunch for fractionation and lignin removal. *Sustainability* 13, 6757.

Domínguez de María, P., 2014. Recent trends in (ligno)cellulose dissolution using neoteric solvents: Switchable, distillable and bio-based ionic liquids. *J. Chem. Technol. Biotechnol.* 89, 11–18.

Dong, H., Bao, J., 2010. Metabolism: Biofuel via biodetoxification. *Nat. Chem. Biol.* 6, 316–318.

Dong, W., Xue, M., Zhang, Y., Xin, F., Wei, C., Zhang, W., Wu, H., Ma, J., Jiang, M., 2017. Characterization of a β-glucosidase from *Paenibacillus* species and its application for succinic acid production from sugarcane bagasse hydrolysate. *Bioresour. Technol.* 241, 309–316.

Du, C., Li, Y., Xiang, R., Yuan, W., 2022. Formate dehydrogenase improves the resistance to formic acid and acetic acid simultaneously in *Saccharomyces cerevisiae. Int. J Mol. Sci.* 23, 3406.

Elgharbawy, A.A., Alam, M.Z., Moniruzzaman, M., Goto, M., 2016. Ionic liquid pretreatment as emerging approaches for enhanced enzymatic hydrolysis of lignocellulosic biomass. *Biochem. Eng. J.* 109. 252–267.

Espro, C., Gumina, B., Szumelda, T., Paone, E., Mauriello, F., 2018. Catalytic transfer hydrogenolysis as an effective tool for the reductive upgrading of cellulose, hemicellulose, lignin, and their derived molecules. *Catalysts.* 8, 313.

Ezeji, T., Qureshi, N., Blaschek, H.P., 2007. Butanol production from agricultural residues: Impact of degradation products on *Clostridium beijerinckii* growth and butanol fermentation. *Biotechnol. Bioeng.* 97, 1460–1469.

Fadimu, G.J., Farahnaky, A., Gill, H., Tuyen, T., 2022. Influence of ultrasonic pretreatment on structural properties and biological activities of lupin protein hydrolysate. *Int. J. Food Sci. Technol.* 57, 1729–1738.

Favaro, L., Basaglia, M., Trento, A., Van Rensburg, E., García-Aparicio, M., Van Zyl, W.H., Casella, S., 2013. Exploring grape marc as a trove for new thermotolerant and inhibitor-tolerant *Saccharomyces cerevisiae* strains for second-generation bioethanol production. *Biotechnol. Biofuels.* 6, 1–14.

Ferreira, S., Gil, N., Queiroz, J.A., Duarte, A.P., Domingues, F.C., 2011. An evaluation of the potential of Acacia dealbata as raw material for bioethanol production. *Bioresour. Technol.* 102, 4766–4773.

Financie, R., Moniruzzaman, M., Uemura, Y., 2016. Enhanced enzymatic delignification of oil palm biomass with ionic liquid pretreatment. *Biochem. Eng. J.* 110, 1–7.

Fitzgerald, D.J., Stratford, M., Gasson, M.J., Ueckert, J., Bos, A., Narbad, A., 2004. Mode of antimicrobial action of vanillin against Escherichia coli, Lactobacillus plantarum, and Listeria innocua. *J. Appl. Microbiol.* 97, 104–113.

Florencio, C., Badino, A.C., Farinas, C.S., 2016. Soybean protein as a cost-effective lignin-blocking additive for the saccharification of sugarcane bagasse. *Bioresour. Technol.* 221, 172–180.

Fonseca B.G., Moutta R.O., Ferraz F.O., Vieira, E. R., Nogueira, A. S., Baratella, B.F., Rodrigues, L.C., Zhang, H.R., da Silva, S.S., 2011. Biological detoxification of different hemicellulosic hydrolysates using *Issatchenkia occidentalis CCTCC M 206097 yeast. J. Ind. Microbiol. Biot.* 38, 199–207.

Ghazali, N. F., Razak, N. D. A., 2021. Recovery of saccharides from lignocellulosic hydrolysates using nanofiltration membranes: A review. *Food Bioprod. Process.* 126, 215–233.

Gírio, F.M., Fonseca, C., Carvalheiro, F., Duarte, L.C., Marques, S., Bogel-Łukasik, R. 2010. Hemicelluloses for fuel ethanol: A review. *Bioresour. Technol.* 101(13), 4775–4800.

Goodey A.R, Tubbs R.S., 1982. Genetic and biochemical analysis of the ability of *Saccharomyces cerevisiae* to decarboxylate cinnamic acid. *J. Bacteriol.* 128, 2615–2620.

Gorsich, S. W., Dien, B.S., Nichols, N.N., Slininger, P.J., Liu, Z.L., Skory, C.D., 2006. Tolerance to furfural-induced stress is associated with pentose phosphate pathway genes *ZWF1, GND1, RPE1*, and *TKL1 in Saccharomyces cerevisiae. Appl. Microbiol. Biot.* 71, 339–349.

Guo, G.L., Chen, W.H., Chen, W.H., Men, L.C., Hwang, W.S., 2008. Characterization of dilute acid pretreatment of silvergrass for ethanol production. *Bioresour. Technol.* 99, 6046–6053.

Guo, P., Mochidzuki, K., Cheng, W., Zhou, M., Gao, H., Zheng, D., Wang, X., Cui, Z., 2011. Effects of different pretreatment strategies on corn stalk acidogenic fermentation using a microbial consortium. *Bioresour. Technol.* 102, 7526–7531.

Gurram, R.N., Datta, S., Lin, Y.J., Snyder, S.W., Menkhaus, T.J., 2011. Removal of enzymatic and fermentation inhibitory compounds from biomass slurries for enhanced biorefinery process efficiencies. *Bioresour. Technol.* 102, 7850–7859.

Hadi, S.M., Rehman, A., 1989. Specificity of the interaction of furfural with DNA. *Mutat. Res. Lett.* 225, 101–106.
Hasunuma, T, Ismail, K.S.K., Nambu, Y., Kondo, A., 2014. Co-expression of TAL1 and ADH1 in recombinant xylose-fermenting *Saccharomyces cerevisiae* improves ethanol production from lignocellulosic hydrolysates in the presence of furfural. *J. Biosci. Bioeng.* 117, 165–169.
He, Y., Zhang, J., Bao, J., 2016. Acceleration of biodetoxification on dilute acid pretreated lignocellulose feedstock by aeration and the consequent ethanol fermentation evaluation. *Biotechnol. Biofuels.* 9, 19.
Hou, Q., Ju, M., Li, W., Liu, L., Chen, Y., Yang, Q., 2017. Pretreatment of lignocellulosic biomass with ionic liquids and ionic liquid-based solvent systems. *Molecules.* 22, 490.
Hu, G., Heitmann, J.A., Rojas, O.J., 2008. Feedstock pretreatment strategies for producing ethanol from wood, bark, and forest residues. *BioResource.* 3, 270–294.
Hu, J., Lin, Y., Zhang, Z., Xiang, T., Mei, Y., Zhao, S., Liang, Y., Peng, N., 2016. High-titer lactic acid production by *Lactobacillus pentosus* FL0421 from corn stover using fed-batch simultaneous saccharification and fermentation. *Bioresour. Technol.* 214, 74–80.
Jaffur, N., Jeetah, P., Kumar, G., 2021. A review on enzymes and pathways for manufacturing polyhydroxybutyrate from lignocellulosic materials. 3 Biotech. 11, 483.
Jin, M.; Lau, M.W.; Balan, V.; Dale, B.E., 2010. Two-step SSCF to convert AFEX-treated switchgrass to ethanol using commercial enzymes and *Saccharomyces cerevisiae 424A (LNH-ST). Bioresour. Technol.* 101, 8171–8178.
Jönsson, L.J., Alriksson, B., Nilvebrant, N.O., 2013. Bioconversion of lignocellulose: Inhibitors and detoxification. *Biotechnol. Biofuels.* 6, 16.
Jönsson, L.J., Martín, C., 2016. Pretreatment of lignocellulose: Formation of inhibitory by-products and strategies for minimizing their effects. *Bioresour. Technol.* 199, 103–112.
Jørgensen, H., Kristensen, J.B., Felby, C., 2007. Enzymatic conversion of lignocellulose into fermentable sugars: Challenges and opportunities. *Biofuels Bioprod. Bioref.* 1, 119–134.
Kim, D., 2018. Physico-chemical conversion of lignocellulose: Inhibitor effects and detoxification strategies: A mini review. *Molecules* 23, 309.
Kim, Y., Mosier, N.S., Ladisch, M.R., 2009. Enzymatic digestion of liquid hot water pretreated hybrid poplar. *Biotechnol. Prog.* 25, 340–348.
Kim, Y., Ximenes, E., Mosier, N.S., Ladisch, M.R., 2011. Soluble inhibitors/deactivators of cellulase enzymes from lignocellulosic biomass. *Enzyme Microb. Technol.* 48, 408–415.
Klinke, H.B., Thomsen, A.B., Ahring, B.K., 2004. Inhibition of ethanol-producing yeast and bacteria by degradation products produced during pre-treatment of biomass. *Appl. Microbiol. Biotechnol.* 66, 10–26.
Ko, J.K., Um, Y., Park, Y.C., Seo, J.H., Kim, K.H., 2015. Compounds inhibiting the bioconversion of hydrothermally pretreated lignocellulose. *Appl. Microbiol. Biotechnol.* 99, 4201–4212.
Kong, X., Defemur, Z., Li, M., Zhang, Q., Li, H., Yue, X., 2021. Effects of combined ultrasonic and grinding pretreatments on anaerobic digestion of vinegar residue: Organic solubilization, hydrolysis, and $CH_4$ production. *Environ. Technol.* 43, 2207–2217.
Koppram, R., Albers, E., Olsson, L., 2012. Evolutionary engineering strategies to enhance tolerance of xylose-utilizing recombinant yeast to inhibitors derived from spruce biomass. *Biotechnol. Biofuels.* 5, 32.
Kumar, V., Yadav, S.K., Kumar, J., Ahluwalia, V., 2020. A critical review on current strategies and trends employed for removal of inhibitors and toxic materials generated during biomass pretreatment. *Bioresour. Technol.* 299, 122633.
Kunasundari, B., Arai, T., Sudesh, K., Hashim, R., Sulaiman, O., Stalin. N.J., Kosugi, A., 2017. Detoxification of sap from felled oil palm trunks for the efficient production of lactic acid. *Appl. Biochem. Biotechnol.* 183, 412–425.
Kurosawa, K., Laser, J., Sinskey, A.J., 2015. Tolerance and adaptive evolution of triacylglycerol-producing *Rhodococcus opacus* to lignocellulose-derived inhibitors. *Biotechnol. Biofuels.* 8, 1–14.
Larsson S., Reimann A., Nilvebrant N.O., Jonsson L.J., 1999. Comparison of different methods for the detoxification of lignocellulose hydrolysates of spruce. *Appl. Biochem. Biotech.* 77, 91–103.
Larsson, S., Cassland, P., Jönsson, L.J., 2001a. Development of a *Saccharomyces cerevisiae* strain with enhanced resistance to phenolic fermentation inhibitors in lignocellulose hydrolysates by heterologous expression of laccase. *Appl. Environ. Microb.* 67, 1163–1170.
Larsson, S., Nivelbrant, N.O., Jönsson, L.J., 2001b. Effect of overexpression of *Saccharomyces cerevisiae* Pad1p on the resistance to phenylacrylic acids and lignocellulose hydrolysates under aerobic and oxygen-limited conditions. *Appl. Microbiol. Biot.* 57, 167–174.
Laureano-Perez, L., Teymouri, F., Alizadeh, H., Dale, B.E., 2005. Understanding factors that limit enzymatic hydrolysis of biomass. *Appl. Biochem. Biotechnol.* 124, 1081–1099.

Liu, G., Sun, J., Zhang, J., Tu, Y., Bao, J., 2015. High titer L-lactic acid production from corn stover with minimum wastewater generation and techno-economic evaluation based on Aspen plus modeling. *Bioresour. Technol.* 198, 803–810.

Liu, H., Liu, X., Jiang, H., Liang, C., Zhang, Z.C., 2021. Enhanced lactic acid production from P2O5-pretreated biomass by domesticated *Pediococcus pentosaceus* without detoxification. *Bioproc. Biosyst. Eng.* 44, 2153–2166.

Liu, Z.L., 2011. Molecular mechanisms of yeast tolerance and in situ detoxification of lignocellulose hydrolysates. *Appl. Microbiol. Biot.* 90, 809–825.

Liu, Z.L., Blaschek, H.P., 2010. Biomass conversion inhibitors and in situ detoxification. In A. A. Vertès, N. Qureshi, H. Yukawa, & H. P. Blaschek (Eds.), *Biomass to biofuels: Strategies for global industries* (pp. 233–259). Blackwell Publishing Ltd. https://doi.org/10.1002/9780470750025.ch12.

Liu, Z.L., Slininger, P.J., Gorsich, S.W., 2005. Enhanced biotransformation of furfural and hydroxymethylfurfural by newly developed ethanologenic yeast strains. *Appl. Appl. Biochem. Biotechnol.* 121–124, 451–460.

López, M.J.; Moreno, J.; Nichols, N.N.; Dien, B.S.; Bothast, R.J., 2004. Isolation of microorganisms for biological detoxification of lignocellulosic hydrolysates. *Appl. Microbiol. Biot.* 64, 125–131.

Lopez, P.C., Peng, C., Arneborg, N., Junicke, H., Gernaey, K. V., 2021. Analysis of the response of the cell membrane of *Saccharomyces cerevisiae* during the detoxification of common lignocellulosic inhibitors. *Sci. Rep.* 11, 6853.

Lu, X., Li, F., Zhou, X., Hu, J., Liu, P., 2022. Biomass, lignocellulolytic enzyme production and lignocellulose degradation patterns by Auricularia auricula during solid state fermentation of corn stalk residues under different pretreatments. *Food Chem.* 384, 132622.

Luo, X., Zeng, B., Zhong, Y., Chen, J., 2022. Production and detoxification of Inhibitors during the destruction of lignocellulose spatial structure. *Bioresources.* 17, 1939–1961.

Maitra, S., Singh, V., 2021. Balancing sugar recovery and inhibitor generation during energycane processing: Coupling cryogenic grinding with hydrothermal pretreatment at low temperatures. *Bioresour. Technol.* 321, 124424.

Minmunin, J., Limpitipanich, P., Promwungkwa, A., 2018. Delignification of Bana grass using sodium hydroxide and ozone. *Waste Biomass Valoriz.* 9, 2099–2105.

Moreno, A.D., Ibarra, D., Alvira, P., Tomas-Pejo, E., Ballesteros, M., 2015. A review of biological delignification and detoxification methods for lignocellulosic bioethanol production. *Crit. Rev. Biotechnol.* 35, 342–354.

Mosier, N., Wyman, C., Dale, B., Elander, R., Lee, Y.Y., Holtzapple, M., Ladisch, M., 2005a. Features of promising technologies for pretreatment of lignocellulosic biomass. *Bioresour. Technol.* 96, 673–686.

Mosier, N.S., Hendrickson, R., Brewer, M., Ho, N., Sedlak, M., Dreshel, R., Welch, G., Dien, B.S., Aden, A., Ladisch, M.R., 2005b. Industrial scale-up of pH-controlled liquid hot water pretreatment of corn fiber for fuel ethanol production. *Appl. Biochem. Biotech.* 125, 77–97.

Nichols, N.N.; Dien, B.S.; Cotta, M.A., 2010. Fermentation of bioenergy crops into ethanol using biological abatement for removal of inhibitors. *Bioresour. Technol.* 101, 7545–7550.

Nichols, N.N.; Hector, R.E.; Saha, B.C.; Frazer, S.E.; Kennedy, G.J., 2014. Biological abatement of inhibitors in rice hull hydrolyzate and fermentation to ethanol using conventional and engineered microbes. *Biomass Bioenerg.* 67, 79–88.

Nichols, N.N., Sharma, L.N., Mowery, R.A., Chambliss, C.K., van Walsum, G.P., Dien, B.S., Iten, L.B., 2008. Fungal metabolism of fermentation inhibitors present in corn stover dilute acid hydrolysate. *Enzyme Microb. Tech.* 42, 624–630.

Nigam, J.N., 2001. Ethanol production from wheat straw hemicellulose hydrolysate by *Pichia stipitis*. *J. Biotechnol.* 87, 17–27.

Okuda, N., Soneura, M., Ninomiya, K., Katakura, Y., Shioya, S., 2008. Biological detoxification of waste house wood hydrolysate using *Ureibacillus thermosphaericus* for bioethanol production. *J. Biosci. Bioeng.* 106, 128–133.

Palmqvist E., Hahn-Hägerdal, B., 2000. Fermentation of lignocellulosic hydrolysates. I: Inhibition and detoxification. *Bioresour. Technol.* 74, 17–24.

Parawira W, Tekere M., 2011. Biotechnological strategies to overcome inhibitors in lignocellulose hydrolysates for ethanol production: Review. *Crit. Rev. Biotechnol.* 31, 20–31.

Park, Y.C.; Kim, J.S., 2012. Comparison of various alkaline pretreatment methods of lignocellulosic biomass. *Energy.* 47, 31–35.

Qin, L., Li, W.C., Liu, L., Zhu, J.Q., Li, X., B.Z. Li, Yuan, Y.J., 2016. Inhibition of lignin-derived phenolic compounds to cellulase. *Biotechnol. Biofuels.* 9, 70.

Rajendran, K., Drielak, E., Varma, V.S., Muthusamy, S., Kumar, G., 2018. Updates on the pretreatment of lignocellulosic feedstocks for bioenergy production-a review. *Biomass Conv. Biorefin.* 8, 471–483.

Reginatto, V., Antonio, R.V., 2015. Fermentative hydrogen production from agroindustrial lignocellulosic substrates. *Braz. J. Microbiol.* 46, 323–335.

Rios-Gonzalez, L.J., Medina-Morales, M.A., Rodriguez-De la Garza, J.A., Romero-Galarza, A., Davila, M.D., Morales-Martinez, T.K., 2021. Comparison of dilute acid pretreatment of agave assisted by microwave versus ultrasound to enhance enzymatic hydrolysis. *Bioresour. Technol.* 319, 124099.

Rizal, N.F.A.A., Ibrahim, M.F., Zakaria, M.R., Abd-Aziz, S., Yee, P.L., Hassan, M.A., 2018. Pretreatment of oil palm biomass for fermentable sugars production. *Molecules.* 23, 1381.

Roberto, I.C., Lacis, L.S., Barbosa, M.F.S., de Mancilha, I.M., 1991. Utilization of sugar cane bagasse hemicellulosic hydrolysate by Pichia stipitis for the production of ethanol. *Process Biochem.* 26, 15–21.

Rosales-Calderon, O., Arantes, V., 2019. A review on commercial-scale high-value products that can be produced alongside cellulosic ethanol. *Biotechnol. Biofuels* 12, 240.

Satari, B., Karimi, K., 2018. Mucoralean fungi for sustainable production of bioethanol and biologically active molecules. *Appl. Microbiol. Biotechnol.* 102, 1097–1117.

Schneider H., 1996. Selective removal of acetic acid from hardwood-spent sulfite liquor using a *mutant yeast. Enzyme. Microb. Tech.* 19, 94–98.

Sen, S.M., Binder, J.B., Raines, R.T., Maravelias, C.T., 2012. Conversion of biomass to sugars via ionic liquid hydrolysis: Process synthesis and economic evaluation. *Biofuels Bioprod. Biorefin.* 6, 444–452.

Sendich, E.D., Dale, B.E., Kim, S., 2008a. Comparison of crop and animal simulation options for integration with the biorefinery. *Biomass Bioenerg.* 32, 1162–1174.

Sendich, E.N., Laser, M., Kim, S., Alizadeh, H., Laureano-Perez, L., Dale, B., Lynd, L., 2008b. Recent process improvements for the ammonia fiber expansion (AFEX) process and resulting reductions in minimum ethanol selling price. *Bioresour. Technol.* 99, 8429–8435.

Silva, J.P., Carneiro, L.M., Roberto, I.C., 2013. Treatment of rice straw hemicellulosic hydrolysates with advanced oxidative processes: A new and promising detoxification method to improve the bioconversion process. *Biotechnol. Biofuels.* 6, 23.

Sjulander, N., Kikas, T., 2020. Origin, impact, and control of lignocellulosic inhibitors in bioethanol production: A review. *Energies.* 13, 4751.

Sodre, V., Vilela, N., Tramontina, R., Squina, F. M., 2021. Microorganisms as bioabatement agents in biomass to bioproducts applications. *Biomass Bioenerg* 151, 106161.

Solowski, G., Konkol, I., Cenian, A., 2020. Production of hydrogen and methane from lignocellulose waste by fermentation. A review of chemical pretreatment for enhancing the efficiency of the digestion process. *J. Clean. Prod.* 267, 121721.

Sun, Y., Li, X., Wu, L., Li, Y., Li, F., Xiu, Z., Tong, Y., 2021. The advanced performance of microbial consortium for simultaneous utilization of glucose and xylose to produce lactic acid directly from dilute sulfuric acid pretreated corn stover. *Biotechnol. Biofuels* 14, 233

Suryadi, H., Judono, J.J., Putri, M.R., Eclessia, A.D., Ulhaq, J.M., Agustina, D.N., Sumiati, T., 2022. Biodelignification of lignocellulose using ligninolytic enzymes from white-rot fungi. *Heliyon.* 8, e08865.

Taherzadeh, M.J., Gustafsson, L., Niklasson, C., Lidén, G., 1999. Conversion of furfural in aerobic and anaerobic batch fermentation of glucose by Saccharomyces cerevisiae. *J. Biosci. Bioeng.* 87, 169–174.

Taherzadeh, M.J., Karimi, K., 2008. Pretreatment of lignocellulosic wastes to improve ethanol and biogas production: A review. *Int. J. Mole. Sci.* 9, 1621–1651.

Talebnia, F., Karakashev, D., Angelidaki, I., 2010, Production of bioethanol from wheat straw: An overview on pretreatment, hydrolysis and fermentation. *Bioresour. Technol.* 101, 4744–4753.

Teymouri, F., Laureano-Perez, L., Alizadeh, H., Dale, B.E., 2005. Optimization of the ammonia fiber explosion (AFEX) treatment parameters for enzymatic hydrolysis of corn stover. *Bioresour. Technol.* 96, 2014–2018.

Thomsen M.H., Thygesen A., Thomsen A.B., 2009. Identification and characterization of fermentation inhibitors formed during hydrothermal treatment and following SSF of wheat straw. *Appl. Microbiol. Biot.* 83, 447–455.

Tian, S., Luo, X.L., Yang, X.S., Zhu, J.Y., 2010. Robust cellulosic ethanol production from SPORL-pretreated lodgepole pine using an adapted strain Saccharomyces cerevisiae without detoxification. *Bioresour. Technol.* 101, 8678–8685.

Toquero, C., Bolado, S., 2014. Effect of four pretreatments on enzymatic hydrolysis and ethanol fermentation of wheat straw: Influence of inhibitors and washing. *Bioresour. Technol.* 157, 68–76.

Tramontina, R., Brenelli, L.B., Sodre, V., Franco Cairo, J.P., Travalia, B.M., Egawa, V.Y., Goldbeck, R., Squina, F.M., 2020. Enzymatic removal of inhibitory compounds from lignocellulosic hydrolysates for biomass to bioproducts applications. *World J. Microbiol. Biotechnol.* 36, 166.

Uppugundla, N., da Costa Sousa, L., Chundawat, S.P., Yu, X., Simmons, B., Singh, S., Gao, X., Kumar, R., Wyman, C.E., Dale, B.E., Balan, V., 2014. A comparative study of ethanol production using dilute acid, ionic liquid and AFEX™ pretreated corn stover. *Biotechnol. Biofuels*. 7, 72.

Wang, J., Gao, Q., Zhang, H., Bao, J., 2016. Inhibitor degradation and lipid accumulation potentials of oleaginous yeast *Trichosporon cutaneum* using lignocellulose feedstock. *Bioresour. Technol.* 218, 892–901.

Wang, X., Yomano, L. P., Lee, J. Y., York, S. W., Zheng, H., Mullinnix, M. T., Shanmugam, K. T., Ingram, L. O., 2013. Engineering furfural tolerance in *Escherichia coli* improves the fermentation of lignocellulosic sugars into renewable chemicals. *Proc. Natl. Acad. Sci. USA*. 110, 4021–4026.

Wierckx, N., Koopman, F., Bandounas, L., de Winde, J.H., Ruijssenaars, H.J., 2010. Isolation and characterization of *Cupriavidus basilensis* HMF14 for biological removal of inhibitors from lignocellulosic hydrolysate. *Microb. Biotechnol.* 3, 336–343.

Wu, D., Wei, Z.M., Mohamed, T.A., Zheng, G.R., Qu, F.T., Wang, F., Zhao, Y., Song, C.H., 2022. Lignocellulose biomass bioconversion during composting: Mechanism of action of lignocellulase, pretreatment methods and future perspectives. *Chemosphere* 286, 131635.

Wyman, C.E., Dale, B.E., Elander, R.T., Holtzapple, M., Ladisch, M.R., Lee, Y.Y., 2005. Coordinated development of leading biomass pretreatment technologies. *Bioresour. Technol.* 96, 1959–1966.

Xie, Y., Hu, Q., Feng, G., Jiang, X., Hu, J., He, M., Hu, G., Zhao, S., Liang, Y., Ruan, Z., Peng, N., 2018. Biodetoxification of phenolic inhibitors from lignocellulose pretreatment using *Kurthia huakuii* LAM0618$^T$ and subsequent lactic acid fermentation. *Molecules*. 23, 2626.

Xu, Y.Y., Hua, K.J., Huang, Z., Zhou, P.-P., Wen, J.B., Jin, C., Bao, J., 2022. Cellulosic hydrocarbons production by engineering dual synthesis pathways in Corynebacterium glutamicum. *Biotechnol. Biofuels Bioprod.* 15, 29.

Yu, Y., Feng, Y., Xu, C., Liu, J., Li, D., 2011. Onsite bio-detoxification of steam-exploded corn stover for cellulosic ethanol production. *Bioresour. Technol* 102,5123–5128.

Zakaria, M.R., Fujimoto, S., Hirata, S., Hassan, M.A., 2014. Ball milling pretreatment of oil palm biomass for enhancing enzymatic hydrolysis. *Appl. Biochem. Biotechnol.* 173, 1778–1789.

Zaldivar, J., Martinez, A., Ingram, L.O., 1999. Effect of Selected Aldehydes on the Growth and Fermentation of Ethanologenic *Escherichia coli*. *Biotechnol. Bioeng.* 65, 24–33.

Zhang, B., Zhan, B., Bao, J., 2021. Reframing biorefinery processing chain of corn fiber for cellulosic ethanol production. *Ind. Crop. Prod.* 170, 113791.

Zhang, J., Zhu, Z., Wang, X., Wang, N., Wang, W., Bao, J., 2010. Biodetoxification of toxins generated from lignocellulose pretreatment using a newly isolated fungus, *Amorphotheca resinae* ZN1, and the consequent ethanol fermentation. *Biotechnol. Biofuels*. 3, 26.

Zhang, N., Tian, C., Fu, P., Yuan, Q., Zhang, Y., Li, Z., Yi, W., 2022. The fractionation of corn stalk components by hydrothermal treatment followed by ultrasonic Ethanol Extraction. *Energies*. 15, 2616.

Zheng, Y., Pan, Z., Zhang, R., 2009. Overview of biomass pretreatment for cellulosic ethanol production. *Int. J. Agric. Biol. Eng.* 2, 51–68.

Zhu, J., Yong, Q., Xu, Y., Yu, S., 2011. Detoxification of corn stover prehydrolyzate by trialkylamine extraction to improve the ethanol production with *Pichia stipitis* CBS 5776. *Bioresour. Technol.* 102, 1663–1668.

Zou, L., Ouyang, S., Hu, Y., Zheng, Z., Ouyang, J., 2021. Efficient lactic acid production from dilute acid-pretreated lignocellulosic biomass by a synthetic consortium of engineered *Pseudomonas putida* and *Bacillus coagulans*. *Biotechnol. Biofuels*. 14, 227.

# *Section III*

## *Case Studies*

# 8 Waste Management System Hybridity

## *A Case Study from the Znojmo Region, Czech Republic*

*Michal Plaček*[1], *Gabriela Daniel*[2], *and František Ochrana*[3]
[1]National Museum of Agriculture, Prague, Czech Republic
[2]Department of Economics, Ambis University, Czech Republic
[3]Department of Public and Social Policy, Faculty of Social Sciences, Charles University, Czech Republic

## 8.1 INTRODUCTION

How public services are delivered remains one of the most crucial issues in public management studies. Waste collection is one of the essential public services that are delivered by municipalities (Bel & Sebő, 2020). Several possible arrangements exist for delivering this service, including privatization, intermunicipal cooperation, and mixed delivery (Bel & Sebő, 2020). One possible arrangement is a hybrid model in which public services are performed by organizations operating at the intersection of the market and the public sector (Grossi & Thomasson, 2015). The impetus for creating a hybrid model for the waste management system is the dynamic changes in which waste management finds itself in times of emergency. These changes, which usually have the character of external destabilizing factors, create new conditions for the functioning of waste management. One of these destabilizing factors is the 'time of emergency'. The term 'time of emergency' refers to a set of phenomena that cause fluctuation and instability in a dynamic system (Prigogine & Stengers, 1984) and lead to the risk of losing its functionality. Management is faced with how to effectively respond to these changes.

In this chapter, using the example of the (good) practices of the city of Znojmo, we show that one positive management response to dynamic environmental changes can be the implementation of a hybrid waste management model. We present the hybrid waste management model as a possible ideological-organizational paradigm for responding appropriately to crisis phenomena and a state of emergency.

From a theoretical point of view, this form of organization could be deemed an easy and comfortable solution to the problem of designing an effective organization (Asquer, 2014). This type of delivery should offer benefits by minimizing issues of information asymmetry, moral hazards, and asset specificity (Asquer, 2014). This should also result in decreased transaction costs of contracting and of transferring knowledge from the private sector to the public sector (Asquer, 2014).

The current mainstream literature focuses on various forms of cost efficiency and cost reduction by presenting comparisons among states of service delivery (Bel & Warner, 2008). Cost effectiveness is undoubtedly an essential factor and indicator for examining the cost effects. However, it is a partial view, usually framed by public choice theory. Some studies have introduced a different point of view than the public choice theory, underlining the importance of competition (Plata-Díaz et al., 2014). Grossi and Thomasson (2015) introduced the problem of accountability of hybrid arrangements; Landi and Russo (2022) mentioned the issue of coproduction in waste management; and Lee-Geiller and Kütting (2021) worked with the concept of environmental stewardship.

DOI: 10.1201/9781032676890-11

Our research enriches a previous study by Plaček et al. (2021) that focused on the behaviour of hybrid waste collection arrangements in an imperfect institutional environment. Their deep exploratory case study aimed to understand the strengths and weaknesses of waste management hybridity and the impact of hybridity on the relationship between innovation and accountability. One of their most important results was the revelation that the key benefit of the hybrid arrangement was a stable intersectoral partnership that allowed comprehensive control of the waste management service delivery.

This conclusion is the starting point of this chapter. We investigate how such a stable partnership might react to the external shocks of the COVID-19 pandemic and the Ukraine crisis. These two shock events have the nature of fluctuation factors. The Ukraine crisis may even become a bifurcation factor in subsequent developments. As a result of its influence, more fluctuations have recently threatened to appear, such as an energy crisis. The Czech Republic does not currently have an adequate response and strategy for this exogenous shock. In this chapter, we ignore these possible new bifurcation effects. We concentrate on two shock events (the COVID-19 pandemic and the Ukraine crisis) and the reactions of the management of the city of Znojmo, and we consider how to effectively absorb the consequences of these crises and find appropriate responses. The given framework is the conceptual starting point for formulating the research questions.

The primary research questions are as follows:

RQ1: How does the hybrid system manage to absorb external shocks?

RQ2: What role does innovation play in these systems as a tool to mitigate the impact of external shocks?

The research questions led our research in two directions. The first is a 'system status scan'. The indicator is (see RQ1) the degree of absorption of external shocks. We describe this absorption level as a response to shock events and management's ability to manage the crisis. The second research direction is seeking new and appropriate tools to mitigate the impact of external shocks. We see innovation as an instrument of renewal (Schumpeter, 2017) and a means of mitigating the effects of external shocks.

These external shocks are crucial in the evolutionary trajectory of waste collection service delivery. The secondary impacts could lead to austerity policies and fiscal stress or could introduce completely different trajectories based on the innovation of current services, new managerial arrangements, or new forms of coproduction of public goods and services.

The resilience of public service delivery in times of uncertainty is essential for sustainable waste collection and for reaching broader and more strategic objectives, such as sustainable development goals or green deal plans.

Our chapter is important from a methodological point of view because we employ historical institutionalism and a historical-evolutionary approach. We examine the institutional environment and look for promising directions for the further development of hybrid waste management ('possible worlds' – see Leibniz, 1970; Kripke et al., 1980). That framework is complemented with desk research and in-depth structured interviews with politicians and bureaucrats. The aim is to identify critical mechanisms and factors that play a role in absorbing external shocks in hybridity arrangements. We rely on a case study. The case study is a standard research method (Yin, 2009; Stake, 1995; Klonoski, 2013). Examining a single case does not reduce the explanatory power of the problem explained. The conclusions drawn from a case study can be tested (Popper, 2004). We examine a typical case in a typical Central-Eastern European (CEE) country, the Czech Republic, and we assume that the results of our analysis can thus be generalized at least to the CEE region of the European Union (EU) countries.

The principal added value of the chapter lies in investigating new phenomena that have not yet been explored. We draw new theoretical conclusions from this analysis. We see the benefit in analysing new findings that contribute to scientific knowledge that have not yet been sufficiently explored, described, or evaluated in depth. This will generate new hypotheses and assumptions that can be further tested through follow-up studies.

## 8.2 THEORETICAL FOUNDATION

The most significant stream of literature on the subject deals with the issue of waste management privatization and how it affects the cost-of-service delivery (Bel & Warner, 2008). The literature review by Bel and Wagner (2008) is quite challenging for traditional economic theories that advocate privatization, e.g. public choice theories and public property rights theories based on simple arguments that competition should lead to decreasing total costs of waste collection service delivery. According to Bel and Wagner (2008), the savings potential offered by the private sector erodes over the long run compared to public sector delivery. Regarding cost savings, there are different approaches to calculating potential benefits. Park and Hong (2021) used the cost-to-budget ratio to calculate these benefits using the example of South Korea. Their findings revealed that more contracting out results in more spending in the MSW budget per capita rather than in cost savings (Park & Hong, 2021). Bel and Wagner (2008) explained this in terms of the rising importance of the regulatory and institutional environment. They offered an industrial theory as the possible theoretical framework for explanation. The benefits of involving the private sector should include lines in competition for the market as a monopoly production benefiting from economies of scale. The main driver of productivity should be innovation and more flexible work practices (Bel & Wagner, 2008).

Plata-Díaz et al. (2014) examined five alternative forms of delivering municipal waste collection services and tried to explain the reasons and conditions for the public sector to use them. The authors focused on the following arrangements: municipal direct delivery, municipal under contract, a joint production by a public community of several local authorities or a public consortium, a joint production by a community or a consortium with a private firm managing the service, and a joint production by a local or supralocal public company. Plata-Díaz et al. (2014) revealed that contract arrangement and other parts of private operators' participation cost. Another central argument for adopting these structures is fiscal stress (Plata-Díaz et al., 2014). Particular forms of municipal cooperation are preferred in the face of decreasing inter-government funding. The lack of government funding is also an essential factor in choosing a private operator. Quality and political variables have an exciting impact. In the eyes of managers, the private sector cannot outperform the public sector in terms of quality.

In contrast with left-oriented parties, conservative parties most likely involve private operators (Plata-Díaz et al., 2014). The importance of environmental factors, specifically socio-economic and geographic conditions, was confirmed by Pérez-López et al. (2021). According to the results of a study by Pérez-López et al. (2021), environmental conditions strongly influence long-term efficiency. The authors generally reported that public provision of waste management services is better suited to the characteristics of the environment than private provision. 'Provision of the service by private entities is not the optimum management form as regards adaptability to heterogeneous environmental conditions' (Pérez-López et al., 2021, p. 451). 'Forms of service delivery by private concerns are more severely affected by environmental characteristics than are the public management forms' (Pérez-López et al., 2021, p. 451). Campos-Alba et al. (2021) introduced the issue of waste selection into the topic of efficiency provision. According to their results, efficiency is enhanced by waste selection; municipality size is a crucial factor. Higher efficiency is associated with progressive government parties and broader coalitions; no significant association with the gender of the mayor was observed (Campos-Alba et al., 2021, p. 18). Bel and Sebő (2021a) described a new form called 'mixed delivery'. This form lies between full contracting and full re-municipalization. 'This involves the fragmentation of a jurisdiction into several territories, of which at least one is subject to public production and another to private production' (Bel & Sebő, 2021, p. 638). Bel and Sebő (2021) argued that a combined governance model could help manage technological volatility, handle unanticipated conflicts, increase the monitoring of external agents, and reduce opportunism. The local government could act as an organizing hub or regulator of production (Bel & Sebő, 2021). Bel and Sebő (2021) tried to prove this concept in Barcelona. They suggested dividing Barcelona into more zones, which should not lead to decreasing returns of scale, and having a public operator operate a newly added zone to get detailed evidence about cost drivers. In another example using Barcelona, the same authors (Bel & Sebő, 2020) introduced the issue of monitoring the quality of service delivered by a private waste collection company. The public sector is facing a problem of information asymmetry, and quite costly managerial tools are needed to monitor service delivery quality. 'Their results suggest that waste companies devote more effort to provide service quality in the areas closer to areas served by competitors, where quality is more directly comparable with that of competitors because this could influence valuations in future tendering processes' (Bel & Sebő, 2020).

Plaček et al. (2021) introduced a different point of view on hybridity in waste management. It is a new point of view that differs from the methodological approach and is based on a holistic approach. This approach makes it possible to qualitatively examine the institutional environment's role and the fundamental mechanisms and factors influencing the activity of a hybrid company in poor institutional conditions. Plaček et al. (2021) also suggested that despite a significant stream of econometrics studies dedicated to the efficiency of hybrid companies in the waste sector, little is known about how fundamental mechanisms influence accountability, innovativeness, and efficiency. They used a qualitative approach to investigate how a hybrid company works in poor institutional conditions. The results underlined the hybrid company's performance in delivering innovation and stable partnerships with the public sector that can absorb external shocks. The authors pointed out a challenge connected with the hybrid arrangement: problematic accountability. The pressure for accountability remains more on politics than on company management. The company does not fully utilize its innovative potential because innovative initiatives are restricted by profit maximization imperatives and by the pressure of the mother company. Innovation thus becomes an essential factor in the development trajectory of the hybrid company. It is a factor that can generate synergistic effects in the hybrid company's performance in the appropriate institutional environment and, at the same time, act as a catalyst for other innovations.

This conclusion opened new issues for investigating hybridity in particular institutional contexts and under the pressure of external factors like economic crises (Zafra-Gomez et al., 2013), new trends like environmental stewardship (Lee-Geiller & Kütting, 2021), coproduction (Landi & Russo, 2022), gender and social equity (Adib & Mahapatro, 2022), and game-changing events like the COVID-19 pandemic and the war in Ukraine. The war in Ukraine is ongoing, and we urgently need research to investigate its impact.

## 8.3 INSTITUTIONAL CONTEXT

The Czech Republic is highly decentralized, with around 6,258 municipalities (Plaček et al., 2020). Most municipalities have under a thousand inhabitants (Plaček et al., 2020). This situation is one of the reasons for the partial inefficiency of public service delivery by Czech municipalities (Plaček et al., 2020). The main factors associated with inefficiency are economies of scale, in which municipalities do not work at optimum scales, and the lack of accountability pressure from citizens, because citizens are not confronted by their municipalities with the actual price of delivered public services due to the existence of fiscal illusion and information asymmetry (Plaček et al., 2020). There are also problems with the processes of contracting out public service delivery. According to several comparative studies, the Czech Republic lacks the quality of an institutional environment (Plaček et al., 2019), has corruption in public procurement (Plaček et al., 2022a, b), and focuses on formalism and lowest prices rather than on results (Plaček et al., 2020).

Regarding waste collection services, only a few have dealt with the service delivery form, and all of them were quantitatively focused. Soukopová and Klimovský (2016) insisted that the local government serves more as a waste collection services provider than a producer. The authors proved that intermunicipal cooperation could reduce the cost of waste collection service delivery, but they noted that intermunicipal cooperation often overlaps with hybrid structures. The second significant result was that neither public nor private provision was a critical factor in the cost efficiency of waste management structures. The same results regarding the cost reduction effect of municipal cooperation were confirmed by Soukopová and Vaceková (2018). Mikušová Meričková et al. (2022) mapped the development of waste management services over the last 20 years. They identified a completely different trend than in Western countries. Instead of remunicipalization, most waste services are delivered in increasing proportions by contracting out. This form of service delivery offers only marginal savings. The ratio of waste services delivered by municipal cooperation seems unchanged over the years. The main reasons for this are excessive fragmentation, high transaction costs, lack of benchmarking solutions, and low motivation to be efficient. Soukopová et al. (2022) confirmed these results. Intermunicipal cooperation seems to be a cost-effective tool for waste management service delivery; public-private partnership increases the market concentration and has a cost-growing effect. The authors stressed the importance of economies of scale. According to Soukopová et al. (2022), the data indicate the beginning of a remunicipalization trend. Struk and Boďa (2022) showed that the existence of recycling consolidation facilities improves the efficiency of municipal waste management services, but programmes focusing on waste separation and green waste collection do not produce the expected effects.

The COVID-19 and Ukraine crises are examples of external shocks that could cause a complete change in the trajectory of municipal waste management. We, unfortunately, lack evidence on how the COVID-19 and Ukraine crises have affected municipal waste management systems. Still, we are aware of some challenges and consequences for local governments, and we can identify some evidences of how local governments have reacted. Plaček et al. (2021) recognized two basic municipal reaction strategies: passive and active. A passive strategy means the municipality relies on backup from the central government; also, the municipality is focused on savings instead of on its economic policy. The second option is to use an active strategy. Municipalities developed strategic initiatives to face the COVID-19 pandemic; they developed policies and actively invited local actors to cooperate. The authors concluded that both strategies could have different results and consequences in the short and long term. The passive strategy focuses more on the long run but could cause severe damage to the city ecosystem. The active strategy offers clear short-term benefits but could be challenging to maintain in the long run.

Regarding the situation in Ukraine, Plaček (2022) insists that local governments are facing an unprecedented period of uncertainty. According to the author, local governments should focus on strategic planning and collateral effects. Collateral effects mean fiscal stress, increasing public debts, and rising inflation. Local governments should critically evaluate their current spending because of the high probability of the need to reallocate money to cover costs associated with migration from Ukraine, extremely high energy prices, and the increasing cost of investment due to inflation.

## 8.4 METHODS AND DATA

### 8.4.1 Methodological Framework

To investigate the impact of shock events on development, we use a historical-evolutionary and retrograde approach. Our concept was inspired by 'universal Darwinism' (Dawkins, 1983), 'institutional evolutionism' (Winter & Nelson, 1982), 'new institutionalism' (Peters, 2019), 'path dependence' (David, 2000, 2007; North, 1990), historical institutionalism (Plaček et al., 2021; Plaček et al., 2022), and retrograde analysis (Smullyan, 1981).

The historical-evolutionary and retrograde approach examines the developmental trajectory of the analysed phenomenon. It examines how past developments influence a given institutional environment. This approach examines how past events (David, 2007; North, 1990) affect the current institutional environment. On the historical-evolutionary trajectory, we find various events $E_i$ (see Figure 8.1). These events are of two types: perturbance (shock events) and positive events (e.g. government intervention). Disorders (shock – see COVID-19; war in Ukraine) cause fluctuations and instability. Positive influences (e.g. effective managerial decisions) dampen disturbances.

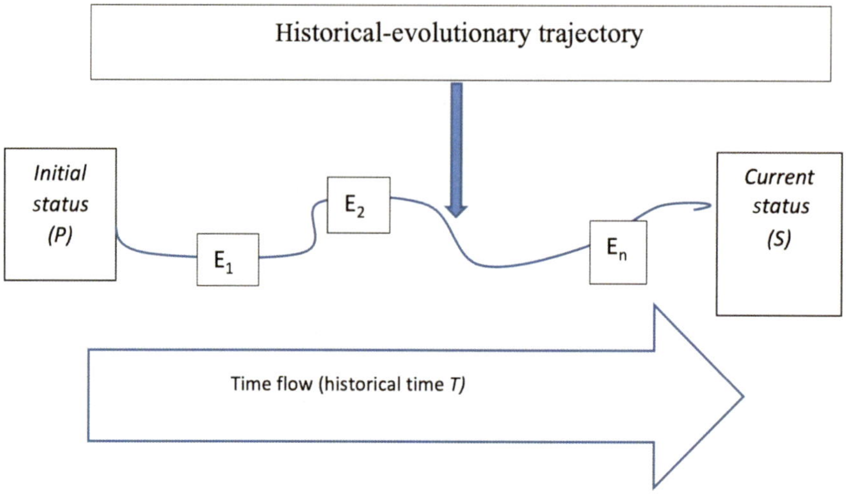


**FIGURE 8.1** Historical-evolutionary trajectories (ontological view).

**Source: Ochrana (2022b).**

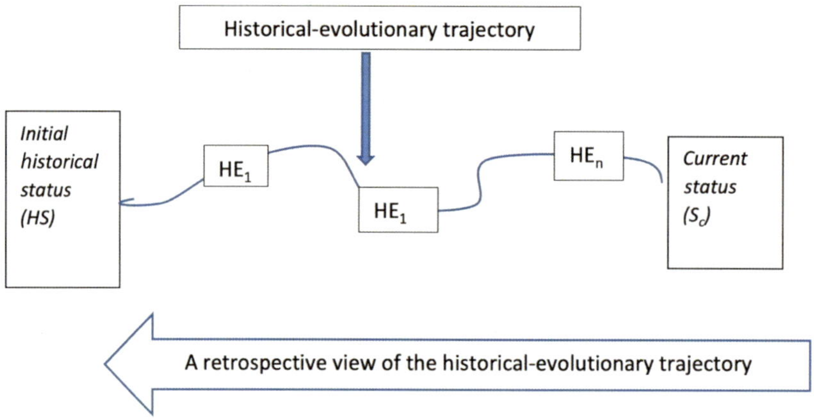


**FIGURE 8.2** Historical-evolutionary trajectories (retrograde view).

**Source: Ochrana (2022b).**

An event $E_i$ takes place at time *T*. It lies on a given historical trajectory defined by the initial state *P* and the current state *S*. The significance and impact of these events are best identified by retrograde analysis.[1]

### 8.4.2 Retrograde Analysis

Retrograde analysis is a retrospective look at past developments (historical trajectory). We can now ask what events happened on a given evolutionary trajectory. How have these historical events (*HEs*) affected the current situation? A retrograde analysis answers this question (see Figure 8.2).

Retrograde analysis is a tool for ex-post examination of a historical-evolutionary trajectory. It is a retrospective epistemological-historical view of the path of social phenomena. A retrograde view reveals the influence of individual HEs on a given evolutionary trajectory. We follow the evolutionary trajectory from the current state (point *S*) back to the beginning (initial historical state of *HS*).

We conducted a retrograde examination of the HEs of *HEi*. We find that these events had different effects on the development. This influence takes various forms: historical nodes, breakpoints, branches, historical turns, embedded history, historical residue, crisis, adaptation, etc. (Ochrana 2022a). COVID-19 and the war in Ukraine are events that we will refer to as a breaking point and a node (historical node).

The node is a milestone on the evolutionary trajectory. A node indicates a significant HE that affected the developmental trajectory. It shows the point on a particular trajectory at which there is potential branching. A breaking point is an event or factor that causes a reversal in the evolutionary trajectory. It is a kind of node with extreme damming and then redirection of current developments. This event 'redirects' the existing evolution and 'reverses' it in another direction. The war in Ukraine is such a breaking point; it redirected the development of EU countries to a different trajectory.[2] We empirically examine one such reaction in the hybridity of the waste management system in Znojmo.

The discovery of the breaking point is of great importance for scientific research, mainly because knowledge of the 'breaking point' phenomenon makes it possible to clarify significant changes in the evolutionary trajectory and identify the start of a new evolutionary route. In retrograde analysis, when we come across these breaking points and determine the branches, we ask ourselves, 'What if?' We create fictional worlds that tell of possible (but unrealized) trajectories. The imaginary worlds are unfulfilled transformations of possibilities into reality. Fictional worlds are the epistemological construction of possible worlds. Possible worlds are new chances and threats for further development.

In the following empirical analysis, we examine how waste management responds to changes in the dynamic environment. The dynamic environment in which waste management works is significantly different than a non-dynamic environment. This difference is related to the essential differences between a dynamic system (which includes society – Prigogine & Stengers, 1984; Luhmann, 1984) and a non-dynamic system. The difference between dynamic systems and non-dynamic systems is shown in Table 8.1. The comparison of both types of systems uses the

**TABLE 8.1**
**Comparison of Non-Dynamic and Dynamic Systems**

| | Type of System and Its Attributes | |
|---|---|---|
| **Comparison Criterion** | **Non-Dynamic System (e.g., Pendulum)** | **Dynamic System (e.g., Society)** |
| System state from the point of view of homeostasis | Balanced system (equilibrium) | Unbalanced system (society is an unbalanced system that changes over time) |
| The 'strength' of the system | Stable system (this system oscillates around an equilibrium state) | Unstable system (society is therefore potentially vulnerable. External shocks cause instability) |
| The 'resilience' of the system in terms of the tendency to return to its original state | Reversibility ('self-healing' system) | Variable state (the dynamic system fluctuates. Therefore, there is a risk of losing system functionality and of irreversibility and dissipation of the system) |
| Response to external stimuli | Response to external stimuli is linear. The behaviour of a non-dynamic system is reliably predictable (see Newton's laws) | The behaviour of dynamic social systems is significantly influenced by sudden emerging phenomena such as pandemics and wars. Response to external stimuli is non-linear. The behaviour of a dynamic system is difficult to predict |

*Source:* Ochrana (2015, p. 39).

following comparative criteria: the system's state, the system strength, the system resistance, and response to external stimuli.

A non-dynamic system is a closed system. It is characterized by a tendency towards equilibrium. If its equilibrium state is disturbed within a certain period of non-equilibrium, the system equilibrium restores itself. Dynamic systems are characterized by instability. The evolution of these systems is determined by the limit cycle. Dynamic systems may fluctuate (Prigogine & Stengers, 1984; Mira, 2000). Fluctuation is a 'pervasive' system disorder that is the attributive property of the system. Provided the system is under dynamic equilibrium, it remains stable. However, if the dynamic system is out of balance, a random phenomenon, the so-called 'butterfly effect' (Lorenz, 1972), can lead to bifurcation. This butterfly effect was in play with COVID-19. The first infected person in Wuhan played the role of the butterfly. This infected person was at the beginning of a chain of dynamic events or fluctuations. We see that fluctuation plays the role of the final initiator, causing a dynamic system to become unstable. The result of this fluctuation was the COVID-19 pandemic. Similarly, the war in Ukraine started many reactions in the EU countries and created many responses in EU countries, including migration, which, in turn, triggered a chain of dynamic events in the Czech Republic.

The case of the hybrid company FCC Znojmo, s.r.o., that we examined is also part of a set of dynamic systems. We empirically examine how it reacts to external stimuli, how it reacts over time, and what its 'resilience' is.

### 8.4.3 Subject of Analysis

The subject of our study is FCC Znojmo, s.r.o., a typical example of a hybrid company. It is a business company in which FCC Environment CEE owns a 49% share, 50% is owned by the Znojmo municipality (public sector), and FCC Únanov, s.r.o., owns 1%. The company belongs to the concern FCC ČR. The concern belongs to the Spanish company Fomento de Construcciones y Contratas, SA (FCC). It is a crucial European firm working in waste management. In the Czech Republic, the consortium has about 20 firms. This is a unique situation in which the private sector owns 50% and the public sector owns the same amount. This setting creates dilemmas and peculiarities that are worth investigating deeply.

FCC Znojmo provides the collection of municipal and sorted waste, including biowaste, bulk trash, and hazardous waste. It includes cleaning and maintaining roads for municipalities and companies in the Znojmo region. Citizens of the city of Znojmo can dispose of massive domestic and hazardous waste, used electrical appliances, etc., in collection yards on Dobšická Street and in Přímětice. The company FCC Znojmo, s.r.o., operates the ROC (regional waste centre), including a sorting line.

According to the latest annual report (2020), the company has net assets of about CZK 64.367 million. The revenue is about CZK 143 million. The profit after taxes is about CZK 14.365 million. The company has 95 regular employees and five employees in executive positions. The governance structure consists of a statutory authority and a supervisory board. The statutory authority consists of two executives nominated by the Znojmo municipality and FCC, s.r.o. The supervisory board includes four members; the city appoints two members, and two are set by FCC, s.r.o., and FCC Únanov, s.r.o. The mayor represents the main shareholder, the Znojmo municipality.

For our analysis, we used a combination of desk research and in-depth structured interviews. The desk research was focused on the FCC annual reports for the years 2010–2021; we analysed mainly the managerial foreword, which summarizes the main events of the previous year, and the financial statements: balance sheet, loss and profit statement, and cash flow statement. This information is publicly available through the official register of the Ministry of Justice. For information about the city of Znojmo and its budget spending, we used the official portal of the Ministry of Finance, called Monitor (https://monitor.statnipokladna.cz/). We examined several documents. For this, we used content analysis. We used retrograde analysis to explore the trajectory and behaviour of FCC.

**TABLE 8.2**
**Recruited Respondents**

| Respondent Code | Position | Responsibility | Experience | Education |
|---|---|---|---|---|
| R1 | Mayor | City investment, financial management, representative of the city at FCC assembly meeting | More than 10 years | Higher |
| R2 | Vice mayor | Tourism | More than 5 years | Higher |
| R3 | Vice mayor | Traffic | More than 10 years | Higher |
| R4 | City secretary | Director of the city office, execution of state administration and self-government | More than 10 years | Higher |
| R5 | Head of the waste department | Responsible for city waste management, waste control, waste analytics, and reporting | More than 10 years | Higher–Ph.D. |
| R6 | Head of the financial department | Budgeting, controlling, financial analysis, cost saving | More than 10 years | Higher |

*Source:* Authors.

In retrospect, we looked for the genesis and development of the events of the past year and their impact on waste management.

The second method was a structured interview with politicians and responsible bureaucrats. During the structured interviews, we followed, from a retrograde perspective, the development trajectory of the events of the past year and their impact on waste management. We used a purposive sampling method to obtain relevant information. The critical criteria for selecting respondents were their competencies and responsibilities connected with waste management and their experience working in the public sector. Table 8.2 presents the characteristics of the recruited respondents.

Figure 8.3 shows the interview questions and their connection with this chapter's central research questions.

## 8.5 FINDINGS

In this section, we present answers to the research questions.

### 8.5.1 RQ1: How Does the Hybrid System Manage to Absorb External Shocks?

First, we reviewed the financial performance of FCC from 2010 to 2021 and data about Znojmo's public expenditure on waste management: a retrograde view. Table 8.3 shows a comparison of fundamental performance indicators of the two systems.

What did the retrograde view of the developmental trajectory reveal? What were the events and effects? The city's performance indicators showed that waste management system expenditure as a percentage of the total budget and current expenditure remained stable.

The fee that citizens pay for waste collection increased from 400 to 600 CZK; there was also an overall increase in public expenditures during the COVID-19 and Ukraine crises. Local governments received compensation transfers from the central government. There is a rising trend in spending on the waste system per citizen. In 2021, this expenditure reached the highest value for the observed period; this was also the main reason for increased local fees. Generally, the system remained stable.

FCC went, perhaps surprisingly, in the opposite direction: its operational profit increased from CZK 18,315 thousand to CZK 30,258 thousand; this shift was caused by increased demands from

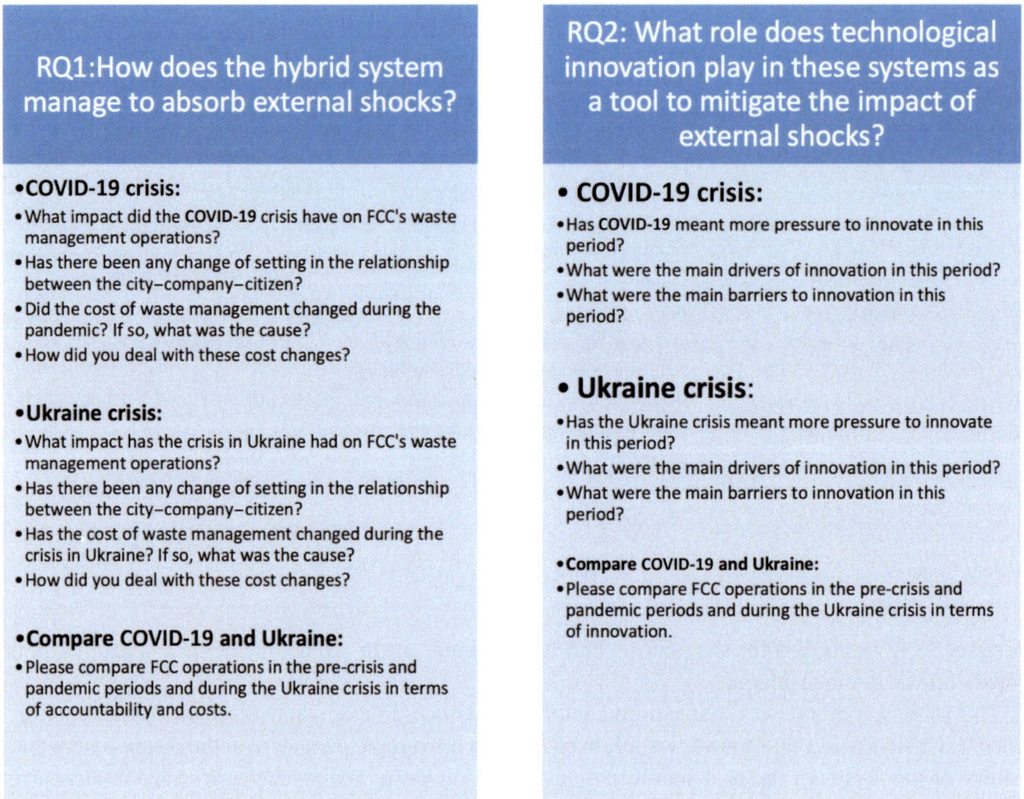


**FIGURE 8.3** Interview questions and their links to principal research questions.

**Source: Authors.**

the city of Znojmo during COVID-19 and by the introduction of a new service for the disposal of medical supplies. The operational profit was the highest for the whole observed period. We can also see improving ROA, ROE, and ROS indicators during the COVID-19 year.

Respondents R2 through R6 (the majority) confirmed no significant break in services during the COVID-19 period; R1 claimed that, 'The company made a lot more money than planned; the amount of waste increased; people used delivery services and the internet; extra collections and bin counts were added; this increased income substantially so there was an increase in revenue; new services were added from waste disposal for testing centres; additional income was not accounted for; and the price of plastics increased'.

Regarding the accountability arrangement, we obtained mixed results. Some respondents (R3, R5, R6) claimed that no significant changes occurred, but other respondents (R1, R2, R4) claimed that FCC management was under increased pressure due to the emergency. Respondent R4 described the situation: 'In the context of the pandemic, the company found itself under more pressure from the city and had to respond more flexibly. There was an emergency and more pressure on the speed of the foreclosure'. All the respondents (R1-R6) confirmed that waste system costs had increased. They reported three main reasons: the change in consumer behaviour during the pandemic, the introduction of new services, and the increasing prices of inputs.

Unfortunately, we have insufficient data for analysing situations connected with the Ukraine crisis, and we must rely on the results of structured interviews. Most respondents (R1, R3, R4, R5, R6) claimed that no significant trending changes occurred during the initial phase of the Ukraine crisis. An opposing opinion was presented only by respondent R2: 'Demand for the purchase of

**TABLE 8.3**
**Comparison of Znojmo and FCC Performance Indicators**

| Category | Indicator | 2010 | 2011 | 2012 | 2013 | 2014 | 2015 | 2016 | 2017 | 2018 | 2019 | 2020 | 2021 | 2022 (05.2022) |
|---|---|---|---|---|---|---|---|---|---|---|---|---|---|---|
| City financial performance | Payment from citizen (in CZK) | 500 | 500 | 500 | 500 | 400 | 400 | 400 | 400 | 400 | 400 | 400 | 600 | 600 |
| | Expenditure per citizen (thousands CZK) | 1,140 | 1,329 | 1,399 | 1,197 | 1,345 | 1,243 | 1,306 | 1,296 | 1,361 | 1,358 | 1,397 | 1,490 | 0,583 |
| | % of total budget | 3.42% | 4.96% | 6.41% | 5.00% | 4.46% | 5.94% | 6.17% | 5.77% | 4.83% | 4.79% | 4.50% | 4.90% | 4.55% |
| | % of current expenditures | 4.51% | 6.01% | 8.59% | 6.30% | 5.65% | 6.64% | 7.06% | 6.42% | 6.27% | 5.69% | 5.45% | 5.80% | 5.42% |
| FCC key financial performance indicators | Operating profit (thousands CZK) | 196,64 | 25,554 | 19,380 | 14,716 | 17,343 | 17,713 | 19,038 | 18,705 | 15,697 | 18,303 | 18,315 | 30,258 | NA |
| | ROA (rentability on assets) | 0.3807 | 0.4141 | 0.3632 | 0.333 | 0.3574 | 0.3813 | 0.3889 | 0.3672 | 0.3016 | 0.3698 | 0.2845 | 0.4190 | NA |
| | ROE (rentability on equity) | 0.8450 | 0.7262 | 0.6671 | 0.7513 | 0.8155 | 0.8303 | 0.8596 | 0.8441 | 0.7295 | 0.8872 | 0.7392 | 0.831 | NA |
| | ROS (return on sales) | 0.1433 | 0.1501 | 0.1158 | 0.10784 | 0.1313 | 0.1334 | 0.1345 | 0.1305 | 0.1024 | 0.1217 | 0.1268 | 0.1762 | NA |

*Source:* Plaček et al. (2021).

sorted waste increased exponentially, increasing the company's revenue'. Respondent R3 confirmed some positive effects: 'The company can profit from Ukrainian workers because it has long-term understaffing'.

Regarding the possibility of changing the accountability arrangement, respondents R3, R4, R5, and R6 insisted that there was no change in the relationship between the city, company, and citizens because the firm has a limited portfolio of activities and is not associated with the Ukraine crisis. Respondents R1 and R2 had different opinions, indicating the issue of fiscal stress as a secondary impact of the Ukraine crisis. Respondent R1 claimed: 'There is more pressure on the company as there are more refugee residents and more children in schools; it takes away from the budget to provide services for residents'.

All the respondents (R1-R6) claimed that the crisis in Ukraine would increase the cost of the waste management system. They identified two main drivers: increasing the prices of inputs and an increasing number of Ukrainian refugees. In the context of this theory, the COVID-19 pandemic and the crisis in Ukraine represent breaking points for companies. These events 'redirected' the development of the waste management system in different directions and on different evolutionary trajectories. This was confirmed by the respondents' answers when they were asked to compare two reference states of the company – before the COVID-19 pandemic and the Ukraine crisis and after these events appeared on the evolutionary trajectory as breaking points.

We asked the respondents to describe the differences between the COVID-19 and Ukraine crises in terms of accountability and financial costs. Three respondents (R1, R3, R6) claimed that the system was under greater pressure during the COVID-19 pandemic than in the Ukraine crisis because all actors were forced to react very quickly during the pandemic, and the FCC company was directly involved because its portfolio of activities included the disposal of medical supplies. All respondents (R1-R6) agreed that the Ukraine crisis would have a more significant impact on system costs than COVID-19 due to increased prices of inputs like fuel, labour, etc.

The final answer to the stated research questions was that a hybrid waste management system can deal with various external shocks of varying natures and perform as a stable arrangement. It allows adverse effects to be distributed in four directions. The first direction is the coverage of the shock by the public sector only; the second direction is the transmission to the citizens; the third direction is the transmission to the firm; and the fourth direction means finding the optimal combination of all three. The results also showed that the hybrid arrangement can perform as an essential actor; for example, during the COVID-19 pandemic, the FCC introduced new services and used the crisis as an opportunity. FCC rapidly increased its profit, and the city's share of the profit could be used to cover social expenditures. It turns out that the hybrid arrangement has the adaptive potential to quickly cope with events that have the nature of destabilizing phenomena. The hybrid arrangement demonstrates a perfect adaptation strategy if a breaking point event appears on the trajectory. Through retrograde analysis, we discovered a fascinating insight regarding accountability. There are some potential problems with accountability arrangements. The financial performance indicators show that citizens paying local fees for waste collection cover less than half of the actual cost of waste collection. Citizens are not confronted with the actual price of this service and are not motivated to ask FCC management and politicians to be accountable. Although politicians claim they feel under more pressure from citizens, they did not introduce any transmission mechanism. Znojmo has not implemented any performance indicators and has no complaints from citizens.

### 8.5.2 RQ2: What Role Does Innovation Play in These Systems as a Tool to Mitigate the Impact of External Shocks?

In this section, we deal with the impact of innovation on the company's development trajectory. We examine the role of innovation as a factor mitigating the effects of external shocks or whether innovation can be a factor that plays the role of a breaking point. We examine the evolutionary

**TABLE 8.4**
**List of FCC Innovations during 2020–2021**

| Year | Innovation |
|---|---|
| 2020 | New cleaning technology against COVID-19 – high-pressure cleaner using hot steam to ensure disinfection of public spaces and urban furniture |
| 2020 | Special waste containers for used masks |
| 2020 | A new service for hospitals and healthcare facilities to dispose of contaminated medical supplies |
| 2020 | New shift rotation system |
| 2021 | Digitalized the system of registration of served waste bins using RFID codes |
| 2021 | Linking container collection records to invoicing |
| 2021 | Introduction of compression technology for waste with a higher compression capacity |
| 2021 | Implementation of a PET bottle perforator that allows for more compact packaging and saves space and money on waste transportation |
| 2021 | Introduction of a vibrator that treats paper waste before pressing and thus ensures greater efficiency of the pressing process |
| 2021 | Switch to the use of chemical spreading material |
| 2021 | Connecting payment terminals with a mobile app |

*Source:* FCC (2020, 2021).

trajectory of innovation and identify which events drove the innovation. We start with a review of innovations during the critical years 2020 and 2021 using information from the FCC annual reports. Table 8.4 shows the list of innovations during that period.

The main driver of innovation in 2020 was the COVID-19 pandemic. The FCC focused on the disposal of materials contaminated with COVID-19. The year 2021 was marked by an acceleration of innovation in two directions. We use retrograde analysis to reveal the two development directions. The first direction was digitalizing internal processes, and the second was the purchase of new waste treatment technologies. The impact of COVID-19 is visible here, as the company tried to respond to possible labour shortages and changes in the labour market.

All the respondents in the structured interview (R1-R6) confirmed that COVID-19 was the main driver of innovation in the given period. The respondents agreed about the critical initiator of innovation. Respondents R4 and R5 insisted that the pressure for innovation came from Znojmo, and the relaxation of rules during the state of emergency enabled a flexible response. Respondents R1, R2, R3, and R6 claimed that the initiator of the innovation was the company itself, which thereby solved the problems associated with COVID-19. We also asked the respondents to identify the main barriers to innovation. Respondents R1, R3, and R4 saw the main reason for complicated decision-making as being on the side of Znojmo. This reason was followed by the complex legislation and the lack of qualified workers. Respondents R2, R5, and R6 saw problems on the FCC side; the firm was not motivated to implement innovations or to focus on profit maximization.

Regarding the situation in Ukraine, there is not sufficient information from FCC's annual reports; we must rely on the results of the structured interviews. The respondents (R1-R6) saw the main trigger for innovation as rising input prices, not the influx of refugees. Respondents also confirmed that it is not yet possible to assess the situation's impact. We do not yet have enough information. This corresponds to the concept of retrograde analysis. We can effectively apply retrograde analysis only when we have a sufficiently long historical path with relevant events to explore.

We asked the respondents to compare the COVID-19 situation and the Ukraine crisis. Respondents R2, R5, and R6 did not see any significant difference between the two events. They were sceptical about the utilization of firm innovation potential. Respondents R1 and R3 deemed COVID-19 to be a more critical innovation driver. Respondent R4 predicted that the main driver of innovations would be the rising cost of inputs.

The final answer to our research question was that innovation could help absorb the crisis events directly connected to FCC's core business. The arrangement between the public and private sectors might complicate the effective execution of innovation. Decisions about innovation face a complicated decision-making structure on the public sector side and for-profit pressure. From the frequent conflicting views among the respondents, we can glean that the city does not have complete information and control over innovation processes in the firm. Regulation also plays a critical role; innovations were implemented more quickly during the pandemic when the rules were loosened during the state of emergency.

## 8.6 CONCLUSIONS AND PERSPECTIVES

According to our results, a hybrid arrangement that could be stable in normal circumstances could be also deemed very stable during emergencies because the public sector has more options for dealing with external shocks. The main instrument for coping with external shock is the city budget. We also remain sceptical about accountability arrangements in hybrid structures. The emergency times did not lead to significant changes. The most crucial difference during the crisis was the utilization of innovative potential; an emergency directly connected to the firm's core business could be an essential innovation driver. The firm could use the crisis to introduce new products and services. One motive could be two sources of pressure: the firm's internal pressure to maintain its profits and the public sector, where firms perform as actors in emergency solutions. The barriers to innovations are not only the profit imperative but also a lack of clear innovation policy on the side of the public sector and a lack of competent workers.

This research opens new directions for future research, in both theoretical and empirical approaches. We tried to show the possibility of using retrograde analysis in the investigation of social phenomena in terms of waste management and to show the evolution of the hybridity of waste management systems in times of emergency. We recognize the limits of our investigation. The single case study describes how some mechanisms work in specific environments and contexts. Understanding how the hybrid arrangement works in a different institutional context is essential. The research should also focus on comparing the performance of hybrid forms with other arrangements such as contracting out, full privatization, and in-house production. The current situation also opens space for correlation analysis, revealing the differences between emergency times and 'normal' conditions. A hybrid organization seems to have the potential to be sufficiently resilient against vulnerability. This is evidenced by the response to external stimuli so far. However, the stated finding does not mean that the hybrid organization is a perfect creation that can adequately respond to all destabilizing influences. It is necessary to constantly look for proactive strategies at the management level. It is desirable to predict possible development trajectories and possible events that may occur in the future. Retrograde analysis of previous developments can be a tool to anticipate these unexpected events. Based on the investigation of the historical trajectory, we can discover historical nodes, breakpoints, branches, historical turns, embedded history, imprints of the past, and historical residue and create a hybrid organization as a 'learning dynamic system'. This is also an opportunity and a challenge for further research.

## NOTES

1 We borrowed the term 'retrograde analysis' from chess theory (Smullyan, 1981). The term 'retrograde analysis' indicates a technique for analysing compositional chess problems. The ideological basis of retrograde analysis is logical deduction. A retrograde solution to the problem is to explain the history of the starting position.

2 In the Czech Republic, this redirection has been so intensely perceived by the public that the Prime Minister of the Czech Republic, Petr Fiala, has been nicknamed 'Ukrainian Prime Minister' by a certain part of the public.

## REFERENCES


Adib, A., & Mahapatro, M. (2022). Private sector involvement in waste management of metropolises: Insights from Dhaka city. *Waste Management*, *142*, 143–151.

Asquer, A. (2014). Explaining partial privatization of public service provision: The emergence of mixed ownership water firms in Italy (1994–2009). *Annals of Public and Cooperative Economics*, *85*, 11–30.

Bel, G., & Sebő, M. (2020). Introducing and enhancing competition to improve the delivery of local services of solid waste collection. *Waste Management*, *118*, 637–646.

Bel, G., & Sebő, M. (2021). Watch your neighbor: Strategic competition in waste collection and service quality. *Waste Management*, *127*, 63–72.

Bel, G., & Warner, M. (2008). Does privatization of solid waste and water services reduce costs? A review of empirical studies. *Resources, Conservation and Recycling*, *52*(12), 1337–1348.

Campos-Alba, C. M., Garrido-Rodríguez, J. C., Plata-Díaz, A. M., & Pérez-López, G. (2021). The selective collection of municipal solid waste and other factors determining cost efficiency. An analysis of service provision by Spanish municipalities. *Waste Management*, *134*, 11–20.

David, P. A. (2000). Path dependence, its critics and the quest for 'historical economics'. In P. Garrouste & S. Ioannides (Eds.), *Market, Competition and Institutions* (pp. 15–40). Edward Elgar Publishing.

David, P. A. (2007). Path dependence: A foundational concept for historical social science. *Cliometrica*, *1*(2), 91–114.

Dawkins, R. (1983). Universal Darwinism. In D. S. Bendall (Ed.), *Evolution: From Molecules to Men* (pp. 403–425). Cambridge University Press.

FCC. (2020). *Annual report 2020*. FCC Group. https://www.fcc-group.eu/en/annual-reports.

FCC. (2021). *Annual report 2021*. FCC Group. https://www.fcc-group.eu/en/annual-reports.

Grossi, G., & Thomasson, A. (2015). Bridging the accountability gap in hybrid organizations: The case of Copenhagen Malmö Port. *International Review of Administrative Sciences*, *81*(3), 604–620.

Klonoski, R. (2013). The case for case studies: Deriving theory from evidence. *Journal of Business Case Studies*, *9*(3), 261–266.

Kripke, S., Davidson, D., & Harman, G. (1980). *Naming and Necessity*. Harvard University Press.

Landi, S., & Russo, S. (2022). Co-production 'thinking' and performance implications in the case of separate waste collection. *Public Management Review*, *24*(2), 301–325.

Lee-Geiller, S., & Kütting, G. (2021). From management to stewardship: A comparative case study of waste governance in New York City and Seoul metropolitan city. *Resources, Conservation and Recycling*, *164*, 105110.

Leibniz, G. W. (1970). Monadológia. (Monadology). In Š. Felber and E. Varossová (Eds.), *Antológia z diel filozofov* (*Anthology of Works of Philosophers*) (pp. 339–351). Epocha.

Lorenz, E. (1972). Predictability: Does the flap of a butterfly's wing in Brazil set off a tornado in Texas? In E. N. Lorenz (Ed.), *The Essence of Chaos* (pp. 181–184). University of Washington Press.

Luhmann, N. (1984). *Soziale Systeme: Grundriss Einer Allgemeinen Theorie*. Suhrkamp.

Mikušová Meričková, B., Soukopová, J., Šumpíková, M., & Křápek, M. (2022). Municipal Solid Waste Management in the Czech Republic and in Slovakia. *NISPAcee Journal of Public Administration and Policy*, *15*(1), 89–112.

Mira, C. (2000). Chaos and fractal properties induced by noninvertibility of models in the form of maps. *Chaos, Solitons & Fractals*, *11*(1–3), 251–262.

North, D. C. (1990). *Institutions, Institutional Change and Economic Performance*. Cambridge University Press.

Nowak, M., & Highfield, R. (2013). *Kooperative Intelligenz. Das Erfolgsgeheimnis der Evolution*. C. H. Beck.

Ochrana, F. (2015). *Methodology of Social Sciences*. Karolinum.

Ochrana, F. (2022a). Historical-evolutionary and retrograde approach to the study of social phenomena and public administration. In Paper Presented at the *30th NISPAcee Annual Conference*, Bucharest, Romania, June 2–June 4.

Ochrana, F. (2022b). *Paradigma, Výzkumný Program a Přístupy ve Vědeckém Zkoumání* (*Paradigm, Research Programme and Approaches in Scientific Research*). Karolinum.

Park, S., & Hong, S. (2021). Assessing the effects of contracting out services on cost in managing municipal solid waste in South Korea. *Waste Management*, *136*, 76–82.

Pérez-López, G., Prior, D., & Zafra-Gómez, J. L. (2021). Modelling environmental constraints on the efficiency of management forms for public service delivery. *Waste Management*, *126*, 443–453.

Peters, B. G. (2019). *Institutional Theory in Political Science: The New Institutionalism*. Edward Elgar Publishing.

Plaček, M. (2022). *Současná Krize na Ukrajině z Pohledu Managementu Obcí. (The Current Crisis in Ukraine from the Point of View of Municipal Management).* [Policy Brief]. Faculty of Social Sciences. Charles University.

Plaček, M., Ochrana, F., Půček, M. J., & Nemec, J. (2020a). Fiscal decentralization reforms and local government efficiency: An introduction. In M. J. Půček, F. Ochrana, & J. Nemec (Eds.), *Fiscal Decentralization Reforms* (pp. 1–49). Springer.

Plaček, M., Ochrana, F., Schmidt, M., Nemec, J., & Půček, M. (2020b). The factors causing delays in public procurement: The Czech Republic versus the UK. *Public Money & Management, 40*(2), 131–139.

Plaček, M., Půček, M., & Ochrana, F. (2019). Identifying corruption risk: A comparison of Bulgaria and the Czech Republic. *Journal of Comparative Policy Analysis: Research and Practice, 21*(4), 366–384.

Plaček, M., Špaček, D., & Ochrana, F. (2020c). Public leadership and strategies of Czech municipalities during the COVID-19 pandemic–municipal activism vs municipal passivism. *International Journal of Public Leadership, 17*, 108–117.

Plaček, M., Vaceková, G., Svidroňová, M. M., Nemec, J., & Korimová, G. (2021). The evolutionary trajectory of social enterprises in the Czech Republic and Slovakia. *Public Management Review, 23*(5), 775–794.

Plaček, M., Vaceková, G., Valentinov, V., & Ochrana, F. (2022). Historical institutionalism: A tool for researching the nonprofit sector in times of pandemic. *Innovation: The European Journal of Social Science Research, 37*, 926–940.

Plaček, M., Valentinov, V., Ochrana, F., Vaceková, G., Langr, I., & Půček, M. (2022). The evolving perceptions of corruption in Czechoslovakia and the Czech Republic. *Governance, 35*(2), 651–659.

Plaček, M., Valentinov, V., Vaceková, G., Ochrana, F., Čurda, S., & Anam, C. (2021). The hybridity of waste management: An in-depth exploratory case study from the Znojmo municipality. *Waste Management, 126*, 433–442.

Plata-Díaz, A. M., Zafra-Gómez, J. L., Pérez-López, G., & López-Hernández, A. M. (2014). Alternative management structures for municipal waste collection services: The influence of economic and political factors. *Waste Management, 34*(11), 1967–1976.

Popper, K. R. (2004). *The Logic of Scientific Discovery*. Taylor and Francis.

Prigogine, Y., & Stengers, I. (1984). *Order out of Chaos: Man's New Dialogue with Nature*. Bantam Books.

Schumpeter, J. A. (2017). *The Theory of Economic Development: An Inquiry into Profits, Capital, Credit, Interest, and the Business Cycle*. Routledge.

Smullyan, R. M. (1981). *The Chess Mysteries of the Arabian Knights*. Knopf.

Soukopová, J., & Klimovský, D. (2016). Local governments and local waste management in the Czech Republic: Producers or providers? *NISPAcee Journal of Public Administration and Policy, 9*(2), 217–237.

Soukopová, J., & Vaceková, G. (2018). Internal factors of intermunicipal cooperation: What matters most and why? *Local Government Studies, 44*(1), 105–126.

Soukopová, J., Mikušová Meričková, B., Nemec, J., & Šumpíková, M. (2022). Institutional factors determining costs of municipal waste management in the Czech Republic. *Waste Management, 144*, 527–532.

Stake, R. E. (1995). *The Art of Case Study Research*. Sage.

Struk, M., & Boďa, M. (2022). Factors influencing performance in municipal solid waste management–A case study of Czech municipalities. *Waste Management, 139*, 227–249.

Winter, S. G., & Nelson, R. R. (1982). *An Evolutionary Theory of Economic Change*. Harvard University Press.

Yin, R. K. (2009). *Case Study Research: Design and Methods* (Vol. 5). Sage.

Zafra-Gómez, J. L., Prior, D., Díaz, A. M. P., & López-Hernández, A. M. (2013). Reducing costs in times of crisis: Delivery forms in small and medium sized local governments' waste management services. *Public Administration, 91*(1), 51–68.

# 9 Potential of Indigenous Earthworms for Agro-Waste Management

## *A Case Study in India*

*Payal Garg, Santosh Satya, and Satyawati Sharma*
Centre for Rural Development and Technology,
Indian Institute of Technology, New Delhi, India

## 9.1 INTRODUCTION

There is a significant amount of organic waste generated from agricultural activities, dairy farms, animal shelters, and household activities. Putrefying waste is generally disposed of without proper management. Considering the prevailing situation of climate change, it becomes very crucial to protect the environment from further degradation by adopting judicious recycling technologies for a variety of organic waste. Vermicomposting is a modern technique for harnessing the potential of micro-ecosystems for effective organic waste utilization with the help of earthworms, resulting in the production of superior-quality organic manure for the agriculture-horticulture sector. It helps to avoid environmental pollution due to the widespread use of biodegradable waste in the rural sector and the expenditure required for waste treatment and disposal. Composted waste/vermicast has a higher quality of manure that can be used to feed our "nutrient-organic matter-hungry" soils (Ismail et al., 2003). There are around 4,400 different species of earthworms found worldwide. However, only a few species of these earthworms are appropriate for vermicomposting. Exotic species used for vermicomposting in India include *Eisenia fetida*, *Eudrilus eugeniae*, and *Perionyx excavatus*. Some demerits associated with the large-scale use of exotic species have been reported. Exotic earthworms may cause environmental risks to our native biodiversity (Ismail, 1994). Considering this, it is necessary to conduct laboratory and field studies to better understand the potential of indigenous earthworm species for agro-waste management and ecological sustainability.

## 9.2 ABUNDANCE OF INDIGENOUS/LOCAL EARTHWORMS—AN OVERVIEW

Julka (1986) found 3,000 species of earthworms around the world. In India, there are approximately 384 species of earthworms, with the possibility that some of them could be used for vermicomposting. India is rich in earthworm resources, and about 8% of the total known species of earthworms inhabit Indian soils. To identify appropriate worm species, Shanthi et al. (1993) investigated the vermicomposting of vegetable waste with three commonly available species of earthworms: *Metaphire posthuma*, *Eisenia* species, and *Perionyx excavatus*. They concluded that *P. excavatus* is the best species for vegetable waste vermicomposting, and the ideal temperature and moisture for worms to survive are 20°C–40°C and 20%–80%, respectively.

Singh (1997) conducted a thorough survey in the Varanasi region to identify the most common Indian earthworm species that could be used for decomposition and soil fertility improvement.

DOI: 10.1201/9781032676890-12

For the first time in the Varanasi region, he identified eleven different species of earthworm. *Octochaetona surensis*, *Dichogaster bolaui*, *Eutyphoeus nicholsoni*, *Lampito mauritii*, and *Eutyphoeus incommodus* are among the species that could be used for soil fertility management. Garg and Bhardwaj (2001) found that vermicompost (VC) prepared with Parthenium and Lantana organic wastes using *Amynthas alexandri*, an indigenous species of the Jaipur area, was rich in fertilizing qualities. Scown and Baker (2006) conducted a comparative study to investigate the effect of livestock dung (sheep, cattle, and horse) on the abundance of exotic (*Aporrectodea trapezoides*, *Microscolex dubius*, and *Microscolex phosphoreus*) and native (*Spenceriella macleayi* and *Spenceriella bywongensis*) earthworms in the grassland of southeastern Australia. They discovered that different species of earthworms responded differently to different types of dung, but horse dung was the preferred substrate for both types of earthworms. Parthasarathy (2007) found that *Lampito mauritii* could be used as an alternative species for vermiculture and vermicomposting due to its high hatching rate, high hatchling production, and tolerance to a wide range of environmental conditions.

Garg et al. (2009) and Suthar (2009) investigated the potential of *Allolobophora parva* for vermicomposting practices and concluded that *A. parva* grows well in various wastes and can be used for long-term land restoration. Paul (2017) studied the diversity of indigenous earthworms in the Sivaganga District of Tamil Nadu. Results showed that indigenous earthworms (*Perionyx excavates*, *Lampito mauritii*, and *D. pellucida pallida*) produced VC efficiently. Although some studies on a few indigenous species have been conducted in India (Suthar, 2007a), an in-depth study of their potential for agro-horticulture waste management through vermicomposting is required. In addition, the quality of VC thus prepared needs to be evaluated for different crop production.

## 9.3 INDIGENOUS EARTHWORM—ECOLOGICAL SIGNIFICANCE

There has been a substantial amount of R&D and field-level studies on vermicomposting using exotic earthworm species. Given the long-term environmental implications of the widespread use of exotic earthworm species, long ago, it was recognized that there was a need to investigate the potential of indigenous earthworms for rural waste vermicomposting (Ismail, 1994; Santosh et al., 1993). Ismail (1994) emphasized that soils contain a variety of earthworm species and that there is no need to import or transfer earthworms from other locations. There are numerous benefits to using indigenous earthworm species. They can be collected locally without any monetary expenditure (Ismail, 1996). Indigenous earthworms are well adapted to their surroundings. Sharma (2017) reported that vermicomposting of organic waste by indigenous earthworms could be done throughout the year, though the VC production potential of indigenous earthworm species varies depending on the species used and the season of the year.

Yusoff et al. (2019) found that the indigenous earthworms would aid local farmers in the composting process because they were inexpensive, and the generated compost was used as organic fertilizer. Organic farm waste can also be managed by allowing indigenous earthworms to decompose it more quickly. As a result, using local or native species would help in achieving ecological security, which is an important aspect of waste recycling in sustainable agriculture. In terms of their inherent characteristics, indigenous earthworm species are able to adapt easily to local geo-climatic conditions (Ismail, 1995), thus promoting eco-interactions. They are also effective soil improvers, contributing to natural soil fertility through physical, chemical, and biological enhancements.

## 9.4 SOLID WASTE MANAGEMENT BY VERMICOMPOSTING—RESEARCH STATUS AND KNOWLEDGE GAP

Vermicomposting has gained prominence in India for mitigating the pollution risks posed by our large amount of organic waste, as well as for reducing the demand for chemical fertilizers (CFs) (Santra and Bhowmik, 2001). Significant research on vermicomposting (both exotic and indigenous earthworms) of various types of waste has been done. This includes sewage sludge (Khwairakpam and Bhargava, 2009); paper waste (Arancon et al., 2008); guar gum industrial waste and biogas sludge (Singh and Sharma, 2002); mango leaf litter and crop residues (Gajalkshmi et al., 2005; Suthar, 2007b); vegetable waste, cow dung, poultry waste, and forestry leaf litter (Jadia and Fulekar, 2008; Muthukumaravel et al., 2008; Manna et al., 2003); and municipal solid waste (Kaviraj and Sharma, 2003; Ghosh, 2004). These findings revealed that limited experiments with native species on various organic substrates have been conducted. Also, research studies on the effect of bioformulations and bioinoculants on earthworm species multiplication, growth, and quality of compost are highly limited. Knowledge of the microorganisms found in the soil and the earthworm gut, as well as their role in the decomposition of organic matter, is very important. More research on heavy metal uptake, elimination, and effects on earthworm growth and reproduction, composting duration, VC quality, and finally field trials on various crops is clearly needed to promote the development of vermiculture and vermicomposting using indigenous earthworm species.

## 9.5 R&D WORK ON INDIGENOUS EARTHWORM (*A. PARVA*) AT IIT DELHI

### 9.5.1 Collection and Inventory of Indigenous Earthworms in the Western UP

During the rainy season, earthworms were collected from a depth of 10–30 cm of soil to create an inventory of indigenous earthworms found in the Bijnor District (Western UP). Earthworms were collected by digging and hand-sorting methods developed by Ismail (1994). Earthworms were collected in a polyethylene bag along with some moist soil from the same site. The bags, after labelling, were taken to the laboratory for the preservation of earthworms to conduct further taxonomic studies.

The preservation of earthworms has three steps: narcotizing, fixation, and storage. The worms were placed in a wide-open flat-bottom container with little water and then were narcotized by gradually adding a 10% solution of formalin in water. The earthworms stopped moving after 10–20 minutes and became completely narcotized. After that, these narcotized worms were placed on a thin layer of cotton, dipped in fixative solution (15% formalin) according to their size, in a straight erect position on a flat dish for 24 hours. The fixed earthworms were then transferred to suitable size labelled glass tubes, filled with 10% formalin solution. These tubes were stored in an inverted position in a big-mouth jar with cotton. This jar was taken to the Zoological Survey of India (ZSI), Solan, for species identification. There were a total of eight indigenous species prevailing in this region, and *Allolobophora parva* (*A. parva*) was the most widely distributed, and hence, it was selected for further study (Table 9.1).

### 9.5.2 Enhancing the Multiplication of Indigenous Earthworms

The rate of multiplication of earthworms determines their suitability for vermicomposting and vermiculture applications. Certain organic growth promoters and microorganisms (bioinoculants), according to the literature, reduce the time period of decomposition and promote the growth of earthworms. Murthy (1987) reported that incorporating some herbal preparations into the compost in minute amounts brought certain cosmic forces into the compost. These forces stimulate the potential and strengthen the action of compost. Biodynamic preparations (502–507) are known to break down organic material rapidly and also accelerate the activity of earthworms in a compost heap (Proctor, 1989).

The rates of reproduction and multiplication of earthworms depend on species and a variety of other factors. Based on the results of the experiment, *A. parva* grows slower than *E. fetida*; thus, different organic growth promoters (homeo-solutions SSY1, BD 506, 507) and bioinoculant (*Azotobacter chroococcum*) were chosen for this purpose. According to the results of the culturing experiment, homeo-formulation (SSY1) gave very interesting results in enhancing earthworm

**TABLE 9.1**
**Inventory of Indigenous Earthworm Species in the Study Area (Western UP)**

| Taxonomic Status (Family and Species) | Ecological Category | Soil Characteristics pH | EC (mmhos/cm) | TOC (%) | TN (%) | Habitat |
|---|---|---|---|---|---|---|
| Lumbricidae | | | | | | |
| *Allolobophora parva* | Epigeic | 7.5 | 4.2 | 1.20 | 0.52 | The soil around the roots of plants, rural sewage system |
| Megascolecidae | | | | | | |
| *Lampito mauritii* | Anecics | 7.2 | 4.50 | 1.70 | 0.53 | Grasslands, forest, crop field, compost pit, domestic garbage, rural sewage system, found in slightly alkaline soil and soil usually with high organic matter |
| *Metaphire posthuma* | Endogeic | 8.3 | 3.40 | 2.30 | 0.22 | Subsoil at 10–20 cm depth in sandy loam soil with high organic content, grasslands, lawns, kitchen garden |
| *Metaphire houlleti* | Endogeic | 6.3 | 3.00 | 1.60 | 0.21 | Compost heaps, plantation croplands, fallow land |
| Octochaetidae | | | | | | |
| *Dichogaster bolaui* | Epigeic | 8.0 | 3.50 | 4.20 | 0.71 | Top 5 cm soil with high organic matter |
| *Octochaetona beatrix* | Epigeic | 8.0 | 1.20 | 1.10 | 0.21 | Sandy loam and clay loam soils with low organic matter in lawns and grasslands |
| *Eutyphoeus incommodus* | Epigeic | 6.9 | 1.40 | 1.40 | 0.43 | Alluvial soil, agriculture fields, plant nurseries, gardens, orchards, and grasslands |
| Ocnerodrilidae | | | | | | |
| *Ocnerodrilus occidentalis* | Epigeic | 8.0 | 4.30 | 1.80 | 0.54 | Found in a wide range of moist habitats in alkaline sandy loam and clay loam soils especially in lowland and upland pastures, crop fields, compost pits, and domestic sewage |

multiplication when compared to BD formulation and bioinoculant (*Azotobacter chroococcum*). It was found that 1.0 ml of homeo-solution/2 kg of the substrate was optimal for increasing the multiplication of both earthworm species (*A. parva* and *E. fetida*).

### 9.5.3 Experimental Design for Vermicomposting Using Agro-Horticulture Waste

Two types of agro-horticulture wastes mixed leaf litter mixtures (sheesham [*Dalbergia sissoo*], jamun [*Syzygium cumini*], and neem [*Azadirachta indica*]) and *Putranjiva roxburghii* were used for the experiment. Experiments were set up in the micromodel complex, at IIT Delhi. A total of 2 kg of substrate (agro-horticulture waste, soil, and cow dung in 3:3:4) was taken in each plastic culture pot separately, and then, the optimum concentration of homeo-solution (1 ml/2 kg) was added to the substrate. Ten mature earthworms were added to each pot containing 2 kg of substrate.

The untreated substrate served as a control. Experiments were carried out in triplicate for 45 days. The moisture content of the substrate was kept between 60% and 70%.

#### 9.5.3.1 Chemical Properties of Vermicompost Prepared with Mixed Leaf Litter Mixture and Horticulture Waste (*Putranjiva roxburghii*) by Both Earthworm Species

The physico-chemical characteristics of mixed leaf litter (*sheesham*, *jamun*, and *neem*) and *Putranjiva roxburghii* VC produced by both earthworm species (*A. parva* and *E. fetida*) were compared. The substrates were studied by determining various parameters, viz. pH, EC, TOC, TN, TK, and TP (Tables 9.2 and 9.3). The pH of the VC in the treated substrate changed from acidic to neutral, but there was little change in the control. According to Mahaly et al. (2018), the near-neutral pH of VC can be attributed to the secretion of $NH4+$ ions, which reduces the pool of $H+$ ions, possibly due to the buffering nature of humic substances. EC increased significantly in both the treated substrates for *A. parva* and *E. fetida* over the EC of the raw substrate. When compared to the initial value of the substrate, the final product had a lower organic carbon content. The reduction in organic carbon with *A. parva* (54%) was statistically at par with *E. fetida* (50%) in both substrates. Chatterjee et al. (2015) found a decrease in TOC in legume substrate VC.

**TABLE 9.2**
**Physico-Chemical Characteristics of Vermicompost (Mixed Leaf Litter) Prepared by *A. parva* and *E. fetida* Earthworms**

| | Treatments | | | |
|---|---|---|---|---|
| Parameters | Substrate + 1 mL SSY1 + *A. parva* | Substrate + 1 mL SSY1 + *E. fetida* | Control (Substrate Alone) | CD at 5% |
| pH | 7.4±0.05 | 7.3±0.06 | 6.4±0.15 | 0.46 |
| EC (mmhos/cm) | 0.65±0.02 | 0.56±0.01 | 0.36±0.01 | 0.09 |
| Total organic carbon (%) | 13.67±0.16 | 14.89±0.17 | 27.53±0.57 | 1.30 |
| Total N (%) | 1.20±0.03 | 0.95±0.02 | 0.69±0.02 | 0.05 |
| Total P (%) | 0.57±0.05 | 0.50±0.03 | 0.29±0.02 | 0.09 |
| Total K (%) | 0.58±0.01 | 0.47±0.03 | 0.34±0.02 | 0.05 |

*Note:* All values are the mean of three replicates ± standard deviation.
Substrate = mixed leaf litter mixture + soil + cow dung (3:3:4) + 1 mL SSY1 (homeo-solution).

**TABLE 9.3**
**Physico-Chemical Characteristics of Vermicompost (*P. roxburghii*) Prepared by *A. parva* and *E. fetida* Earthworms**

| | Treatments | | | |
|---|---|---|---|---|
| Parameters | Substrate + 1 mL SSY1 + *A. parva* | Substrate + 1 mL SSY1 + *E. fetida* | Control (Substrate Alone) | CD at 5% |
| pH | 7.4±0.10 | 7.2±0.26 | 6.8±0.10 | 0.41 |
| EC (mmhos/cm) | 0.58±0.01 | 0.56±0.04 | 0.27±0.07 | 0.02 |
| Total organic carbon (%) | 15.70±0.32 | 15.72±0.14 | 33.97±0.15 | 2.42 |
| Total N (%) | 0.78±0.05 | 0.76±0.09 | 0.48±0.02 | 0.12 |
| Total P (%) | 0.52±0.05 | 0.50±0.03 | 0.29±0.02 | 0.09 |
| Total K (%) | 0.52±0.06 | 0.48±0.14 | 0.30±0.05 | 0.22 |

*Note:* All values are the mean of three replicates ± standard deviation.
Substrate = *Putranjiva roxburghii* + soil + cow dung (3:3:4) +1 ml SSY1 (homeo-solution).

The nutrient content (NPK) was considerably higher in the VC prepared by *A. parva than in E. fetida*. In comparison with the control (0.69% and 0.48%), *A. parva* (1.2% and 0.78%), and *E. fetida* (0.95% and 0.76%), VC had significantly higher N content (Tables 9.2 and 9.3). According to Suthar (2007a), the amount of nitrogen accumulated in earthworm casts indicates the total N content of the substrate used for worm feeding. In general, the literature reveals that the quality of VC is determined by the type of waste and the initial nutrient content of the waste.

Total phosphorus (TP) content was considerably higher in the case of *A. parva* (96% and 79%) than in the case of *E. fetida* (72%) in both substrates, with respect to the control. According to Aalok and Tripathi (2010), earthworm gut phosphatase contributes to the release of P in available form, and further P release may occur via P-solubilizing microorganisms in worm casts. In mixed leaf litter substrate, TK content was significantly higher (0.58%) with *A. parva* VC compared to *E. fetida* (0.47%) than with the control (0.34%), whereas the difference between *A. parva* and *E. fetida* in *P. roxburghii* VC was not statistically significant. Pathma and Sakthivel (2012) found that organic matter that passes through the gut of earthworms during vermicomposting increases the amount of P and exchangeable K nutrients significantly.

Vermicomposting of homeo-solution-treated substrates accelerates mineralization and, as a result, enriches the end product with more available forms of plant nutrients. Better adaptability and earlier acclimatization of indigenous species to the nature of the substrate might have been the reason for better results with native species.

### 9.5.4 Indigenous Earthworms and Heavy Metal Tolerance

As a result of the improper management of waste coming out from the metal industry, heavy metals such as Cu, Zn, Cr, Cd, and Pb are widely distributed in soils. While most organisms in soil tolerate essential trace elements at certain concentrations and exposure times, this can have an effect on their abundance, diversity, and distribution (Lukkari et al., 2004). Hence, an assessment of heavy metal effects on soil organisms is required in order to comprehend the environmental risk of contamination (Malley et al., 2006). A distinguishing feature of earthworms has recently received more attention. Many studies have shown that earthworms can alter the availability, uptake, and accumulation of heavy metals by passing and accumulating toxic metals through their body tissues (Turgay et al., 2011). Furthermore, earthworms are easy to handle and measure various life-cycle parameters such as growth and reproduction, metal accumulation and excretion, and biochemical responses.

There are few reports available on earthworm toxicity data. There have been no systematic studies on the uptake and elimination characteristics of heavy metals in indigenous/local earthworm species reported. Even the relationship between body metal concentration and the effect on indigenous earthworm growth and reproduction has not been studied. In light of this history, the purpose of this study was to compare the effect of different metal concentrations on morphological changes, reproduction, mortality, and total metal accumulation in two different earthworm species, *Eisenia fetida* and *Allolobophora parva*, kept in substrate spiked with different heavy metal concentrations.

#### 9.5.4.1 Experimental Design

Experiments were carried out in triplicate in earthen pots with a capacity of 2 kg each. The substrate (leaf litter, cow dung, and soil [40:30:30, w/w]) was artificially fortified by adding various metal concentrations (dry weight basis). Cu, Cd, Cr, Pb, and Zn were used in the study as metallic nitrate salts. These metals were added to the substrate at 500, 1,500, and 2,500 mg/kg concentrations. The metal concentrations used in the study were chosen based on a bibliographical survey conducted by the authors. During the study, the moisture percentage in the substrate was kept between 50% and 60%. As a control, an untreated substrate was used. Individually, ten healthy earthworms (*Eisenia fetida* and *Allolobophora parva*) were placed in each pot. The duration of the study lasted 45 days.

### 9.5.4.2 Heavy Metal Accumulation in Earthworms

The results show that the concentrations of heavy metals (Cu, Cr, Pb, Zn, and Cd) in both earthworm species were significantly higher than in the control at all doses (Table 9.4). Cu was observed to be the most toxic metal for both earthworm species. None of the copper doses were tolerated by the earthworms. After 45 days, *Eisenia fetida* accumulated heavy metals in the order Cr>Cd>Pb>Zn, whereas *A. parva* accumulated heavy metals in the order Cd>Cr>Pb>Zn. Pb concentrations in *E. fetida* and *A. parva* were initially 0.62 and 0.47 mg/kg, respectively. However, as Pb concentrations increase, the concentration of Pb in both earthworm species increases gradually. In general, *A. parva* accumulated more Pb as compared to *E. fetida*, which may be due to being slightly more sensitive to the presence of lead than *A. parva*. Corp and Morgan (1991) investigated the patterns of metal accumulation (Pb, Zn, Cd, and Cu) in nine epigeic earthworm populations. They discovered that the concentrations of Pb, Zn, and Cd in the "indigenous" worms were significantly higher in most cases than in the "introduced" worms.

The findings also revealed that higher doses of heavy metals affected worm growth. Due to the decrease in reproduction at high concentrations of heavy metals, there were a delay in the completion of the life cycle and a decrease in the total population, resulting in a significant decrease in juveniles and adults, delaying the time for population doubling. Additionally, some morphological changes in both species of earthworm were observed. Zn causes worm sluggishness at 500 mg/kg. At 1,000 mg/kg Cd concentration, the clitella of some earthworms swelled and ruptured. It was concluded that heavy metal concentrations were higher in *Allolobophora parva* than in *Eisenia fetida*. The findings indicated that the earthworm *A. parva* was more adapted to surviving in metal-contaminated environments than the exotic ones (*E. fetida*). Thus, indigenous earthworm species have enormous potential for heavy metal-contaminated soil remediation.

**TABLE 9.4**
**Heavy Metal Concentrations (mg/kg) in *Allolobophora parva* and *Eisenia fetida***

| Metal | Species | Background Concentration of Heavy Metals in Earthworms (mg/kg) | Heavy Metal Concentration in Earthworms on the 45th Day | | |
|---|---|---|---|---|---|
| | | | Spiked Heavy Metal (mg/kg) | | |
| | | | 500 | 1,500 | 2,500 |
| Cu | *Allolobophora parva* | 0.62±0.05 | – | – | – |
| | *Eisenia fetida* | 0.57±0.90 | – | – | – |
| Cr | *Allolobophora parva* | 0.30±0.25 | 416[a]±2.0 | 728[a]±5.8 | 757[a] ±4.5 |
| | *Eisenia fetida* | 0.28±0.13 | 323±3.0 | 617±2.5 | 658±4.0 |
| Pb | *Allolobophora parva* | 0.47±0.04 | 352[a]±0.5 | 490[a]±5.0 | 611[a]±3.0 |
| | *Eisenia fetida* | 0.62±0.03 | 195±1.7 | 246±2.9 | 415±8.0 |
| Zn | *Allolobophora parva* | 23±1.65 | 168±2.4 | – | – |
| | *Eisenia fetida* | 34±0.52 | 155±0.3 | – | – |
| Cd | *Allolobophora parva* | 0.60±0.05 | 368±2.5 | 790±1.0 | – |
| | *Eisenia fetida* | 0.56±0.24 | 216±2.0 | 660±3.2 | – |

*Note:* Results are presented as means of three replicates±SD.
[a] Shows a significant difference ($p<0.05$) in metal concentration between species.
– *Shows mortality.*

## 9.6 FIELD INVESTIGATION OF VERMICOMPOST PRODUCED BY INDIGENOUS AS WELL AS EXOTIC EARTHWORMS ON THE BLACK GRAM (*VIGNA MUNGO*)

The field experiments to assess the response of VC prepared by both earthworm species were carried out in agriculture plots (sandy loam soil, pH 8.5) at the micromodel complex, IIT Delhi, in 2005. The experiments were carried out in a randomized block design (RBD) with a total of six treatments: a combination of two seed varieties, T-9 and indigenous, and three treatments: control (T1), *A. parva* VC (T2), and *E. fetida* VC (T3) (Table 9.5). By calculating the equivalent amount of both types of VC, the recommended dose of N was applied to the soil. The results indicated that the effect of different treatments on total yield in both varieties was in the following order: T2>T3>T1 (Table 9.5). *A. parva* VC (T2) produced a significantly higher yield. This was 19.08 and 15.75 q/ha in both seed varieties in comparison with *E. fetida* VC (T3) and control (T1). However, the control of indigenous variety also yielded a significantly higher yield (8.33 q/ha) compared to the control of T-9 (6.47 q/ha). These findings supported Rana's (2004) conclusion that indigenous *Vigna mungo* seeds produced a higher grain yield (4.2 q/ha) with organic manure than T-9 (3.4 q/ha), which also reported that even in the control, the indigenous seeds gave a statistically higher yield (3.1 q/ha) than the improved variety (2.3 q/ha).

It was noted that a significantly higher grain yield (8%–10%) was obtained in the case of *A. parva* VC as compared to *E. fetida* VC (Table 9.5). The difference was significant for both varieties in terms of grain yield. VC application at 3 tonnes/hectare produced higher chickpea dry matter (19.78 gm/plant), pods/plant (27.38), leaf-area index (1.57), straw yields (3.81 tonnes/ha), and seed (2.35 tonnes/ha) according to Jat and Ahlawat (2004). Jeyabal and Kuppuswamy (2001) found that adding VC to the soil increased rice stalk growth and soil fertility. They reported that with an integrated application of 50% VC, 50% CF, and biofertilizers, rice and legume grain yields were 6.25 and 0.51 tonnes/ha, respectively. These yields were 12.2% and 19.9% higher, respectively, than those obtained with only CF. Javed and Panwar (2013) discovered that VC had the highest seed germination percentage (88%) and CF had the lowest (64%) for urad crops (*Vigna mungo*).

Similarly, in terms of biomass yield, the indigenous variety was clearly superior to the improved (T-9) variety in T2 (14.83 and 11.35 q/ha) in comparison with T3 (12.63 and 10.63 q/ha). The results of our study are corroborated by Rana (2004), who reported significantly higher biomass yields of the indigenous variety (22 q/ha) than the improved variety (13.9 q/ha) of *Vigna mungo*. Under control plots, the indigenous variety gave more straw yield (6.5 q/ha) as compared to the improved one (4.7 q/ha).

**TABLE 9.5**
**Effect of Two Types of Vermicompost on Urad Crop (*Vigna mungo*)**

| Variety | Treatments | Total Yield (q/ha) | Grain Yield (q/ha) | Straw Yield (q/ha) |
|---|---|---|---|---|
| Indigenous | T1 control | 8.33±0.14 | 1.83±0.05 | 6.50±0.35 |
| | T2 *A. parva* (vermicompost) | 19.08±0.25 | 4.25±0.09 | 14.83±1.10 |
| | T3 *E. fetida* (vermicompost) | 16.53±0.19 | 3.90±0.07 | 12.63±0.18 |
| T-9 | | | | |
| | T1 control | 6.47±0.35 | 1.77±0.05 | 4.70±0.23 |
| | T2 *A. parva* (vermicompost) | 15.75±0.29 | 4.40±0.05 | 11.35±0.66 |
| | T3 *E. fetida* (vermicompost) | 14.63±0.50 | 4.00±0.07 | 10.63±0.61 |
| CD at 5% | | | | |
| Variety | | 0.31 | 0.08 | 0.59 |
| Treatment | | 0.38 | 0.10 | 0.73 |

All values are the mean of three replicates ± standard deviation.

The difference was statistically significant between both varieties of seeds. VC prepared by native earthworms (*A. parva*) produced significantly higher yields for both varieties of seeds, according to the study. The data on biomass yield corroborate the theory of Shiva (1992), and the research findings of Arya (1995) reported for the wheat crop that indigenous varieties yield higher biomass than improved ones.

Agarwal et al. (2003) investigated the effects of VC, farmyard manure (FYM), and CFs on the growth and yield of wheat (*Triticum aestivum* L. var. HD2643) and found that the use of VC in all treatments increased total biomass production and yield of the wheat crop over the control. Materechera (2002) discovered that VC-grown maize had significantly higher plant height, stem diameter, and dry weight, as well as higher nutrient uptake and lower fertilizer response than the control.

## 9.7 NURSERY RAISING OF INDIGENOUS EARTHWORM SPECIES AND OUTREACH ACTIVITIES

As a result of the above findings, nursery raising of native earthworm species (*A. parva*) and VC units were put up on the farms of Mr. Sheesh Ram (Village Garhi) and Mr. Prashant Mahrishi (Village Khari). Since it was a new beginning for the indigenous earthworm species under field conditions, a few problems were faced. These problems were noted and discussed with experts, especially Dr. Sultan Ismail and experienced farmers. The results indicated that indigenous earthworm species can be employed successfully for composting agro-horticulture waste on a larger scale. Hence, a training workshop on "Organic farming utilizing native earthworm species" was also organized along with JVT (NGO), Govindpur, at the farm of Mr. Sheesh Ram Ji. Farmers discussed various aspects of indigenous earthworm collection and handling, as well as vermicomposting technology, at this workshop. A few farmers also shared their experiences related to this technology using exotic species of earthworms. Afterward, participants were given a demonstration on nursery-raising earthworms and vermicomposting. After attending the training workshop, people's perceptions about using indigenous earthworm species in manure production and organic farming changed.

As part of a DST-sponsored project, a technical manual on "local species of earthworms for vermicomposting and vermiwash" was also published. This manual's goal was to explain how to encourage the growth of indigenous earthworm species. The bioeconomic aspect of the technology package for rural prosperity is also included.

## 9.8 CONCLUSIONS AND PERSPECTIVES

This study highlighted the potential of indigenous earthworm species, particularly *Allolobophora parva*, for enhancing vermicomposting and soil fertility. The results of vermicomposting experiments clearly showed that the NPK content of *A. parva* VC was either higher or at par with that of *E. fetida* VC in most cases. Experiment results on the relatively heavy metal accumulation and tolerance of native earthworm species also revealed an intriguing fact: *A. parva* exhibited greater heavy metal tolerance, suggesting its suitability for remediating contaminated soils. Field investigations showed that VC produced by *A. parva* significantly improved yields of black gram (*Vigna mungo*) compared to both control and exotic species. Overall, the findings of the study have undoubtedly contributed to our understanding of the potential and significance of indigenous earthworms in terms of eco-sustainability and bioeconomic feasibility for achieving rural prosperity. This must be emphasized by connecting the "waste management and recycling system" to sustainable agriculture.

Utilizing indigenous earthworm species would be an ecologically sound and sustainable method for managing and efficiently utilizing agro-horticultural waste. Indigenous earthworms have certain merits over exotic species, and their fast multiplication can be successfully achieved by adding some new inoculants/formulations. Nursery of indigenous earthworms can be raised on locally available biomass by simple techniques at the farm level. Additionally, the successful establishment of nurseries for *A. parva* and subsequent outreach activities underscored the importance of educating farmers about the benefits of using native earthworms for composting agro-horticulture waste. Once the optimum earthworm population becomes available, it can be used for VC production at a large scale. The training workshops not only facilitated knowledge transfer but also positively shifted farmers' perceptions regarding indigenous earthworms. As a result, a technical manual on promoting local earthworm species for vermicomposting was developed, emphasizing the bioeconomic advantages of this technology. The study also highlighted the importance, potential, and scope of using indigenous earthworm species (*A. parva*) in sustainable agriculture and pollution control (including heavy metals). Such studies should be carried out in various agro-climatic regions, and an inventory of indigenous earthworm species should be made in order to select suitable species for vermicomposting of various organic waste.

## ACKNOWLEDGEMENTS

We gratefully acknowledge Dr. J.M. Julka (Solan, Himachal Pradesh) and Dr. Sultan Ismail (Chennai) for their insightful suggestions. We are grateful to Dr. Ransingh Arya (JVT Bijnor) and Mr. Sheeshram Ji (a progressive farmer) for making this innovative effort successful.

## REFERENCES

Aalok, A., and A.K. Tripathi. 2010. Composting-vermicomposting of different types of leaves using earthworm species *Eisenia fetida*. *Dyn. Soil Dyn. Plant* 4: 139–144.

Agrawal, S.B., A. Singh, G. Dwivedi. 2003. Effect of vermicompost, farm yard manure and chemical fertilizers on growth and yield of wheat (*Triticum aestivum L.* var HD 2643). *Plant Arch.* 3: 9–14.

Arancon, N.Q., C.A. Edwards, A. Babenko, J. Cannon, P. Galvis, D. M. James. 2008. Influences of vermicompost, produced by earthworms and microorganisms from cattle manure, food waste, and paper waste, on the germination, growth, and flowering of petunias in the greenhouse. *Appl. Soil Ecol.* 39: 91–99.

Arya, R.S. 1995. Synergetic agriculture system and development. Ph.D. Thesis. Indian Institute of Technology, Delhi.

Chatterjee, R., S. Bandyopadhyay, J. C Jana. 2015. Evaluation of vegetable wastes recycled for vermicomposting and its response on yield and quality of carrot (*Daucus carota L.*). *Int. J. Recycl. Org. Waste. Agric.* 3: 1–7.

Corp, N., and A.J. Morgan. 1991. Accumulation of heavy metals from polluted soils by the earthworm *Lumbricus rubellus*: Can laboratory exposure of 'control' worms reduce biomonitoring problem? *Environ. Pollut.* 74: 39–52.

Gajalakshmi, S., E.V. Ramasamy, S.A. Abbasi. 2005. Composting–vermicomposting of leaf litter ensuing from the trees of mango (*Mangifera indica*). *Bioresour. Technol.* 96: 1057–1061.

Garg, K., and N. Bhardwaj. 2001. Production of vermicompost through vermin-technology using two obnoxious weeds. *J. Environ. Pollut.* 8: 49–52.

Garg, P., S. Satya, S. Satyawati. 2009. Effect of heavy metal supplementation on local (*Allolobophora parva*) and exotic (*Eisenia fetida*) earthworm species: A Comparative Study. *J. Environ. Sci. Health A* 44: 1025–1032.

Ghosh, C. 2004. Integrated vermi-pisciculture-an alternative option for recycling of solid municipal waste in rural India. *Bioresour. Technol.* 93: 71–75.

Ismail, S.A. 1994. The use of local species of earthworms. In: *B. N. Ganguly* (Ed.), *Agriculture in Changing Villages* (pp. 27–37). Chennai: University of Madras.

Ismail, S.A. 1995. Earthworms in soil fertility management. In: P. K. Thampan (Ed.), *Organic Agriculture* (pp. 77–100). Chennai: Peekay Tree Crops Development.

Ismail, S.A.1996. *Vermitech (Vermicompost and Vermiwash)* (Vol. 10). Chennai, India: Ajju's Wormery.

Ismail, S.H., P. Joshi, A. Grace. 2003. The waste in your dustbin is scarring the environment – The technology of composting. *Adv. Biotechnol.* 5: 30–34.

Jadia, C., and M. fulekar. 2008. Vermicomposting of vegetable waste: A bio-physicochemical process based on the hydro-operating bioreactor. *Af. J. Biotechnol.* 7: 20.

Jat, R.S., and I.P.S. Ahlawat. 2004. Effect of vermicompost, biofertilizer and phosphorus on growth, yield, and nutrient uptake by gram (*Cicer arietinum*) and their residual effect on fodder maize (*Zea mays*). *Indian J. Agric. Sci.* 74: 359–361.
Javed, S., and A. Panwar. 2013. Effect of biofertilizer, vermicompost, and chemical fertilizer on different biochemical parameters of *Glycine max* and *Vigna mungo*. *Recent Res. Sci. Technol.* 5: 40–44.
Jeyabal, A., and G. Kuppuswamy. 2001. Recycling of organic wastes for the production of vermicompost and its response in rice–legume cropping system and soil fertility. *Eur. J. Agron.* 15: 153–170.
Julka, J.M. 1986. Earthworm resources of India. In R. C. Dash, B. K. Senapati, and P. C. Mishra (Eds.), *Vermicomposting, Part B: Worms and Vermicomposting* (pp. 1–7). Proceedings of the National Seminar on Organic Waste Utilization. Burla: Five Star Printing Press.
Kaviraj, and S. Sharma. 2003. Municipal solid waste management through vermicomposting employing exotic and local species of earthworms. *Bioresour. Technol.* 90: 169–173.
Khwairakpam, M., and R. Bhargava. 2009. Vermitechnology for sewage sludge recycling. *J. Hazard. Mater.* 161: 948–954.
Lukkari, T., M. Taavitsainen, M. Soimasuo, A. Oikari, and J. Haimi. 2004. Biomarker responses of the earthworm *Aporrectodea tuberculata* to copper and zinc exposure: Differences between populations with and without earlier metal exposure. *Environ. Pollut.* 129: 377–386.
Mahaly, M., K.A. Senthilkumar, C. Kaliyaperumal, N. Karupannan. 2018. Vermicomposting of distillery sludge waste with tea leaf residues. *Sustain. Environ. Res.* 28: 223–227.
Malley, C., J. Nair, and G. Ho. 2006. Impact of heavy metals on enzymatic activity of substrate and on composting worms *Eisenia fetida*. *Bioresour. Technol.* 97: 1498–1502.
Manna, M.C., S. Jha, P.K. Ghosh, C.L. Acharya, 2003. Comparative efficacy of three epigeic earthworms under different deciduous forest litters decomposition. *Bioresour. Technol.* 88: 197–206.
Materechera, S.A. 2002. Nutrient availability and maize growth in a soil amended with earthworm cast from a South African indigenous species. *Bioresour. Technol.* 84: 97–101.
Murthi, G.S.R. 1987. *Homeopathy in Agriculture*. New Delhi, India: B. Jain Publishers.
Muthukumaravel, K., A. Amsath, M. Sukumaran. 2008. Vermicomposting of vegetable wastes using cow dung. *J. Chem.* 5: 810–813.
Parthasarathi, K. 2007. The life cycle of *Lampito mauritii (*Kinberg) in comparison with *Eudrilus* eugeniae (Kinberg) cultured on different substrates. *J. Environ. Biol.* 28: 803–812.
Pathma, J. and N. Sakthivel. 2012. Microbial diversity of vermicompost bacteria that exhibit useful agricultural traits and waste management potential. *Springer*Plus 1: 26. https://doi.org/10.1186/2193-1801-1-26
Paul, J. 2017. Biodiversity and vermicomposting potential of indigenous earthworm resources of Sivagangai District. *Int. J. Adv. Res. Biol. Science*. 4: 173–180.
Proctor, P. 1989. The Biodynamic preparations. In: New Zealand Biodynamic Association (Ed.), *Biodynamics; New directions for farming and gardening in New Zealand* (pp. 106–129). Auckland: Random House.
Rana, V. 2004. Potential of indigenous seeds (Vigna mungo) and plant nutrient recycling through biogas technology for sustainable agriculture. Ph.D Thesis. Indian Institute of Technology, New Delhi.
Santosh, S., R.S. Arya, P. Vasudevan. 1993. *Towards Sustainable Agriculture (Booklet)*. Delhi: IIT Delhi.
Santra, S.K., and Bowmik, K.L. 2001. Vermiculture and development of agriculture. *Yojna* 45: 43–45.
Scown, J., G., and Baker. 2006. The influence of livestock dung on the abundance of exotic and native earthworms in grassland in south-eastern Australia. *Eur. J. Soil Biol.* 42: 310–315.
Shanthi, N.R., R.V. Bhoyar, A.D. Bhide. 1993. Vermicomposting of vegetable wastes. *Compost. Sci. Util.* 1: 27–30.
Sharma, D. 2017. Parthenium weed management by vermicomposting using local earthworm species of Jammu. *Int. J. Appl. Res.* 6: 1170–1171.
Shiva, V. 1992. *The Violence of the Green Revolution: Third World Agriculture, Ecology and Politics*. Goa, India: The Other India Press.
Singh, A., and S. Sharma. 2002. Composting of a crop residue through treatment with microorganisms and subsequent vermicomposting. *Bioresour. Technol.* 85: 107–111.
Singh, J. 1997. Habitat preference of selected Indian earthworm species and their efficiency in the reduction of organic materials. *Soil Biol. Biochem.* 29: 585–588.
Suthar, S. 2007a. Nutrient changes and biodynamics of epigeic earthworm *Perionyx excavatus* (Perrier) during recycling of some agriculture wastes. *Bioresour. Technol.* 98: 1608–1614.
Suthar, S. 2007b. Vermicomposting potential of *Perionyx sansibaricus* (Perrier) in different waste resources. *Bioresour. Technol.* 98: 1231–1237.
Suthar, S. 2009. Bioremediation of agricultural wastes through Vermicomposting. *Bioremediation*, 13: 21–28.

Turgay O, C., R. Kizikaya, A. Namli, S.C. Cetin. 2011. Detoxification of heavy metals using earthworms. In I. Sherameti and A. Varma (Eds.), *Detoxification of Heavy Metals* (pp. 407–421). New York: Springer-Verlag Heidelberg.

Yusoff A., S. B. Lobb, C. Ibrahim. 2019. The potential of local earthworms in vermicomposting goat manure (gm) and spent coconut flakes (SCF). *J. Undergrad. Res.* 1: 19–28.

# 10 Institutional Factors Determining Efficiency of Waste Management Services in the Czech Republic and Slovakia – A Case Study

*Beata Mikusova Merickova*[1], *Juraj Nemec*[2], *Jana Soukopova*[3], *and Kristina Murinova*[4]

[1]Faculty of Economics, Department of Public Economics and Regional Development, Matej Bel University Banska Bystrica, Slovakia

[2]Faculty of Economics and Administration, Department of Public Economics, Czech Republic and Matej Bel University, Faculty of Economics, Department of Finance and Accounting, Masaryk University Brno, Banska Bystrica, Slovakia

[3]Faculty of Economics and Administration, Department of Public Economics, Masaryk University Brno, Czech Republic

[4]Faculty of Economics, Department of Public Economics and Regional Development, Matej Bel University Banska Bystrica, Slovakia

## 10.1 INTRODUCTION

Municipal solid waste (MSW) management is a key local public service, and its costs represent a substantial part of local budgets. Municipalities have to decide on the most efficient way to provide this service in all countries. This question is specifically critical in highly fragmented countries with an excessive number of small municipalities, such as the Czech Republic and Slovakia, where the provision of waste collection and disposal services is the responsibility of local self-governments (Soukopová & Klimovský, 2016). Current literature and practice on contracting out and local public service provision methods show that the impact of contracting out on the efficiency and quality of privately provided public services is country and sector specific, not least because transaction costs appear to be an important bias. Therefore, this literature not only compares the internal and external production of waste management services, but also examines several institutional factors determining these costs in depth.

Due to the increasingly fewer positive findings related to the contracting out of MSW services, most recent books even propose the need for "re-municipalization" or "public ownership" (Albalate et al., 2021a, 2021b; Voorn, 2021). For example, the volume edited by Koprić et al. (2018, p. 335) concludes that "Ideological beliefs that NPM-related mechanisms such as contracting out, outsourcing, public–private partnership, privatization, and others will deliver better solutions to users may or may not be supported by hard data on efficiency and user satisfaction. This depends on the type of service, the size of the municipality, the national legal and institutional contexts, the level and nature of competition in specific local service sectors, local regulatory and managerial capacity,

DOI: 10.1201/9781032676890-13

and other variables. Furthermore, contrary to ideological beliefs, the privatization and implementation of NPM-driven measures can result in price increases, loss of investment, and other significant detrimental long-term effects, whether these are the consequences of model inappropriateness, regulatory deficiencies, or implementation inconsistencies." Kishimoto et al. (2020) mentioned that "(Re)municipalization is one of the key strategies to address the challenges of our time, including pandemics such as COVID-19." Their book identifies more than 1,400 successful (re)municipalization cases involving more than 2,400 cities in 58 countries.

The aim of this chapter is to contribute to an informed discussion about the efficiency of MSW services. Primarily based on own data collected through primary field research, this chapter examines trends in municipal MSW collection and disposal for the Czech Republic and Slovakia in terms of developments over the last 20 years, focusing on the extent of contracting out and the results from contracting out. In addition, it addresses selected factors that determine the results from contracting out, namely the quality of contract management, market concentration, and inter-municipal cooperation (IMC) in the provision of this service.

The flow of the text is standard with some peculiarities. The first section presents a literature review. Then, the methodology section is delivered separately for each of the four analytical lines (scope of contracting out and its results, contract management, market concentration, and IMC) of research to make the text flow more understandable. The final sections summarize the results and provide conclusions and perspectives.

## 10.2 THEORETICAL BACKGROUND

The issue of contracting out local services has been the subject of academic research for many years, and many existing studies focus on or cover the area of waste management and disposal. Some articles merely present arguments as to why public ownership should be viewed positively or negatively (Monteduro, 2014). EU institutions, national governments, and local public authorities are increasingly paying attention to cooperation with the private sector. Questions have been raised about the extent to which these partnerships will promote the public interest and defend collective interests. International institutions, policymakers, and academics have taken an interest in this issue; in particular, the EU has suggested implementing different types of public–private partnership arrangements in the provision of public services. However, most advanced papers in academic research on this issue provide more than a normative perspective. Instead, they seek to assess the situation from several viewpoints, addressing not only the question of whether to contract out, but what process contracting out should follow or what factors are involved in the contracting out process and their results (Agiamoh, 2020; Soukopová et al., 2014; Schoute et al., 2021; Bel & Rosell, 2016).

As contracting out of public service production is based on a contractual relationship – a contract between a public institution and an external entity – many studies focus on the concepts of contracts and of contracting out of public services (Cooper, 2003; Weghmann & Hall, 2021). Books and articles on the topic highlight the tools that managers need to create, operate, terminate, and transform these contracts amidst a complex web of intergovernmental relationships.

Many studies focus on the problem of comparing the cost-effectiveness of internalization and contracting out perceived from different perspectives. The authors attempt to determine whether and to what extent there are differences in performance, particularly economic performance, between public and private production of local services. Domberger and Jensen (1997) showed that private production of local services led to the efficient provision of a number of government services. However, more recent evidence on the cost efficiency of private provision is rather ambiguous and more mixed (Bel & Costas, 2006; Bel & Mur, 2009; Bel et al., 2010; Benito et al., 2015; Callan & Thomas, 2001; Dijkgraaf & Gradus, 2007; Gradus et al., 2016; Ohlsson, 2003; Schoute et al., 2021). This simplistic view of contracting out public sector services is limited to addressing the "make or buy" question and does not answer why contracting out public sector services does not increase efficiency to the extent expected.

To conclude this section, it is worth mentioning that the topic of contracting out local services has already been covered to some extent by national sources in the two countries studied, in many cases by authors of this chapter. One of the most comprehensive analyses was presented in the dissertation of Majlingová (2005), focusing on waste collection and disposal services. Majlingová used a representative sample of 213 municipalities and mapped several aspects of waste management at the local level, including contracting out efficiency. Two other important monographs in Slovakia were prepared by Balážová (2006) with the support of Transparency International Slovakia. Finally, in the Czech Republic, an analysis by Pavel (2006) documented the limited transparency and inefficiency of local companies responsible for the provision of local services. Other relevant studies are Balážová (2006), Majlingová (2005), Koprič et al. (2018), Nemec et al. (2020), and Soukopová and Klimovský (2016).

### 10.2.1 Contractual Relations and Contract Management

The perception of the problem of contracting out activities related to the performance of state functions in the economy, influenced by the principal–agent theory, sees the contract as a certain agreement between two entities with different interests – the principal and the agent (Andersen et al., 2008). In our understanding, the principal is the branch of government that has a legislatively defined obligation to provide a given service – let us also call it the service provider, who bears the political responsibility for deciding what sources will finance the service (public vs. private sources) and what form of service production (internal vs. external) will be chosen.

The agent is the specific service producer (it can be a private, non-profit, or public organization, whose management is not directly influenced by the principal). We deliberately refer to management influence because often times commercial companies with a majority principal ownership stake have been automatically mandated to provide a particular public service outside of the public procurement regime, which means that it is an internalized, "in house" way of providing the service. In our view, at least in such cases, this is a violation of the ethics of public procurement. Both the principal and the agent pursue their own objectives, which need not (and usually do not) align (DeHoog, 1990; Sclar, 2000; Bel & Gradus, 2018; Andersen et al., 2008). The mismatch between the goals of the principal (service provider/service purchaser) and the agent (service producer/supplier) is the cause of the two basic problems associated with contracting out that we have described in the previous text: (1) the moral hazard or hidden activity problem and (2) the problem of misselection or hidden information (Arrow, 1985; Pratt & Zeckhauser, 1986).

Therefore, the outcome or success of contracting out depends on the extent to which the aforementioned problems, or risks, can be eliminated. In this sense, several authors mention the factors or determinants of the benefits of contracting out and public sector services that depend on the quality of contract management (these factors are tied to the contracting organization, so we will understand them as internal):

- *the degree of competition in being awarded a public contract* – can be simplistically assessed according to the procurement procedure used (Savas, 1987; Green, 2002; Hodge, 2000; Brown et al., 2015; Bel & Fageda, 2011),
- ex ante *evaluation of a bidder for a public contract* (Romzek & Johnston, 2002; Yescombe & Farquharson, 2018),
- a *clearly defined subject of procurement* (Brown et al., 2015),
- *the extent and intensity of monitoring of external production* (Lavery, 1999; Brown & Potoski, 2003; Johansson et al., 2016),
- *penalties for non-compliance with contract terms* (DeHoog, 1990; Brown et al., 2015; Johansson et al., 2016),

- the *contracting authority's knowledge and experience in contract management* (DeHoog, 1990; Romzek & Johnston, 2002; Yescombe & Farquharson, 2018),
- the *expertise of the contracting authority on the technical parameters of the procured service* (Yescombe & Farquharson, 2018). These are factors derived from the quality of the management of the contracting out process.

In addition to the conventional view of contracting out from a principal–agent theory perspective, where the relationship between the service provider, the purchaser of the service, and the producer, the service supplier, is considered managerial or controlling (due to the divergence of the parties' objectives and their efforts to enforce them) in the literature (DeHoog, 1990; Williamson, 1986; Sclar; 2000; Smith, 1996; Bennett & Ferlie, 1996; Johansson et al., 2016; Rendon, 2012; Rose-Ackerman & Palifka, 2016), we also find a more recent perspective that sees the relationship as a partnership based on flexible collaboration. In this conception, other intrinsic factors stand out as determinants of the benefits of contracting out:

- the frequency of communication between the service provider and the service producer (Behn & Kant, 1999; DeHoog, 1990; Rendon, 2012; Johansson et al., 2016),
- collaborative problem solving (Cooper, 2003; Sclar, 2000; Rendon, 2012; Johansson et al., 2016),
- mutual trust, shared values, and a move away from penalties to negotiation (Rose-Ackerman & Palifka, 2016).

Suppose we talk about the factors influencing the outcome or benefits of contracting out in the public sector. In that case, we cannot forget the *political support for pluralization of ownership forms* in the exercise of state functions, or privatization in the public sector, understood in this sense as creating a certain competitive environment in the public sector. Several authors (Romzek & Johnstone, 2002; Marra & Carlei, 2014; Johansson et al., 2016; Koprič et al., 2018; Petersen et al., 2018) consider this factor to be important. However, assessing or quantifying this factor in order to empirically verify its impact on the benefits of contracting out is challenging in our setting. Therefore, some authors have attempted to evaluate this factor in terms of the creation and development of the legislative, economic, and organizational preconditions necessary for successful contracting out of public sector services (Yescombe & Farquharson, 2018).

### 10.2.2 Market Concentration

More recent economic literature also addresses factors independent/semi-independent of the contracting institution (principal), such as financial constraints and political and legal institutions as determinants of government decisions to contract out (Megginson et al., 2000), market constraints to contracting out – high transaction costs (Brown & Potoski, 2003; Johansson, 2008), hidden costs (Porcher, 2017), or an unstructured market that manifests itself in market concentration (Bel & Costas, 2006; Gradus et al., 2018).

Market concentration can be identified as a significant market disruption in contracting out public services. Studies (Gradus et al., 2018) increasingly point to the fact that a more concentrated market can lead to higher market prices in terms of contracting out. This can be justified by the fact that the provision of public services to municipalities is a specific area in which there may be weak competition and high concentration due to the nature of the service (as evidenced by the high HHI index). The competitive environment as a factor in the efficiency of service provision may impact higher spending by municipalities or economies of scale. According to some authors, the competitive environment is an even more important factor than the type of ownership. We agree with the results of the research (Soukopová & Vaceková, 2018) that states competitiveness has a high impact on local public service prices. Therefore, when providing a public service by contracting out, more attention should be paid to the public procurement process (Nemec et al., 2020) to ensure maximum participation of market players.

The competitive environment can be viewed dichotomously. Firstly, as competition for the market, where one assesses how the market is developing over time, what conditions are set for entry into the industry, or whether barriers (administrative or legislative regulation by the government) are set for the entry of new entities into the industry. Second, how competition in the market that replicates traditional market approaches focuses on internal processes. Within this view, we can monitor the volume of total production or the marginal market shares of the entities. Both internal and external factors are result-oriented in terms of the transformation process. For services, however, it is important to consider the quality-of-service provision by individual entities.

When looking at the provision of local public services at the municipal level, the serviceability of the area needs to be considered. As far as the service area is concerned (Di Foggia & Beccarello, 2021), high market concentration may be present as the area shrinks. This factor is considered in the calculation of the Herfindahl–Hirshman index (HHI). The effect of market concentration on the price of the service and increasing costs is demonstrated by the increasing HHI index.

Applying this approach in practice was the goal of the authors Gradus et al. (2018). The authors use special indices to measure Dutch market concentration in the provision of a local public MSW collection service. The study was conducted over the period 2002–2014 and shows that the waste collection market is highly concentrated, which increases the cost of private collection. During the period under review, 25 and later 16 entities operated in this market. Although, in the initial period, the largest entity had a contractual relationship with 87 municipalities for the provision of waste collection services, by 2014 the share of municipalities had decreased to 37. The price for the provision of waste collection services has increased over time, since 2006 mainly due to the introduction of a compensation fund. From the HHI measurement results s, we find that if the market area was 50 km, the market concentration was 4,400 in 2014; if the area was 30 km, the market concentration was 6,660 in 2014. If the geographic area was expanded to 70 km, the concentration in 2014 was 3,700.

A different approach to market concentrations in local public service provision is offered by Demuth et al. (2022). This approach is based on reverse privatization, which aims to take control of service provision by municipalities. The analysis of reverse privatization was followed in a study over the period 2003–2015 in German municipalities, and the authors used the concentration ratio (CR) index and the HHI index. The results show that municipalities are more likely to switch to traditional local public service provision (in sourcing) in response to a concentrated market. In doing so, municipalities are also replicating their neighbor's service provision, for example, if a neighboring municipality applies in sourcing, it is likely to be used by its neighboring municipalities. The authors argue that this is an induced vertical connection between municipalities. However, such behavior can lead to a reverse privatization spiral, which is associated with a high market share of municipalities, the exit of small private entities, and an increase in the dominance of large private entities, further increasing market concentration. Demuth et al. (2022) point out that the effective outcome of such cyclicality is highly questionable.

### 10.2.3 Inter-Municipal Cooperation

As a form of decentralized and democratized ideology for municipalities, IMC plays a key role in advancing governance at the local and municipal level. Local government structures in many parts of the world exhibit municipal fragmentation making it difficult for central and/ or federal governments to deal with resource allocation effectively and efficiently. For example, Matějová et al. (2014) posit that Central and Eastern European countries have some of the most territorially fragmented municipalities in the context of European local government. This presents a challenge for the provision of public services at the local level. Collaboration across the spheres of each nation's economic fabric presents a great opportunity for local government authorities to gauge proactive means to create engagement between municipalities to deal with the challenges of the economic crisis and recent financial constraints facing the public sector.

As a territorial policy engrained in structural reformation (Hulst & van Montfort, 2007), IMC has been used as a propeller of efficiency, as many scholars (Swianiewicz & Teles, 2019) have noted, to the woes of municipal fragmentation to reduce the associated costs with administrative compliance. However, it also ensures significant economies of scale and scope (Southwick, 2012; Niaounakis & Blank, 2017). Bel and Warner (2015) discussed that IMC analysis has centered on waste collection, water, electricity, and gas (Garrone et al., 2013). However, the effect of IMC has contextual connotations and the type of public service which the cooperative municipal governance arrangement is oriented toward in the production and provision of services.

There exist differences in opinions in the literature on the reasons that municipalities consider when cooperating with each another. Nevertheless, this form of cooperative arrangement has been around for some time, gaining popularity among scholars and morphing into different forms in terms of the shape and level of institutionalization as well as the extent of engagement, as is the case in America (Warner, 2006). In the context of the European Union, West (2007) discussed how French municipalities have coped with and thwarted the lack of scale at the municipal level, where collaborative arrangements have been organized in a voluntary form to provide the public services. In a similar vein, Fedele and Moini (2007) discussed the different forms of Italian IMCs from an institutional and organizational perspective to show the difficulties and advantages of municipal collaboration, whether voluntary agreements or forced collaboration as a consequence of local government structural reforms. Elsewhere in Finland, Haveri and Airaksinen (2007) discussed modern and old IMC arrangements in Finland's public administration context.

From the above, it is clear that there are diverse institutional arrangements for IMC and the forms and shape depend on the ultimate goal of the cooperating entities. More important are the forms as discussed by Fedele and Moini (2007), the dichotomy of inter-municipal cooperation or collaboration. Starting from the Italian case, municipalities collaborate in the provision of services through agreements and conventions, such as joint activities backed by regional policy contracts. On the other hand, cooperation arrangements persist on consortia and unions through norms and regulations. This brings more formalities in inter-municipal relations instituted that are based on service provision.

Within the public service, waste management has come to the fore as a result of climate change and policy changes relating to sustainability objectives. To date, many reforms have been implemented to ensure waste collection and management efficiency (Gradus et al., 2018; Bel & Sebő, 2020). Bel and Sebő (2021), in their analysis of strategic competition in waste collection and quality of service, distinguished between two overlapping strands of literature on the current topic. Waste management has taken a new turn due to inefficiencies within public-dominated production, which called for more market-oriented solutions to introduce competition among waste management firms to compete for municipal waste service delivery contracts; however, the main challenge has been the case of cost-effectiveness. Lack of monitoring and undesirable contract outcomes led to the cancellation of many contracts, culminating in the excesses of market or government failure.

This opening to secure the goal of efficiency and cost-effectiveness ushered in IMC as a novel model that provides policymakers with the right mix of contracting out and re-municipalizing. It provides them with the means to consolidate resources to take proper control while ensuring efficiency and cost-effectiveness on the one hand, and to create a cooperative platform for municipalities to provide services to benefit their residents on the other (Spicer, 2015). Municipalities in collaborative arrangements can efficiently allocate scarce resources in the provision of local services through cohesive and strategic planning in funding inter-municipal collaboration in service provision, which is a requirement of any IMC framework.

## 10.3 SCALE AND EFFICIENCY OF CONTRACTING OUT MUNICIPAL SOLID WASTE SERVICES

This section of the analysis seeks to summarize trends in terms of the proportion of contracted MSW management and the efficiency of contracted-out MSW services compared to in-house production. The contracting out rate is assessed using a simple indicator: the proportion of municipalities that outsource MSW collection and disposal. The core methodological problem with this indicator is the fact that different research samples used different weighting approaches in relation to the size of municipalities – for this reason, the pictures provided by the selected samples are not fully comparable, although they are sufficient to show general trends.

Contracting out efficiency is predominantly assessed via the cost-per-resident indicator. The supporting indicator is the cost-per-unit of output (one ton of waste generated); this indicator is only calculated for selected years, depending on the availability of the necessary data. Two main problems limit the possibility to accurately compare contracting out efficiency. One problem is that the costs with contracting out have to be higher due to the existing accounting system (cost center accounting, as full cost accounting is not used at municipal level). Municipalities do not track overheads, or the transaction costs associated with, for example, public procurement or follow-up of external delivery. However, invoiced costs from external entities include direct costs, overheads (including most transaction costs), added value, and, in most cases, also (monopoly or oligopoly) profit. The second problem is the limited reliability of the data collected. Cost data were collected using questionnaires for the Czech Republic at the outset and for all Slovak samples; thus, not all data are necessarily correct. More recent economic data for the Czech Republic are collected from the national information system (ARIS).

## 10.4 METHODOLOGY

For the selected Slovak sample from 2009, 2010, 2015, and 2020, the authors used Best Value Performance Indicators (BVPI) to construct composite efficiency scores based on a combination of three indicators as efficiency benchmarks: the costs of service provision per inhabitant, the unit costs of output, and the quality of service. BVPI is one of the nonparametric multivariate approaches to evaluating the efficiency of Decision-Making Units (DMUs) based on the weighted sum algorithm. Here, we designated the service provision method as DMU. The method makes it possible to express the efficiency of the evaluated DMUs, considering multiple indicators measured in different units. In our analysis, we chose three indicators of evaluation as efficiency benchmarks: costs of service provision per capita, unit costs of service provision – 1 metric ton of waste, and quality of service.

Data on the quality of service were provided by the users, citizens of different municipalities, through a questionnaire. The samples are non-representative, so we accept that our aggregated data are partly preliminary. Citizen satisfaction was used as a measure of local public service quality. The users, the citizens of different municipalities, provided data on service quality through a questionnaire. Citizens rated the quality of local public service on a scale of absolutely satisfied 100%, satisfied 80%, more satisfied than unsatisfied 60%, more unsatisfied than satisfied 40%, unsatisfied 20%, and unsatisfied 0%.

To calculate BVPI, we assigned weights (vj) to the selected indicators using Saaty's (1983) approach as follows: unit costs of service provision per capita – 30%, unit costs of service provision per service outcome – 30%, and quality of service – 40%. We used Saaty's method to set the weights for the indicators with inputs from a panel of ten contracting out experts in both countries. The data presented are from our own original field research, and data produced by other authors have been used for a few years. Table 10.1 for the Czech Republic and Table 10.2 for Slovakia characterize the research samples. In some cases, the research covered a representative sample of municipalities; in others, only an illustrative sample. The response rate is calculated as the proportion of responses received to the total number of municipalities in a given country in a given year.

**TABLE 10.1**
**Samples Analyzed for the Czech Republic**

| Year | Sample | Source |
|---|---|---|
| 2000 | 53 municipalities of different sizes (non-representative sample, response rate 0.85%) | Authors |
| 2007 | 900 municipalities of different sizes (representative sample, response rate 14.5%) | Authors |
| 2011 | 673 municipalities of different sizes (representative sample, response rate 10.85%) | Authors |
| 2013 | 1,300 municipalities of different sizes (representative sample, response rate 20.9%) | Authors |
| 2014a | 102 municipalities of different sizes (representative sample, response rate 1.6%) | Authors |
| 2014b | 53 municipalities of different sizes (non-representative sample, response rate 0.85%) | Authors |
| 2014c | 2,065 municipalities of different size (representative sample, response rate 33.3%) | Authors |
| 2015a | 675 municipalities of different sizes (representative sample, response rate 10.85%) | Authors |
| 2015b | 1,962 municipalities of different sizes (representative sample, response rate 31.6%) | Authors |
| 2019 | 90 municipalities of different sizes (representative sample, response rate 1.4%) | Authors |

**TABLE 10.2**
**Samples Analyzed for Slovakia**

| Year | Research Sample | Sources |
|---|---|---|
| 2001 | 55 municipalities of different sizes (representative sample, response rate 2.24%) | Authors |
| 2004 | 213 municipalities of different sizes (representative sample, response rate 8.69%) | Majlingová (2005) |
| 2005 | 17 municipalities of the same size (non-representative sample, response rate 1.75%) | Authors |
| 2006 | 100 municipalities of different sizes (representative sample, response rate 4.88%) | Authors |
| 2008 | 28 municipalities of different sizes (non-representative sample, response rate 1.14%) | Authors |
| 2009 | 131 municipalities of different sizes (representative sample, response rate 5.3%) | Authors |
| 2010 | 141 municipalities of different sizes (representative sample, response rate 5.7%) | Authors |
| 2015 | 9 municipalities of different sizes (non-representative sample, response rate 0.36%) | Authors |
| 2020 | 115 municipalities of different sizes (representative sample, response rate 4.69%) | Authors |
| 2021 | 54 municipalities of different sizes (representative sample, response rate 4.69%) | Authors |

## 10.5 RESULTS: EFFICIENCY OF CONTRACTED-OUT MUNICIPAL SOLID WASTE SERVICES

The next step is to evaluate the effect of contracting out on the efficiency of service provision. Tables 10.4 and 10.5 construct an index of production efficiency for a service using data on the costs of providing services that rely primarily on contracting out compared to those that rely primarily on internal production. Using the average costs of internalized services as the base (= 100), the index score is constructed by taking the ratio of the cost for outsourced services to the cost for internalized services and normalizing by multiplying by 100. Thus, Table 10.3 presents the results for the research, where unit costs have been calculated, and Table 10.4 presents the data for samples where this indicator has been calculated.

Table 10.5 presents the data for the multivariate evaluation. Based on this and the calculations above, it is impossible to draw firm conclusions or reject the assumption that contracting out

**TABLE 10.3**
**Efficiency of Contracting Out MSW Services (Expenditures per Ton MSW) for Slovakia (In-House Costs = 100%)**

| 2009 | 2010 | 2015 | 2020 | 2021 |
|---|---|---|---|---|
| 109 | 60 | 189 | 146 | 199 |

*Source:* Authors.

**TABLE 10.4**
**Efficiency of Contracting Out MSW Services (Expenditures per Inhabitant, In-House Costs = 100%)**

| Czech Republic | | Slovakia | |
|---|---|---|---|
| 2000 | 76 | 2001 | 94 |
| 2005 | 137 | 2005 | 94 |
| 2007 | 136 | 2006 | 125 |
| 2011 | 115 | 2008 | 184 |
| 2019 | 116 | 2009 | 100 |
| | | 2010 | 138 |
| | | 2015 | 158 |
| | | 2020 | 114 |
| | | 2021 | 79 |

*Source:* Authors.

**TABLE 10.5**
**Efficiency of Contracting Out MSW Collection and Disposal: Multidimensional Evaluation for Slovakia, 2009, 2010, 2015, and 2020 Sample (Maximum Score = 100)**

| Year of Evaluation | | 2009 | 2010 | 2015 | 2020 |
|---|---|---|---|---|---|
| Collection and disposal of MSW | Internalization | 92.13 | 88.20 | 79.67 | 52.72 |
| | Contracting out | 96.58 | 95.99 | 29.29 | 99.20 |

*Source:* Authors.

positively impacts the efficiency of the MSW service provision (the values obtained contain significant extremes). The "jumpy" results are primarily related to the problems with measuring efficiency as described in the methodology. However, they do suggest to some extent that contracting out has the potential to deliver savings. Secondly, given the small percentage of municipalities that provide this service in-house, it is impossible to draw valid conclusions as to whether in-house production would be cheaper.

However, based on the expert opinion obtained from an expert who has been dealing with municipal public contracts in the field of waste management for more than 20 years, it can be concluded that the outsourcing of municipal waste collection by waste collection companies is overpriced. The is due to cartel agreements on market sharing and pricing between the companies concerned and, secondly, the cost inefficiency of providing the service to one or more municipalities in a region that another company largely serves.

The multi-criteria assessment of the efficiency of contracting out services evaluated in the different years of the research carried out appears to be detrimental to contracting out. In these cases, one or more of the assessed indicators (expenditure per citizen/employee, expenditure per performance indicator, quality of service) or several indicators perform worse for contracting out. In addition, we must reiterate that we consider expenditure on contracted-out services from the organization's budgets, not the total costs associated with contracting out, where the transaction costs of contracting out, that is, the costs associated with the process of procurement of the external output and the costs associated with its subsequent monitoring would have to be included.

## 10.6 QUALITY OF CONTRACT MANAGEMENT

This section presents the results of the evaluation of the quality of contract management of MSW services using data only from Slovakia. The internal determinants of contract management quality can be categorized according to three contracting risks: moral hazard/hidden activity, information asymmetry/hidden information, and non-transparency. Since the factors of the mentioned risks are defined as qualitative characteristics, for the research we have transformed them (Saaty's method used) into a quantitative form (see Table 10.6).

### 10.6.1 Calculated Values of the Quality of Contract Management

Table 10.7 illustrates the calculated values of the quality of contract management for the three Slovak samples. The results indicate serious problems in the quality of contract management in MSW collection and disposal service; these problems increase the risks associated with contracting out, limit the efficiency and quality, and should negatively impact the efficiency of contracting out. A specific contract management problem associated with its initial phase is non-competitive tendering. Table 10.8 illustrates the situation in Slovakia (the Czech situation is similar).

Non-transparent procurement procedures can lead to the contracting authority not selecting the most suitable supplier/agent for external service provision. Such cases often occur for subjective reasons on the part of the contracting authority, who does not "want" to select the most suitable bid.

## 10.7 MARKET CONCENTRATION

Limited competition for contracted MSW services may result from low contract management quality by the municipality and a specific monopoly. The first problem is discussed in the previous section; this section deals with market concentration. Our research follows the methodology of previous studies already published for the Czech Republic (such as Soukopová et al., 2022) and uses only Slovak data. We observed 195 municipalities in Central Slovakia. The data we use in the paper are data obtained through primary research, which was conducted in the form of a questionnaire survey. In the first step, we measured the HHI index (see the literature review part). In the second

**TABLE 10.6**
**Quantification of the Outcome Effect Factors of Contracting Out Public Sector Services in Relation to the Risks of Contracting Out**

| Risk of Contracting Out | Factor | Qualitative Character | Quantification |
|---|---|---|---|
| Non-transparency | Degree of competition for the award of a public contract | Public tender | 100 |
| | | Restricted competition | 70 |
| | | Negotiating procedure | 50 |
| | | Price tender | 30 |
| | | Direct award | 0 |
| Hidden information | Ex ante evaluation of the contractor for the public contract | the most economically advantageous offer | 100 |
| | | lowest price | 50 |
| | Do the municipality's staff involved in the contracting out process have sufficient expertise in the technical parameters of the service to be procured | fundamentally agree | 100 |
| | | agree | 50 |
| | | Disagree | 0 |
| | | Fundamentally disagree | 0 |
| Risk of hidden activity | Frequency of monitoring of the output of the contracted service (how often the service is inspected by the municipality) | Regularly | 100 |
| | | On an ongoing basis as required | 50 |
| | | No monitoring | 0 |
| | Duration of the contract – length of the contract | Up to and including 1 year | 100 |
| | | Up to and including 2 years | 70 |
| | | Up to and including 5 years | 30 |
| | | Indefinitely | 0 |
| | The method of payment to the external supplier for the service | Payment for performance | 100 |
| | | Fixed payment+performance, i.e., a combination | 50 |
| | | Fixed payment | 0 |

*Source:* Authors.

**TABLE 10.7**
**Values for the Quality of Contract Management in % (Max=100)**

| | Rate of Competition | Ex Ante Evaluation of Applicant | Expert Knowledge | Monitoring Frequency | Contract Length | Type of Payment to Supplier |
|---|---|---|---|---|---|---|
| 2010 | 42.84 | 67.12 | N.A. | 70.32 | 38.63 | 65.65 |
| 2020 | 46.53 | 50.00 | 48.15 | 67.89 | 34.36 | 71.89 |
| 2021 | 43.25 | 38.75 | 48.33 | 75.30 | 28.57 | 26.90 |

*Source:* Authors.

step, we tried to verify if market concentration has a visible impact on the costs of contracted-out MSW services. The costs of the contracted-out service are monitored by the cost of service per capita and the cost of service per performance indicator (one ton of waste). We used causal analysis to examine the interdependencies (causes and outcomes), that is, how private producers set prices in monopolies and lower market concentration.

**TABLE 10.8**
**Public Procurement Procedures for Contracting Out MSW Services in Slovakia (in %)**

| Procedure | 2001 | 2005 | 2006 | 2008 | 2009 | 2010 | 2015 | 2020 | 2021 |
|---|---|---|---|---|---|---|---|---|---|
| Open procedure | 16 | 17 | 27 | 32 | 17 | 21 | 87 | 24 | 17 |
| Restricted procedure | 5 | 0 | 5 | 3 | 14 | 6 | 9 | 2 | 0 |
| Negotiated procedure | 0 | 13 | 30 | 0 | 7 | 1 | 0 | 14 | 5 |
| Price quote | 0 | 0 | 0 | 25 | 4 | 5 | 0 | 27 | 68 |
| Direct award | 31 | 17 | 38 | 30 | 11 | 5 | 0 | 33 | 11 |
| Municipality did not specify | 48 | 55 | – | 25 | 66 | 62 | 4 | 0 | 0 |

*Source:* Authors.

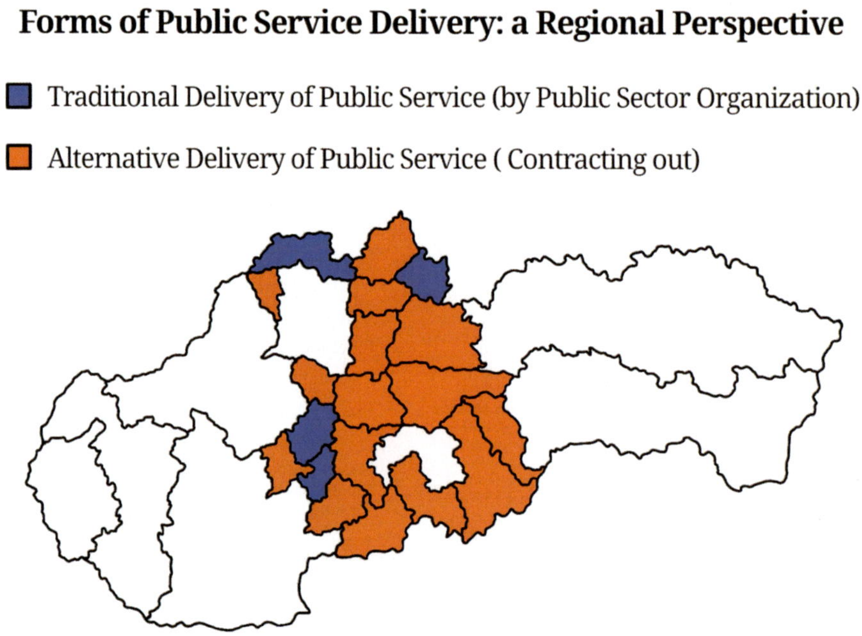


**FIGURE 10.1** NUTS 4 distribution according to the form of provision of municipal solid waste collection and disposal services ($n = 195$).

### 10.7.1 Results

The observed municipalities (Figure 10.1) prefer alternative MSW collection and disposal provision. Only 11.28% use traditional forms of provision of this service. In terms of public–private linkages, we see positively that the remaining 173 municipalities use contracting out for this purpose. In the traditional provision of MSW collection and disposal service, 45.45% of municipalities use technical services established within the municipality or city. 31.81% use the services of an independent organization established by the municipality, and 22.72% of municipalities use public utilities.

When we assess the results of market concentration according to the HHI index (Figure 10.2), in none of the districts did we observe a value lower than 2,500. The average value of the HHI set is 7,260 (statistical deviation 215,361) and is indicated by a line. Based on this result, we rule out the existence of perfect competition in the market or the possible existence of monopolistic competition. On the contrary, the presence of a natural monopoly (HHI = 10,000) was observed in the districts of Revúca, Žiar nad Hronom, Martin, Tvrdošín, and Žilina. In most of these municipalities, there is only one private entity. For example, for the municipalities of Revúca and Martin, it is Brantner

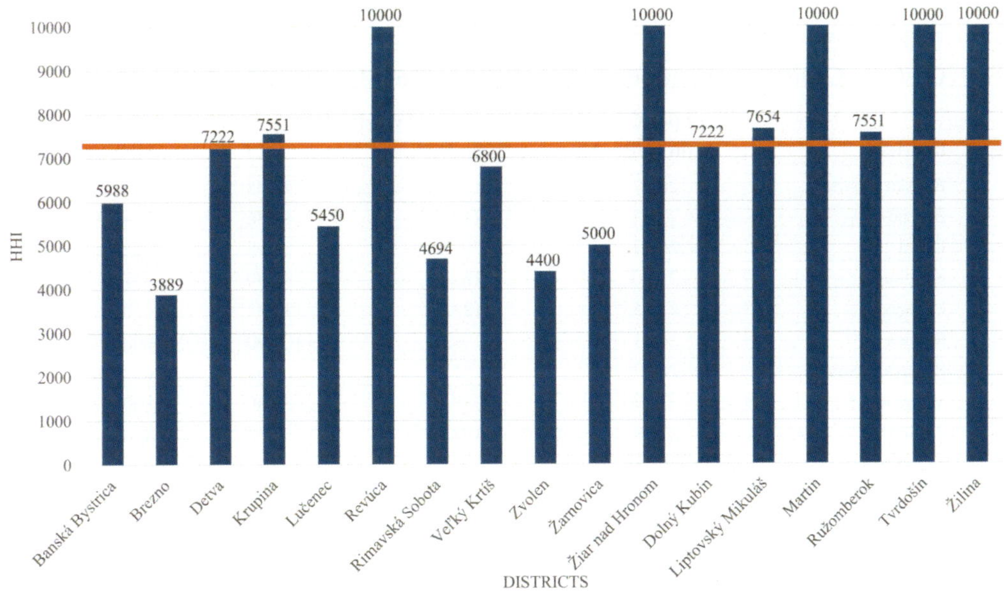


**FIGURE 10.2** The result of market concentration in districts according to the HHI index.

s.r.o. The only private producer for the city of Žilina is T+T a.s. which was established after 1991 following the city's technical services' abolition. Today, it provides waste collection for 365,000 inhabitants and annually processes 40,000 tons of municipal waste in the Žilina Region, which is used to produce alternative fuel for cement plants.

HHI index in the range <6,800.00; the districts of Detva, Krupina, Veľký Krtíš, Dolný Kubín, Liptovský Mikuláš, and Ružomberok report 7,654.32. There are two to three players in these districts. The dominant company is Marius Pedersen, a. s., which has been the largest provider of waste management services in Slovakia since 2003. It currently cooperates with more than 400 municipalities, providing its services to almost 1,000,000 inhabitants and 4,000 companies. HHI index <3,888.89; the districts of Banská Bystrica, Brezno, Lučenec, Rimavská Sobota, Zvolen, and Žarnovica recorded 5,987.65. Rimavská Sobota district shows the lowest market concentration in the observed set. The reason is that there are four organizations providing collection and disposal of MSW in the district. The companies serving an area of 40 km$^2$ include Marius Pedersen, a. s., Brantner Gemer, p. r. o., and Profax, MEPOS s.r.o.. Based on the HHI index values, we can clearly confirm that the market for the provision of MSW collection and disposal services in the municipalities of Central Slovakia is highly concentrated and characterized by a monopoly or oligopoly with a dominant entity.

Table 10.9 shows the resulting values of the cost per capita and cost per ton of waste indicators found as the arithmetic average of the cost per capita and ton of waste in all those municipalities within a single district under study. For comparison, the table also shows the individual HHI and CR indexes.

The lowest values of the first indicator of costs per capita were recorded in the district of Tvrdošín at a level of € 14.32, which is also a district with a monopoly. It is here that we observe the paradox that higher market concentration does not translate into higher prices for services. The reason for this may be the appropriate setting of public procurement, price capping, or supervision of compliance by the founder. Other low values of the first indicator were measured in the districts of Lučenec, Dolný Kubín, Ružomberok, and Žiar nad Hronom, where a monopoly was also recorded. In the Lučenec district, there are two companies competing among the largest players in Central Slovakia, and therefore, we assume that their efforts to gain the upper hand led to lower prices for the district. The highest values of the indicator were in the districts of Banská Bystrica, Detva, Žarnovica, and Martin. Only in the last district was the monopoly of Brantner s.r.o. In the remaining three districts, it was Brantner s.r.o. or Marius Pedersen a.s. Based on the

**TABLE 10.9**
**Cost per Capita and Cost per Ton of Waste in the Monitored Districts**

| Districts | Cost per Capita (in EUR) | Cost per Ton of Waste (in EUR) | CR | HHI |
|---|---|---|---|---|
| Banská Bystrica | 26.57 | 144.84 | 77.38 | 5,987.65 |
| Brezno | 23.62 | 119.76 | 62.36 | 3,888.89 |
| Detva | 26.07 | 202.12 | 84.98 | 7,222.22 |
| Krupina | 23.40 | 160.41 | 86.90 | 7,551.02 |
| Lučenec | 22.25 | 123.05 | 73.82 | 5,450 |
| Revúca | 24.41 | 178.81 | 100.00 | 10,000 |
| Rimavská Sobota | 24.27 | 161.10 | 68.51 | 4,693.88 |
| Veľký Krtíš | 23.42 | 148.98 | 82.46 | 6,800 |
| Zvolen | 23.89 | 192.89 | 66.33 | 4,400 |
| Žarnovica | 26.78 | 127.03 | 70.71 | 5,000 |
| Žiar nad Hronom | 22.74 | 130.14 | 100.00 | 10,000 |
| Dolný Kubín | 18.96 | 168.04 | 84.98 | 7,222.22 |
| Liptovský Mikuláš | 23.32 | 121.50 | 87.49 | 7,654.32 |
| Martin | 26.34 | 152.02 | 100.00 | 10,000 |
| Ružomberok | 22.72 | 102.42 | 86.90 | 7,551.02 |
| Tvrdošín | 14.32 | 137.27 | 100.00 | 10,000 |
| Žilina | 24.09 | 127.52 | 100.00 | 10,000 |

*Source:* Authors.

values found, we observe that the highest prices were recorded in districts where purely private players provided the service. This confirms the negative impact of market concentration on the price of the service.

The lowest cost value per ton of waste was in the Ružomberok district with € 102.42. In addition, low values were recorded in the districts of Lučenec, Liptovský Mikuláš, and Brezno. However, Brezno is the district with the lowest HHI value, and, as in the Lučenec district, the largest part of the market is shared by two private players – Profax and Brantner s.r.o. The highest values of the costs per ton of waste indicator were recorded in the district of Detva, which also had high values for the first indicator. In the district of Zvolen, the market is dominated by the private organization Marius Pedersen a.s. In the case of Central Slovakia, we perceive as a problem area where, despite the different HHI values, all the districts examined are highly concentrated and are therefore difficult to estimate how the situation would change in less concentrated markets. Similarly, the market for MSW collection and disposal services has large players.

## 10.8 INTER-MUNICIPAL COOPERATION

Data provided by already published studies suggest that IMC in the Czech Republic and Slovakia is delivering very good results in terms of the efficiency and quality of the MSW service provision. For example, Soukopová et al. (2022) tested a model of the impact of selected factors on the cost of MSW collection and disposals. The descriptive power of the model was high – more than 90% – and almost all variables (except distance to landfill/incinerator) provided results. Furthermore, the model confirmed that the IMC variable was the second strongest variable in terms of impact on cost.

Despite the positive results, the share of IMC as a form of MSW service provision has not substantially increased over the 20-year period under study in both countries (Soukopová et al., 2022). This situation raised our main research question for this section:

RQ: Why is inter-municipal cooperation not used more frequently for the provision of MSW management services in the Czech Republic and Slovakia?

In order to answer our main research question, we decided to use a qualitative approach, that is, the opinions of experts. In both countries, we approached three well-known experts on the topic – two academic experts and one practitioner. They were asked to answer the following question:

> According to our data collected over the last 20 years, contracting out is the main form of provision of municipal waste management services in the Czech Republic and Slovakia. According to these data, IMC is the most efficient form in many local cases; however, the share of this mode of provision has not visibly increased in recent years. We would like to ask for your opinion on why municipalities do not use IMC more frequently in the provision of waste management services.

This first attempt was expected to serve as the beginning of the standard Delphi method with 10–15 respondents. However, since the responses of this small sample of experts were nearly identical, we decided that the results from this limited survey were fully representative and there was no need to pursue this line of inquiry further.

### 10.8.1 Results

The six experts involved in the qualitative part of our research provided a nearly identical set of three critical purposes: transaction costs (of different types), non-existent benchmarks (no regular comparisons of best solutions), and limited motivation to select the optimal mode of service provision. Transaction costs in all their different forms were the most frequently cited factors. The experts mentioned several barriers related to this dimension. One response directly linked this problem to the limited reliability of any efficiency calculation, as also noted in the methodology section of this paper:

"Without including full transaction costs in the calculations, any comparison of the efficiency of internal and external forms of provision is misleading" (academic expert, Slovakia).

All the experts mentioned the financial and non-financial transaction costs and limits of cooperation. The following responses are representative in relation to transaction costs:

"IMC requires complicated coordination and acceptance by management and leaders. Some municipalities are not ready to cooperate simply because of the relationships between their mayors" (academic expert, Slovakia).

"Municipalities do not trust each other. In particular, smaller municipalities do not trust the larger ones. Therefore, municipal leaders may prefer to tender the service, negotiating with a legally tendered organization to provide the service to avoid having to defend their own solutions" (academic expert, Czech Republic).

"The creation of a new inter-municipal authority for the management of MSW requires significant starting costs and guarantees" (practitioner, Czech Republic).

The fact that there are no existing benchmarks and municipalities do not routinely conduct any in-depth investigation of the most effective solutions was confirmed in the following statement:

"Most municipalities never approach this topic in any systemic way and do not conduct or commission the analysis of the efficiency of different provision options" (academic expert, Slovakia).

The third critical limit – lack of motivation to select the "best" option – was confirmed in several responses, for example: "Municipal leaders are interested in whether the MSW management system is working; they do not want to deal with details or to search for possible improvements. Lobbying and other activities of private waste management companies, who protect their business as best they can, also play a role" (practitioner, Czech Republic).

"If citizens do not show visible dissatisfaction, municipal managers do not see the need to change existing solutions. Political motivation, will, shared information, local capacities – all these elements to push for change are lacking" (practitioner, Slovakia).

In short, our qualitative research into the opinions of experts in this field has identified several reasons for this: high transaction costs, limited motivation of municipalities to choose the most efficient form of provision, ill-functioning benchmarking (successful examples of IMC, so-called cases of good practice, are often not known by other municipalities), and insufficient or absent "testing of services." Other authors have added the additional administrative costs associated with the organization of cooperation (Klimovský, 2014).

## 10.9 CONCLUSIONS AND PERSPECTIVES

Contracting out waste management services is one of the most commonly used approaches to the provision of this service. Under this arrangement, the state is responsible for the service provision, but production is transferred to private bodies. This approach was mainly promoted by the New Public Management (NPM) ideology, which proposed that contracting out should help increase service provision's efficiency through competition. Globally, the issue of contracting out public sector services has attracted the attention of both the professional and lay public for the past three decades because the government of each country spends a significant quantity of public money each year to fund contract out services. In contrast to Western European countries undergoing re-municipalization, the data obtained for Slovakia and the Czech Republic over 20 years specifically show that municipal MSW services are dominated by contracting out, which has increased over the period under study. This fact is probably a reflection of the high rate of fragmentation in municipal structures and the municipalities' low willingness to cooperate in exercising their authority. More than half of Czech municipalities have less than 500 residents, and approximately 78% have less than 1,000; the situation is similar in Slovakia. Therefore, waste management is an integral and indispensable part of local budgets. The data also suggest that the use of external suppliers seems to yield marginal savings that could certainly be substantially larger if the quality of contract management is improved (according to our Slovak samples, it has not improved at all over the last 10 years!). We also show that market concentration may not directly impact the cost of MSW services. This fact suggests that other factors (in particular, the quality of contract management) should be the key determinants of costs. The fact that the share of IMC between service provision modes did not increase significantly over the period under study was investigated via expert opinions. The experts almost unanimously agreed on three critical limits related to the issue of IMC in MSW management delivery – transaction costs of different types, no regular comparisons of best solutions, and limited motivation to select optimal service provision modes. Non-existent benchmarking and limited motivation to select optimal service provision modes are at least partially linked and are highly predictable in countries with limited government accountability and responsibility of politicians and bureaucrats. If citizens do not demand better and more efficient services, local "bosses" are not motivated to seek better options and change.

Our findings suggest that contracting out may have potential, even in areas with high market concentration. However, in Czech and Slovak practice this potential is not exploited due to the low quality of contract management. The perspectives in this situation are quite obvious – the solution is a systematic and regular evaluation of possible provision alternatives when deciding between internal and external production and an effective selection of supplier if external provision seems to be better option, implemented by responsible municipalities.

## ACKNOWLEDGMENTS

The chapter is the result of a research project supported by VEGA 1/0003/24 Inter-municipal cooperation in providing local public services.

## REFERENCES


Agiamoh, R. 2020. From bureaucracy to market? Ongoing reform and performance challenges of solid waste administration in Moscow. *Public Administration Issues* 5: 149–170.

Albalate, D., G. Bel, R. Gradus and E. Reeves. 2021b. Re-municipalization of local public services: Incidence, causes and prospects. *International Review of Administrative Sciences* 87(3): 419–424.

Albalate, D., G. Bel and E. Reeves. 2021a. Government choice between contract termination and contract expiration in re-municipalization: A case of historical recurrence? *International Review of Administrative Sciences* 87(3): 461–479.

Andersen, B., B. Henriksen and I. Spjelkavik. 2008. Benchmarking applications in public sector principal-agent relationships. *Benchmarking: An International Journal* 15(6): 723–741.

Arrow, K. J. 1985. The economics of agency. In *The Structure of Business*, ed. J. Pratt and J. W. Zeckhauser, pp. 37–51. Boston: Harvard Business School Press.

Balážová, E. 2006. *Benchmarking služieb miestnej samosprávy na Slovensku*. Bratislava: Transparency International Slovensko. Adin.

Behn, R. D. and P. A. Kant. 1999. Strategies for avoiding the pitfalls of performance contracting. *Public Productivity & Management Review* 22(4): 470–489.

Bel, G. and A. Costas. 2006. Do public sector reforms get rusty? Local privatization in Spain. *Journal of Policy Reform* 9(1): 1–24.

Bel, G. and R. Gradus. 2018. Privatisation, contracting-out and inter-municipal cooperation: New developments in local public service delivery. *Local Government Studies* 44(1): 11–21.

Bel, G., X. Fageda and M. E. Warner. 2010. Is private production of public services cheaper than public production? A meta-regression analysis of solid waste and water services. *Journal of Policy Analysis and Management* 29(3): 553–577.

Bel, G. and X. Fageda. 2011. Big guys eat big cakes: Firm size and contracting in urban and rural areas. *International Public Management Journal* 14(1): 4–26.

Bel, G., X. Fageda and M. Mur. 2014. Does cooperation reduce service delivery costs? Evidence from residential solid waste services. *Journal of Public Administration Research and Theory* 24(1): 85–107.

Bel., G. and M. Mur. 2009. Intermunicipal cooperation, privatization and waste management costs: Evidence from rural municipalities. *Waste Management* 29(10): 2772–2778.

Bel, G. and J. Rosell. 2016. Public and private production in a mixed delivery system: Regulation, competition and costs. *Journal of Policy Analysis and Management* 35(3): 533–558.

Bel, G. and M. Sebő. 2020. Introducing and enhancing competition to improve delivery of local services of solid waste collection. *Waste Management* 118: 637–646.

Bel, G. and M. Sebő. 2021. Watch your neighbor: Strategic competition in waste collection and service quality. *Waste Management* 127: 63–72.

Bel, G. and M. E. Warner. 2015. Inter-municipal cooperation and costs: Expectations and evidence. *Public Administration* 93(1): 52–67.

Benito, B., M. D. Guillamón and F. Bastida. 2015. Public versus private in municipal services management. *Lex Localis* 13(4): 995–1018.

Bennett, C. and E. Ferlie. 1996. Contracting in theory and in practice: Some evidence from the NHS. *Public Administration* 74(1): 49–66.

Brown, T. L. and M. Potoski. 2003. Contract–management capacity in municipal and county governments. *Public Administration Review* 63(2): 153–164.

Brown, T. L., M. Potoski and D. Van Slyke. 2015. Managing complex contracts: A theoretical approach. *Journal of Public Administration Research and Theory* 26(2): 294–308.

Callan, S. J. and J. M. Thomas. 2001. Economies of scale and scope: A cost analysis of municipal solid waste services. *Land Economics* 77(4): 548–560.

Cooper, P. J. 2003. *Governing by Contract*. Washington D.C.: CQ Press.

Dehoog, R. H. 1990. Competition, negotiation, or cooperation: Three models for service contracting. *Administration & Society* 22(3): 317–340.

Demuth, J., H. W. Friederiszick and S. Reinhold. 2022. Reverse privatization as a reaction to the competitive environment: Evidence from solid waste collection in Germany. *Review of Industrial Organization* 60(2): 217–261.

DiFoggia G. and M. Beccarello. 2021. Market structure of urban waste treatment and disposal: Empirical evidence from the Italian industry. *Sustainability* 13(13): 7412–7424.

Dijkgraaf, E. and R. Gradus. 2007. Collusion in the Dutch waste collection market. *Local Government Studies* 33(4): 573–588.

Domberger, S. and P. Jensen. 1997. Contracting out by the public sector: Theory, evidence, prospects. *Oxford Review of Economic Policy* 13(4): 67–78.

Fedele, M. and G. Moini. 2007. Italy: The changing boundaries of inter-municipal cooperation. In *Inter-Municipal Cooperation in Europe*, ed. R. Hulst and A. van Montfort, pp. 117–138. Dordrecht: Springer.

Garrone, P., L. Grilli and X. Rousseau. 2013. Management discretion and political interference in municipal enterprises. Evidence from Italian utilities. *Local Government Studies* 39(4): 514–540.

Gradus, R., E. Dijkgraaf and M. Schoute. 2016. Is there still collusion in the Dutch waste collection market? *Local Government Studies* 42(5): 689–697.

Gradus, R., M. Schoute and E. Dijkgraaf. 2018. The effects of market concentration on costs of local public services: Empirical evidence from Dutch waste collection. *Local Government Studies* 44(1): 86–104.

Green, J. D. 2002. *Cities and Privatization: Prospects for the New Century*. Upper Saddle River, New Jersey: Prentice Hall.

Haveri, A. and A. Jenni. 2007. Inter-municipal cooperation in Finland: Old traditions and new promises. In *Inter-Municipal Cooperation in Europe*, ed. R. Hulst and A. van Montfort, pp. 39–65. Dordrecht: Springer Netherlands.

Hodge, G. A. 2000. *Privatization: An International Review of Performance*. Boulder, CO: Westview Press.

Hulst, R. and A. van Montfort. 2007. The Netherlands: Cooperation as the only viable strategy. In *Inter-Municipal Cooperation in Europe*, ed. R. Hulst and A. van Montfort, pp. 139–168. Dordrecht: Springer.

Johansson, T. 2008. Municipal contracting out: Governance choices, misalignment and performance in Swedish local government. *Financial Accounting Management* 24(3): 243–264.

Johansson, T., S. Siverbo and C. Camén. 2016. Managing cooperation, coordination, and legitimacy: Control of contracted public services. *Accounting Auditing & Accountability Journal* 29(6): 1012–1037.

Kishimoto, S., L. Steinfort and O. Petitjean. 2020. *The Future Is Public: Towards Democratic Ownership of Public Services*. Amsterdam, Paris: Transnational Institute.

Klimovský, D. 2014. *Inter-Municipal Cooperation in Slovakia: The Case of Regions with Highly Fragmented Municipal Structure*. Novo Mesto: Faculty of Organization Studies.

Kopričˇ, I., H. Wollmann and D. Marcou, D. 2018. *Evaluating Reforms of Local Public and Social Services in Europe*. London: Palgrave McMillan.

Lavery, K. 1999. *Smart Contracting for Local Government Services: Processes and Experience*. Westport, CT: Praeger.

Majlingová, Ľ. 2005. *Spôsoby Zabezpečovania Služieb Odpadového Hospodárstva v SR*. Dissertation Thesis, EF UMB, Banská Bystrica.

Marra, A. and V. Carlei. 2014. Institutional environment and partial privatizations: Policy implications for local government. *Annals of Public and Cooperative Economics* 85(1): 31–52.

Matějová, L., M. Plaček, M. Křápek, M. Půček and F. Ochrana. 2014. Economies of scale–empirical evidence from the Czech Republic. *Procedia Economics and Finance* 12: 403–411.

Megginson, W. L., R. C. Nash, J. M. Netter and A. B. Poulsen. 2000. The choice between private and public markets: Evidence from privatizations. Working Paper, University of Georgia, Athens.

Monteduro, F. 2014. Public-private versus public ownership and economic performance: Evidence from Italian local utilities. *Journal of Management & Governance* 18(1): 29–49.

Nemec, J., J. Soukopová and B. Mikušová Meričková, B. 2020. Local public service delivery arrangements in the Czech Republic and Slovakia. In *Open Government: Concepts, Methodologies, Tools, and Applications*, ed. M. Khosrow-Pour, pp. 1392–1410, Pennsylvania: Information Science Reference.

Niaounakis, T. and J. Blank. 2017. Inter-municipal cooperation, economies of scale and cost efficiency: An application of stochastic frontier analysis to Dutch municipal tax departments. *Local Government Studies* 43(4): 533–554.

Ochrana, F., B. Meričková and J. Nemec. 2008. Introducing benchmarking in the Czech Republic and Slovakia. Processes, problems and lessons. *Public Management Review* 10(5): 673–684.

Ohlsson, H. 2003. Ownership and production costs: Choosing between public production and contracting-out in the case of Swedish refuse collection. *Fiscal Studies* 24(4): 451–476.

Pavel, J. 2006. *Efektivnost a transparentnost obecních obchodních společností*. Prague: Transparency International.

Petersen, O. H., U. Hjemar and K. Vrangbaek. 2018. Is contracting out of public services still the Great Panacea? A systematic review of studies on economic and quality effects from 2000 to 2014. *Social Policy & Administration* 52(1): 130–157.

Porcher, S. 2017. The hidden costs of water provision: New evidence from the relationship between contracting-out and price in French water public services. *Utilities Policy* 48: 166–175.

Pratt, J. W. and R. J. Zeckhauser. 1986. Principals and agents: The structure of business. *Journal of Economic Literature* 24(4): 1788–1789.

Rehfuss, J. A. 1989. *Contracting Out in Government*. San Francisco: Jossey-Bass.

Rendon, R., G. 2012. Public procurement: Public administration and public service perspectives. *Journal of Public Affairs Education* 18(2): 327–348.

Romzek, B. S. and J. M. Johnston. 2002. Effective contract implementation and management: A preliminary model. *Journal of Public Administration Research and Theory* 12(3): 423–453.

Rose-Ackerman, S. and B. J. Palifka. 2016. *Corruption and Government: Causes, Consequences, and Reform*, 2nd ed. Cambridge: Cambridge University Press.

Savas, E. S. 1987. *Privatisation: The Key to Government*. New York: Chatman House.

Sclar, E. D. 2000. You *Don't Always Get What You Pay For: The Economics of Privatization*. Ithaca: Cornell University Press,

Schoute, M., R. Gradus and T. Budding. 2021. Drivers of service delivery modes in Dutch local government: An analysis over time and across domains. *International Review of Administrative Sciences* 87(3): 425–439.

Smith, S. R. 1996. Transforming public services: Contracting for social and health services in the US. *Public Administration* 74(1): 113–27.

Soukopová, J. and D. Klimovský. 2016. Local governments and local waste management in the Czech Republic: Producers or providers? *NISPAcee Journal of Public Administration and Policy* 9(2): 217–237.

Soukopová, J., B. Mikušová-Meričková, J. Nemec and M. Šumpíková. 2022. Institutional factors determining costs of municipal waste management in the Czech Republic. *Waste Management* 144: 527–532.

Soukopová, J., J. Nemec, L. Matějová and M. Struk. 2014. Municipality size and local public services: Do economies of scale exist? *NISPAcee Journal of Public Administration and Policy* 7(2): 151–171.

Soukopová, J. and G. Vaceková. 2018. Internal factors of intermunicipal cooperation: What matters most and why?. *Local Government Studies* 44(1), 105–126.

Southwick, L. 2012. Economies of scale in local government: General government spending. *Ibusiness* 4(03): 265.

Spicer, Z. 2015. Regionalism, municipal organization, and inter-local cooperation in Canada. *Canadian Public Policy* 41(2): 137–150.

Swianiewicz, P., Teles, F. 2019. The institutionalization of inter-municipal arrangements in Europe: Findings from the unusual suspects. *Transylvanian Review of Administrative Sciences* 15(57): 119–136.

Voorn, B. 2021. Country, sector and method effects in studying remunicipalization: A meta-analysis. *International Review of Administrative Sciences* 87(3): 440–460.

Warner, M. E. 2006. Inter-municipal Cooperation in the US: A regional governance solution? *Urban Public Economics Review* 6: 221–239.

Weghmann, V. and Hall, D. 2021. The unsustainable political economy of investor-state dispute settlement mechanisms. *International Review of Administrative Sciences* 87(3): 480–496.

West, K. 2007. Inter-municipal cooperation in France: Incentives, instrumentality and empty shells. In *Inter-Municipal Cooperation in Europe*, ed. R. Hulst and A. van Montfort, pp. 67–90. Dordrecht: Springer.

Williamson, O. E. 1986. *Economic Organization: Firms, Markets, and Policy Control*. New York: New York University Press.

Yescombe, E. R. and E. Farquharson. 2018. Public-sector contract management. In *Public-Private Partnerships for Infrastructure: Principles of Policy & Finance*, ed. E. R. Yescombe, pp. 261–274. Oxford: Butterworth-Heinemann.

# Index

For Product Safety Concerns and Information please contact our
EU representative GPSR@taylorandfrancis.com Taylor & Francis
Verlag GmbH, Kaufingerstraße 24, 80331 München, Germany